BOOK LO

Genetics
for Fish Hatchery
Managers

Second Edition

Genetics
for Fish Hatchery
Managers

Second Edition

Acknowledgments

No book is written without help, and I owe many individuals a great debt for their assistance and guidance. First and foremost, E. W. Shell convinced me that there was a need for this book and convinced me that I could write the book when I began to doubt myself. Fred P. Meyer, Michael C. Wooten, William K. Hershberger, Terry Campbell, Klaus D. Kallman, Stephen P. Malvestutso, Daryl L. Kuhlers, Mary U. Ball, and J. C. Williams reviewed all of or portions of the first edition. Rex A. Dunham and Harold L. Kincaid reviewed portions of the second edition. They made many helpful suggestions and greatly improved the book. Beverly Owen, Ingrid A. L. Smith, Teresa Rodriguez, and Sally Rader took my crudely drawn figures and converted them into excellent illustrations. Stuart M. Tave and Robert P. Romaire helped procure references for both the first and second editions.

I would like to thank everyone who bought the first edition, which made a second edition necessary.

A special thanks is due to Kai Stuart Tave who gave up valuable play time and had to play his Nintendo games with the sound off so that I could finish this book.

I owe my greatest debt to the world's prettiest editor, my lovely wife Katherine, who encouraged me, made suggestions, proofread, criticized, and goaded me into finishing the second edition, and who rescued me from the clutches of the Nightmare.

COPYRIGHT ACKNOWLEDGMENTS. I would like to thank the following publishers and individuals for the use of copyrighted and uncopyrighted material: American Association for the Advancement of Science and Gary H. Thorgaard for Fig. 2.2; Jack Turner for Fig. 3.1; The World Aquaculture Society for Fig. 3.2; T.F.H. Publishers, Inc. for Table 3.7; Genetics Society of America for Table 3.9; Nauka Publishers, Moscow and Valentin S. Kirpichnikov for Fig. 3.4 and Table 3.11; Blackwell Scientific Publishers, Ltd. for Fig. 3.9; American Genetic Association and Jack S. Frankel for Fig. 3.5; New York Zoological Society for Fig. 3.6; Bamidgeh for Fig. 3.7; Abdel El Gamel for Fig. 3.8; Longman Group Ltd. for Fig. 3.10; John Wiley and Sons, Inc. for Figs. 4.3 and 4.4; Elsevier Scientific Publishers for Figs. 4.17 and 4.21 and Tables 4.3 and 4.15; Trygve Gjedrem for Table 4.3; B. Chevassus for Fig. 4.21; G. W. Wohlfarth for Table 4.15; Universities Federation for Animal Welfare for Figs. 4.8 and 4.13; Aquaculture Magazine for Figs. 4.14 and 5.7; Harold L. Kincaid for Figs. 4.15 and 4.16; B.C. Salmon Farmers Association for Figs. 4.18 and 4.19; Jesse A. Chappell for Fig. 4.31 and Table 4.6; Alabama Agricultural Experiment Station, Auburn University for Figs. 4.20 and 4.27 and Tables 4.10, 4.12, and 4.14; The American Fisheries Society for Table 4.7; Prentice-Hall, Inc. for Fig. 4.23; The American Society of Ichthyologists and Herpetologists and William H. LeGrande for Fig. 4.22; The Progressive Fish-Culturist for Figs. 4.28 and 6.1; The Japanese Society for Scientific Fisheries for Figs. 4.29 and 4.30; Washington Sea Grant Program and the University of Washington Press for Figs. 5.6 and 5.10 and Table 5.1; Penny Riggs and C. Larry Chrisman for Fig. 5.9.

CHAPTER 1

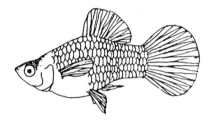

Introduction

1
Introduction

No two fish are identical. The variability that is observed or measured in any population of fish extends to virtually all phenotypes (observable or measurable characteristics): color, number of dorsal fin spines, length, weight, the exact shape of a bone, etc. The idea that man can exploit the variability that exists in a population was recognized eons ago when he began to breed and to domesticate sheep, goats, horses, dogs, and cattle. However, a scientifically sound theory that explained why phenotypic variation exists and also explained how it is inherited did not exist until 1866 when Gregor Mendel published his monograph on the inheritance of phenotypes in peas. Mendel's work was ignored until the twentieth century when it was finally understood and was later amplified and refined into the sciences of genetics and breeding.

The underlying causes of the variations that exist for the phenotypes and the way that the differences are inherited is the science called genetics. Breeding is the applied science of genetics that exploits the heritable component of variation in order to change the population for man's benefit.

Basic breeding concepts have only recently been applied to aquaculture. Fish culture lags far behind other areas of animal husbandry in that fish culturists raise animals that either come from wild stocks or are only

3

a few generations removed from the wild. Except for a few strains of common carp and rainbow trout and some species of tropical fish, domesticated stocks of fish do not exist.

Because aquaculture genetics is comparatively new, but especially because most aquaculturists feel that genetics and geneticists are incomprehensible, genetic aspects of fish husbandry have had comparatively little impact on productivity and profits, except in the ornamental fish industry. However, several breeding programs have been developed and are being used, and they demonstrate that tremendous gains are possible.

Despite the fact that breeding has had minimal impact on improving productivity in the fish farming industry, it is imperative that fish culturists have a good grasp of genetics and breeding principles because they are among the major factors that govern productivity. Animal husbandry is, first and foremost, the management of an animal, and it is impossible to maximize productivity until the biology of that animal is properly understood.

One reason why it is imperative that farmers understand the basic principles governing genetics and breeding is the unfortunate fact that some effort is often needed simply to maintain the status quo. It is also an unfortunate fact that inattention to the genetic aspects of fish husbandry can actually cause productivity to decline as inbreeding or other genetic consequences of mismanagement result in slower growth, decreased viability, decreased disease resistance, and decreased egg production.

To date, most effort in fish culture has been directed toward improved diets, health management, and water quality management. As important as they are, these disciplines deal with the environment in which fish live, and improvements and breakthroughs in these areas simply improve the environment. Breeding deals with the animal itself, and improvements in this area improve the biological potential of the fish. Productivity cannot be optimized if the biological potential of the population is not optimized. Fish that have a faster growth rate, greater dressing percentage, lower food conversion, and greater disease resistance are more economical to raise. Because of this, one of the goals of hatchery management should be to incorporate basic genetic and breeding concepts into routine hatchery management in order to maximize the biological potential of the fish. Until basic breeding concepts are incorporated into everyday hatchery management, fish culture will not emerge into twentieth century farming, and the twenty-first century is nearly upon us.

1

Introduction

No two fish are identical. The variability that is observed or measured in any population of fish extends to virtually all phenotypes (observable or measurable characteristics): color, number of dorsal fin spines, length, weight, the exact shape of a bone, etc. The idea that man can exploit the variability that exists in a population was recognized eons ago when he began to breed and to domesticate sheep, goats, horses, dogs, and cattle. However, a scientifically sound theory that explained why phenotypic variation exists and also explained how it is inherited did not exist until 1866 when Gregor Mendel published his monograph on the inheritance of phenotypes in peas. Mendel's work was ignored until the twentieth century when it was finally understood and was later amplified and refined into the sciences of genetics and breeding.

The underlying causes of the variations that exist for the phenotypes and the way that the differences are inherited is the science called genetics. Breeding is the applied science of genetics that exploits the heritable component of variation in order to change the population for man's benefit.

Basic breeding concepts have only recently been applied to aquaculture. Fish culture lags far behind other areas of animal husbandry in that fish culturists raise animals that either come from wild stocks or are only

3

a few generations removed from the wild. Except for a few strains of common carp and rainbow trout and some species of tropical fish, domesticated stocks of fish do not exist.

Because aquaculture genetics is comparatively new, but especially because most aquaculturists feel that genetics and geneticists are incomprehensible, genetic aspects of fish husbandry have had comparatively little impact on productivity and profits, except in the ornamental fish industry. However, several breeding programs have been developed and are being used, and they demonstrate that tremendous gains are possible.

Despite the fact that breeding has had minimal impact on improving productivity in the fish farming industry, it is imperative that fish culturists have a good grasp of genetics and breeding principles because they are among the major factors that govern productivity. Animal husbandry is, first and foremost, the management of an animal, and it is impossible to maximize productivity until the biology of that animal is properly understood.

One reason why it is imperative that farmers understand the basic principles governing genetics and breeding is the unfortunate fact that some effort is often needed simply to maintain the status quo. It is also an unfortunate fact that inattention to the genetic aspects of fish husbandry can actually cause productivity to decline as inbreeding or other genetic consequences of mismanagement result in slower growth, decreased viability, decreased disease resistance, and decreased egg production.

To date, most effort in fish culture has been directed toward improved diets, health management, and water quality management. As important as they are, these disciplines deal with the environment in which fish live, and improvements and breakthroughs in these areas simply improve the environment. Breeding deals with the animal itself, and improvements in this area improve the biological potential of the fish. Productivity cannot be optimized if the biological potential of the population is not optimized. Fish that have a faster growth rate, greater dressing percentage, lower food conversion, and greater disease resistance are more economical to raise. Because of this, one of the goals of hatchery management should be to incorporate basic genetic and breeding concepts into routine hatchery management in order to maximize the biological potential of the fish. Until basic breeding concepts are incorporated into everyday hatchery management, fish culture will not emerge into twentieth century farming, and the twenty-first century is nearly upon us.

CHAPTER 2

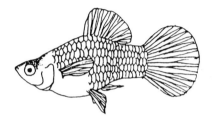

Review of
Basic Genetics

Genetics
for Fish Hatchery
Managers

Second Edition

Douglas Tave

An AVI Book
Published by Van Nostrand Reinhold
New York

An AVI Book
(AVI is an imprint of Van Nostrand Reinhold)

Copyright © 1993 by Van Nostrand Reinhold

Library of Congress Catalog Card Number 92-24563
ISBN 0-442-00417-6

Printed in the United States of America.

Van Nostrand Reinhold
115 Fifth Avenue
New York, New York 10003

Chapman and Hall
2-6 Boundary Row
London, SE1 8HN, England

Thomas Nelson Australia
102 Dodds Street
South Melbourne 3205
Victoria, Australia

Nelson Canada
1120 Birchmount Road
Scarborough, Ontario M1K 5G4, Canada

16 15 14 13 12 11 10 9 8 7 6 5 4 3 2 1

Library of Congress Cataloging-in-Publication Data

Tave, Douglas.
    Genetics for fish hatchery managers / Douglas Tave.—2nd ed.
        p. cm.
    An AVI book.
    Includes bibliographical references and index.
    ISBN 0-442-00417-6
    1. Fishes—Breeding. 2. Fishes—Genetics. I. Title.
SH155.5.T38   1992
639.3—dc20   92-24563
                                                         CIP

This book is for
my parents who provided the V_G
and for
Katherine who provided the V_E

Contents

Contents

Preface

Since the first edition of this book was published in 1986, several books on fish genetics and the proceedings of several symposia on fish genetics have been published. Despite the geometric increase in monographs on fish genetics, this book remains the only one that is devoted to the genetic aspects of hatchery management and that is written for non-geneticists. That, coupled with the warm reception given to the first edition, convinced me to update the book and to create a second edition. The creation of the second edition enabled me to add some topics that were not covered in the first edition.

This book makes genetics and the basic ideas of animal breeding accessible to fish farmers, hatchery managers, or anyone who raises fish. The genetic aspects of fish husbandry have received far less attention than nutrition, health management, or water quality management, but breeding is as important, if not more important, than other aspects of fish culture because the genetics of a population defines its potential.

Unfortunately, most aquaculturists view genetics as an exotic disease, because geneticists tend to explain their discipline in an obtuse manner using 50¢ words and $3 jargon. This book tries to bridge the gap between the geneticist and the aquaculturist by presenting topics and ideas as simply as possible and by explaining how the ideas and concepts can be used to increase productivity. The first purpose of the book is to convince those who raise fish that genetics and basic animal breeding concepts should become an everyday part of hatchery management. The second purpose is to convince aquaculturists that genetics is not incomprehensible—and it is not; only geneticists are. The final purpose is to convince hatchery managers that productivity depends, to a great extent, on breeding practices and how the population is managed genetically. The final purpose is the crux of the book, because hatchery managers, aquaculturists, and tropical fish hobbyists practice selection every time they decide which fish will be spawned and which will not. Consequently, it behooves all who work with fish to understand the consequences of their actions and understand how their actions affect productivity.

Because this book was written for nongeneticists, the topics are explained in a simple manner without pages of qualifying comments and without a long list of "exceptions to the rule." An example of this is the discussion about sex determination. Sex determination is a very complicated subject, and this topic could easily encompass a 100-page chapter; but because most aquaculturists need to know only the rudiments of sex determination, this topic was abridged to just a few pages.

The book was kept purposely short so that it would be a readable length. As such, it is not a complete text on fish genetics.

I tried to cite the original reference when giving examples about a particular form of inheritance, but some studies are almost impossible to obtain and, when that is so, I cite a more easily obtainable paper so that anyone who wishes to pursue the inheritance of a particular phenotype can quickly find a foothold in the literature. No attempt was made to make this book an all inclusive review of fish genetics.

The book is divided into six chapters: an introduction; a brief review about how genes are transmitted from generation to generation, coupled with a brief discussion about sex determination; the genetics and management of qualitative phenotypes, such as body color; the genetics and management of quantitative phenotypes, such as weight; non-traditional genetics, such as the production of sex-reversed broodstock, chromosomal manipulation, and genetic engineering; and a final chapter on how to incorporate the ideas that have been discussed into broodstock management.

In addition, there are two appendixes and a glossary. Appendix A lists the scientific names of fishes cited in the book. Appendix B is a brief discussion on how to calculate the mean, variance, standard deviation, coefficient of variation, and range of a quantitative phenotype.

2

Review of Basic Genetics

Before I discuss the genetics of fish stock management and explain how genetic variation can be exploited in order to improve productivity, it is necessary to digress and capsulize how the genetic material is transmitted from generation to generation via the gametes (sperm and eggs). This abridged discussion on genetics will cover only information that shows how the genetic information passes from parents to offspring. This brief digression is necessary because the ideas and processes that are discussed in this chapter are used to help explain some of the concepts that are developed in Chapters 3, 4, and 5.

This chapter is divided into three sections: genes and chromosomes; meiosis; and sex determination. These three sections briefly discuss a few of the important ideas and processes of these subjects. The vast majority of information about the basic laws of genetics, particularly that dealing with molecular genetics and cytogenetics, will not be discussed. Those wishing information about these areas or those desiring a more complete discussion about basic genetics can find it in most introductory genetics textbooks.

GENES AND CHROMOSOMES

The basic unit of inheritance is the gene or locus (plural: loci). Technically, the word locus refers to the location of a gene on a chromosome, but colloquially the terms "gene" and "locus" are used interchangeably, and I shall use them synonymously in this book (gene = locus). The gene is the genetic unit that contains the blueprint or biological code for the production of a phenotype. A gene is a linear array of very specific subunits and is but a small segment of a much larger molecule called deoxyribonucleic acid (DNA); see Fig. 2.1.

Genes can occur in one or more forms. Each form of a specific gene is called an allele. In a population, any gene may have only one allele or may have over a dozen. A gene with only one allele is said to be monomorphic. A gene with two or more common [frequency $(f) \geq 0.01$] alleles is said to be polymorphic. For example, the P gene in the platyfish, which controls tail spot patterns, has nine alleles (Kallman 1975); see Fig. 3.6 on p. 66.

Different alleles have slightly different sequences of paired bases. These differences produce different chemical messages which, in turn, produce variations of a particular phenotype. The different phenotypes, the variances that the phenotypes exhibit in a population, and the way that they are controlled genetically must be understood before breeding programs can be used to improve productivity. These subjects will be developed in Chapters 3 and 4.

The DNA, and thus the genes, forms structures called chromosomes which are located in the nucleus of a cell. The number of chromosomes varies from species to species but remains constant within a species. There are some exceptions to this rule, but generally the number is constant within a species. In most fish, chromosomes occur in pairs, and organisms in which they do are called diploids ($2N$). One chromosome of each pair is from the fish's father, while the other chromosome is from the fish's mother.

There are some exceptions to the rule that fish are diploids. Tetraploidy, a doubling of the diploid number to the $4N$ state, has been a major driving force in the evolution of fishes. Some fishes, notably the catostomids (suckers) (Ferris 1984) and the salmonids (salmon and trout) (Allendorf and Thorgaard 1984), evolved via tetraploidy, i.e., the original $2N$ number of chromosomes doubled to $4N$. However, for practical fish culture work these fish behave like diploids. Some fish have even become triploids ($3N$) during their evolution (Schultz 1967); the occurrence of natural populations of triploid fish is extremely rare and need not concern us. All fish will be considered diploids throughout this

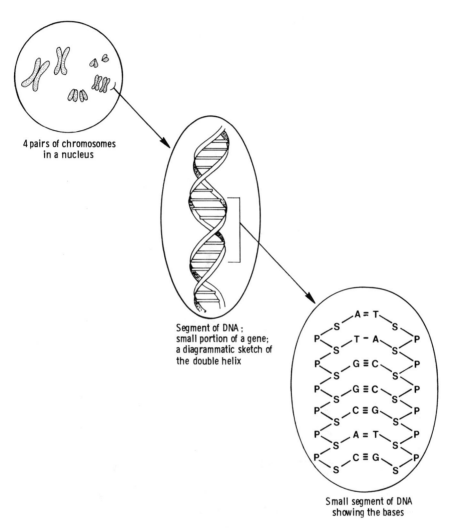

4 pairs of chromosomes
in a nucleus

Segment of DNA;
small portion of a gene;
a diagrammatic sketch of
the double helix

Small segment of DNA
showing the bases

Figure 2.1 Schematic drawing of chromosomes and deoxyribonucleic acid (DNA). Genes are located on chromosomes. A gene is a linear array of very specific subunits and is a small segment of a much larger molecule called DNA. A gene is composed of a series of paired bases [adenine (A) paired with thymine (T) and guanine (G) paired with cytosine (C)] arranged in a specific linear order and joined by a double backbone of deoxyribose sugar (S) and phosphorus (P). Each gene has a unique linear sequence of paired bases. The overall structure of DNA is called a double helix.

book. Good reviews about chromosome numbers in fish are found in Gyldenholm and Scheel (1971), Denton (1973), Gold *et al.* (1980), and Sola *et al.* (1981).

Because fish are diploids, not only do the chromosomes occur in pairs, but each gene in an individual also occurs in pairs (exceptions to

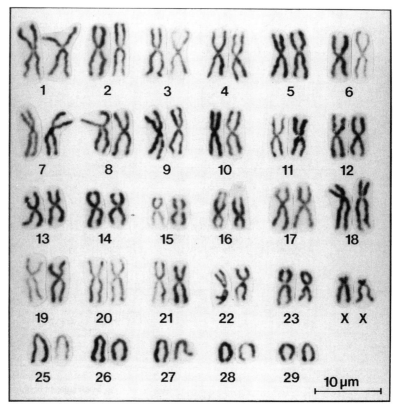

FEMALE

Figure 2.2　Karyotypes of male and female rainbow trout. Rainbow trout have 58 chromosomes (29 pairs). Twenty-eight pairs (56 chromosomes) are autosomes: pairs 1-23 and 25-29. The chromosomes in pair 24 are the sex chromosomes. The two chromosomes in pair 24 in female rainbow trout are morphologically identical, while those in the male are a mismatched pair. This pattern produces what is called the XY sex-determining system. Males are said to be XY and are heterogametic, while females are said to be XX and are homogametic.

Source: G. H. Thorgaard (1977). Heteromorphic sex chromosomes in male rainbow trout. Science 196, 900–902.

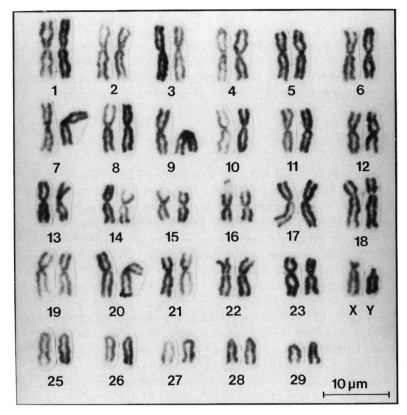

MALE

Figure 2.2 (*continued*)

this rule are some of the genes that are located on the sex chromosomes). Thus, while a population may contain two, three, four, etc. alleles per locus, each fish can have only one or two alleles at a given locus. Individuals that have only one type of allele at a given locus, i.e., the pair are identical, are homozygous at that locus. Individuals that have two different alleles at a given locus are heterozygous at that locus.

Fish culturists can now alter ploidy level by manipulating eggs, sperm, and zygotes and can create haploids (*N*), triploids, tetraploids, and fish whose chromosomes come either only from the mother (gynogens) or only from the father (androgens). Chromosomal manipulation is an active area of research and is the basis for some fish farming industries. This topic will be discussed in Chapter 5.

There are two basic types of chromosomes: autosomes and sex chromosomes. Sex chromosomes are those chromosomes that usually determine an individual's sex, and the pair(s) of sex chromosomes is often morphologically different in the two sexes (Fig. 2.2). Sex chromosomes are difficult to detect in fish. In a review of the karyotypes of 810 bony fishes, Sola *et al.* (1981) found that sex chromosomes could be identified in only 29 species (3.6%). Autosomes are the other pairs of chromosomes and are morphologically the same in males and females.

MEIOSIS

Although every cell that has a nucleus contains the full complement of an individual's chromosomes, those that exist in the primary gametocytes are the most important genetically, because the eggs and the sperm that will carry the genes and create the next generation are produced from the primary gametocytes. To produce the gametes, primary gametocytes undergo a reduction division called meiosis. Meiosis differs from mitosis (cell division of somatic or nonsex cells) in that the end result of meiosis is haploid (N) gametes that contain only one chromosome from each chromosome pair, whereas mitosis produces two daughter cells, each of which contains the normal 2N complement of paired chromosomes. Meiosis is schematically outlined in Fig. 2.3.

Many important processes occur during meiosis, but I shall concentrate on only three. The first important process occurs during the initial phase of meiosis. During this phase, each chromosome replicates itself, and the replicated homologues then pair. The bundles of four units are called tetrads. The chromosomes are elongated and drawn out rather than compact, and the chromosomes that comprise each tetrad twist around each other. Typically, two or more of the units that form a tetrad break, and pieces from different homologues rejoin. When this occurs, genetic material is transferred from one chromosome to another. This process is called crossing over. Crossing over is a very important aspect of population genetics because it reshuffles the genes and creates new combinations of genes in the gametes. Crossing over greatly increases genetic and phenotypic variance in a population.

The second important process is the reduction division which reduces the chromosome complement from the diploid state (2N) to the haploid state (N). During this division, all chromosome pairs separate, with one chromosome of the pair going to each of the two secondary spermatocytes or to the secondary oocyte and the first polar body. The reduction division does not separate the halves of the replicated chromosomes (remember, each chromosome is double); it separates the homologous chromosomes that form a pair, i.e., the replicated chromo-

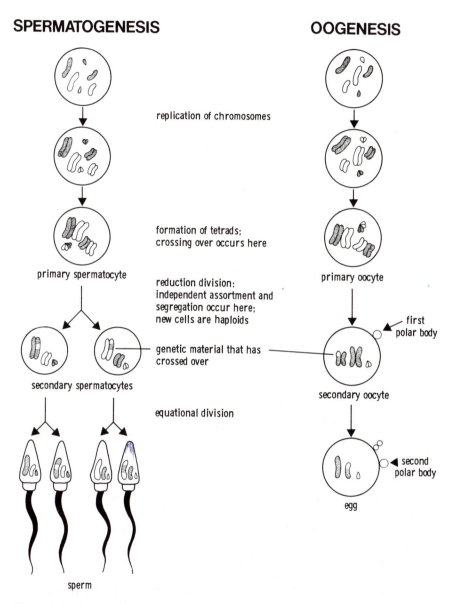

SPERMATOGENESIS

OOGENESIS

replication of chromosomes

formation of tetrads;
crossing over occurs here

primary spermatocyte

primary oocyte

reduction division;
independent assortment and
segregation occur here;
new cells are haploids

first
polar body

genetic material that has
crossed over

secondary spermatocytes

secondary oocyte

equational division

second
polar body

egg

sperm

Figure 2.3 Schematic and abridged diagram of meiosis: stippled are maternal chromosomes; white are paternal chromosomes.

somes from the father go to one cell while those from the mother go to the other. The directions in which the replicated maternal and paternal chromosomes of each pair go is totally random, and each pair is divided independently of all other pairs. This random division of the maternal and paternal chromosomes from each pair is very important because it greatly increases gametic and genotypic variability which, in turn, increase phenotypic variation.

Because the reduction division reduces the chromosome complement from the diploid to the haploid state, the secondary spermatocytes and the secondary oocyte are haploid. The reduction of the diploid complement of chromosomes to the haploid state is necessary so that diploids will be created during fertilization. If reduction to the haploid state did not occur, the number of chromosomes would double each generation, and it would not be long until fish were blobs of DNA.

The reduction division and the random separation of the chromosomes that form each pair are two of the most important aspects of genetics. These two events are named in honor of Mendel. Mendel's First Law is the Law of Segregation: A pair of alleles and the pair of chromosomes on which the alleles (the gene) exist are separated during meiosis. Mendel's Second Law is the Law of Independent Assortment: A pair of alleles and the pair of chromosomes on which the alleles (the gene) are located segregate independently from all other pairs, i.e., they move to the secondary spermatocytes or to the secondary oocyte and first polar body in a random manner (linked genes are an exception to this rule; this will be discussed in Chapter 3).

Segregation and independent assortment are two of the most important biological processes because of their contribution to genetic variance and ultimately to phenotypic variance. If these processes did not occur, a parent's genotype would be transmitted intact to each offspring and variation among individuals would be minimal. The only variability among individuals would be among families or that caused by mutation.

Segregation and independent assortment reshuffle an individual's genome, and this maximizes genotypic variability. If these two processes did not occur, an individual could produce only one kind of gamete—its genotype. Segregation and independent assortment increase tremendously the possible kinds of gametes that an individual can produce. The number of possible kinds of gametes that can be produced is determined by using the following formula:

$$\text{possible kinds of gametes} = 2^{(\text{no. of heterozygous genes})} \qquad (2.1)$$

A fish with 10 heterozygous genes would produce:

$$\text{possible kinds of gametes} = 2^{10}$$
$$\text{possible kinds of gametes} = 1024$$

If a fish has just one heterozygous gene per chromosome pair (most fish have far more than one per chromosome pair), the number of possible gametes that can be produced can become astronomical. For example, channel catfish have 29 pairs of chromosomes (LeGrande *et al.* 1984). A channel catfish that has only one heterozygous gene per chromosome pair can produce over half a billion possible kinds of gametes:

$$\text{possible kinds of gametes} = 2^{29}$$
$$\text{possible kinds of gametes} = 536,870,910$$

The final process that occurs during meiosis is the equational division. Here, the replicated halves of each chromosome separate and go either into one of the four sperm cells or to the one egg or the second polar body. The second polar body is not produced until the egg is fertilized. The gametes are haploid, because each contains only one chromosome from each pair. For example, a fish that has 20 chromosomes (10 pairs of chromosomes) produces gametes that have 10 chromosomes.

SEX DETERMINATION

Methods of sex determination are known for relatively few species of fish, but there are nine known systems in fish, and sex is controlled by sex chromosomes in eight of these systems. The sex chromosomes of some species are morphologically distinct and can be identified (Fig. 2.2). But those in other species are not morphologically distinct and are inferred from sex-reversal studies (Jalabert *et al.* 1974; Guerrero 1975; Hunter *et al.* 1982, 1983; K. B. Davis *et al.* 1990), hybridization experiments (F. Y. Chen 1969), chromosomal manipulation (Stanley 1976B; Mirza and Shelton 1988), or analysis of sex-linked phenotypes (Aida 1921; Winge 1922).

Understanding the way sex is determined is not trivial. Once this is known, it can be used to produce monosex populations through hormonal sex reversal and/or chromosomal manipulation. These topics will be covered in Chapter 5.

The most common system that has been detected in fish is the XY sex-determining system, which is also the system that exists in humans. Most of the important aquacultured species have this system of sex determination (Table 2.1).

Table 2.1   Examples of Nine Systems of Sex Determination

Species	System	Reference
Channel catfish	XY	K.B. Davis et al. (1990)
Rainbow trout	XY	Thorgaard (1977)
Lake trout	XY	Phillips and Ihssen (1985)
Coho salmon	XY	Hunter et al. (1982)
Chinook salmon	XY	Hunter et al. (1983)
Sockeye salmon	XY	Thorgaard (1978)
Tilapia nilotica	XY	Jalabert et al. (1974)
Tilapia mossambica	XY	F.Y. Chen (1969)
Common carp	XY	Nagy et al. (1981)
Silver carp	XY	Mirza and Shelton (1988)
Bighead carp	XY	W.L. Shelton (1990)
Grass carp	XY	Stanley (1976B)
Goldfish	XY	Yamamoto and Kajishima (1968)
Guppy	XY	Winge (1922)
Medaka	XY	Aida (1921)
Tilapia aurea	WZ	Guerrero (1975)
Tilapia hornorum	WZ	F.Y. Chen (1969)
Mosquitofish	WZ	T.R. Chen and Ebeling (1968)
Japanese eel	WZ	Park and Kang (1979)
Congor eel	WZ	Park and Kang (1979)
Platyfish	WXY	Gordon (1946)
Dollar hatchetfish	XO	T.R. Chen (1969)
Dwarf gourami	ZO	Rishi (1976)
Catarina pupfish[2]	$X_1X_1X_2X_2/X_1X_2Y$	Uyeno and Miller (1971)
Filefish	$X_1X_1X_2X_2/X_1X_2Y$	Murofushi et al. (1980)
Freshwater gobi	$X_1X_1X_2X_2/X_1X_2Y$	Pezold (1984)
Virolito	ZZ/ZW_1W_2	Filho et al. (1980)
Hoplias sp. from Aripuña River, Brazil	XY_1Y_2/XX	Bertolo et al. (1983)
Swordtail	autosomal	Kosswig (1964)
Blue poecilia	autosomal	Kosswig (1964)

[a] Species described by Miller and Walters (1972).

In the XY sex-determining system, the sex chromosomes in females are identical, while those in males are a mismatched pair. The sex chromosomes in females are named "X." Males also contain an X chromosome; the non-matching sex chromosome, found only in males, is named "Y." Consequently, females are said to be XX, while males are said to be XY. Because the pair is identical in females, they are called "homogametic" since they can produce only one kind of gamete—all eggs have an X chromosome. Because the pair is not identical in males, they are called "heterogametic" since they can produce two kinds of gametes—half the sperm carry an X chromosome and half carry a Y chromosome. Figure 2.2 shows the karyotype of a male (XY) and a female (XX) rainbow trout.

Regardless of sex-determining system, the heterogametic sex is the sex that usually determines sex of the offspring. In the XY sex-determining system, the Y chromosome is the chromosome that determines sex. In reality, it may be only one or two genes that determine sex. Two possible sex-determining genes have been found on the Y chromosome in humans, so it is likely that such genes exist in fish.

Because the Y chromosome is the chromosome that determines sex, the male is the parent that determines sex in species that have the XY sex-determining system: offspring that receive his Y chromosome become males, while those that receive his X chromosome become females:

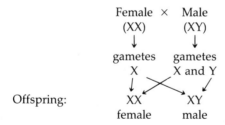

The second system is the WZ sex-determining system. This system is the mirror image of the XY sex-determining system. In this system, the sex chromosomes in the male form an identical pair, while those in the female form a mismatched pair. In this system, males are said to be "ZZ," while females are said to be "WZ." In the WZ sex-determining system, males are the homogametic sex, while females are the heterogametic sex.

In the WZ sex-determining system, the W chromosome is the chromosome that determines sex. Consequently, it is the female that determines the sex of her offspring for species that have the WZ sex-determining system: offspring that receive her W chromosome become females, while those that receive her Z chromosome become males:

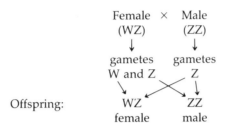

The third, fourth, and fifth sex-determining systems have multiple sex chromosomes, and all three are variations on either the XY or the WZ sex-determining systems. The third sex-determining system has multiple X chromosomes. In this system, females are $X_1X_1X_2X_2$ and males are X_1X_2Y. In this system of sex determination, females are the homogametic sex, and males are the heterogametic sex. Consequently, the male determines the sex of his offspring: those that receive his X_1X_2 chromosomes become females, while those that receive his Y chromosome become males.

The fourth system has multiple W chromosomes. In this sex-determining system, males are ZZ and females are ZW_1W_2. Males are the homogametic sex, and females are the heterogametic sex. As a result, the female determines the sex of her offspring: those that receive her W_1W_2 chromosomes become females, while those that receive her Z chromosome become males.

The fifth system has multiple Y chromosomes. In this system, males are XY_1Y_2 and females are XX. In this system, males are the heterogametic sex, and females are the homogametic sex. Thus, a male determines the sex of his offspring: those that receive his Y_1Y_2 chromosomes become males, while those that receive his X chromosome become females.

The sixth system is the WXY sex-determining system. This is a variant on the XY sex-determining system. The W chromosome is a modified X chromosome that can block the male-determining ability of the Y chromosome (D. Nakamura *et al.* 1984). Thus, XY and YY fish are males, while XX, WX, and WY fish are females. Both males and females can be either homogametic or heterogametic. Additionally, either parent can determine the sex of its offspring, depending on its sex chromosomes.

The seventh and eighth systems are those where only one sex chromosome exists: the XO and ZO systems (O is the symbol for no chromosome). The XO sex-determining system is a variant of the XY sex-determining system. In the XO sex-determining system, females are XX and males are XO. Females are the homogametic sex, while males are the heterogametic sex. Consequently, a male determines the sex of his offspring: those that receive his X chromosome become females, while those that receive no sex chromosome become males.

The ZO sex-determining system is a variant of the WZ sex-determining system. In the ZO sex-determining system, females are ZO and males are ZZ. Here, females are the heterogametic sex, while males are the homogametic sex. In the ZO sex-determining system, a female determines the sex of her offspring: those that receive her Z chromosome become males, while those that receive no sex chromosome become females.

When species have the $X_1X_1X_2X_2/X_1X_2Y$, ZW_1W_2/ZZ, XY_1Y_2/XX, XO, and ZO sex-determining systems, the number of chromosomes is not constant within a species. Males have one more chromosome than females in the XY_1Y_2/XX and ZO sex-determining systems, while females have one more chromosome in the $X_1X_1X_2X_2/X_1X_2Y$, ZW_1W_2/ZZ, and XO sex-determining systems.

Sex determination in fish that have sex chromosomes is not really as simple and as regimented as I have described. Although sex in these species is controlled by the sex chromosomes, an individual's sex can also be influenced or controlled by autosomal sex-influencing or sex-modifying genes (Kosswig 1964). The influence that these sex-influencing or sex-modifying genes have on sex determination may become important and can become a source of frustration in the production of monosex populations.

For example, the major goal of tilapia breeding projects is to produce monosex populations to prevent reproduction during grow-out. Hybridization is one technique that has been used to try and produce all-male populations (more about this in Chapter 4). The use of this technique in producing all-male populations is based on a simple system of sex determination, one controlled solely by sex chromosomes. Unfortunately, tilapia have autosomal sex-influencing or sex-modifying genes (Avtalion and Hammerman 1978; Hammerman and Avtalion 1979; Majumdar and McAndrew 1983; W. L. Shelton *et al.* 1983; Mair *et al.* 1991A, 1991B; Wohlfarth and Wedekind 1991), and these genes turn some fish that should be males into females. In addition, crossing over between the sex chromosomes can alter expected sex ratios in tilapia (Avtalion and Don 1990).

The final method of sex determination is not controlled by sex chromosomes, but is controlled autosomally. Some species of fish do not have sex chromosomes. In these species, sex is determined by the number of male or female genes that are located on the autosomes.

Examples of the nine different sex determining systems are shown in Table 2.1. Kosswig (1964), Yamamoto (1969B), and Kallman (1984) provide good reviews of sex-determination in fish.

Although sex determination is primarily under genetic control, environmental factors such as temperature, photoperiod, salinity, and crowding can help determine sex in fish. Mair *et al.* (1990) found that temperature could influence sex determination in tilapia, which adds another layer of complexity and confusion to the production of all-male populations of tilapia. Chan and Yeung (1983) provide a good review of this subject. The ability to control sex determination by manipulating environmental vectors has tremendous implications for fish culture. The use of hormones to produce monosex populations for grow-out has become a reliable technology, and it is being used at hatcheries around the world. This topic will be discussed in Chapter 5.

CHAPTER 3

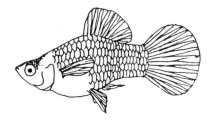

Genetics of
Qualitative
Phenotypes

3

Genetics of Qualitative Phenotypes

Although geneticists and breeders work with and try to manipulate and exploit a fish's genes, they can only be seen or measured indirectly through a fish's phenotypes. Therefore, it is the phenotypes which are analyzed and studied, and the variation that exists for each phenotype is of special concern. Consequently, in order to exploit the biological potential of a population, it is imperative that the phenotypic variance be analyzed and understood.

There are two basic types of phenotypic variation: qualitative variation and quantitative variation. Although both kinds ultimately depend on the same basic biological unit—the gene—each has to be analyzed and exploited by different approaches. Quantitative phenotypes are those that are measured, such as length and weight. The genetics of these phenotypes will be discussed in Chapter 4. This chapter will discuss the genetics and exploitation of qualitative phenotypes in fish.

Qualitative phenotypes are those that most people understand because they are the ones that are taught in basic biology and in basic genetics; they are the characteristics that most people associate with genetic variance. Qualitative phenotypes can also be called the either/or phenotypes: a fish either expresses one phenotype or it expresses another. Individuals fall into one of two or more discrete categories; there is no gradation joining the phenotypes into a continuum. Examples of such phenotypes are

albino vs normally pigmented channel catfish
golden vs palomino vs normally pigmented rainbow trout
blue vs normally pigmented common carp
veil tail vs round tail guppies
saddleback vs normal *Tilapia aurea*
spotted vs unspotted platyfish
line vs leather vs mirror vs scaled common carp
gold vs bronze vs black *Tilapia mossambica*
red-fleshed vs white-fleshed chinook salmon

Because individuals fall into discrete categories, the distribution of individuals in these categories forms certain ratios: e.g., $3:1$; $1:0$; $1:1$; $9:7$; $100:1$; $9:3:3:1$; etc. The specific ratio that is exhibited in a population depends on several factors which will be discussed later: number of genes needed to produce the phenotype; mode of gene action; and gene frequency.

The genetics of qualitative phenotypes is rather simple and is called Mendelian genetics or classical genetics. Most qualitative phenotypes are controlled by one, two, or three genes. Phenotypic expression depends on the number of genes and the mode of gene action at each locus. This discussion assumes that all genes have more than one allele per locus; if only one allele exists, no genetically heritable variation exists at that locus.

This chapter begins with a description of qualitative phenotypes, explains how they are inherited, and discusses the manner in which the genes express themselves phenotypically. Understanding the genetics that provides the biological blueprint for a particular phenotype makes it possible to determine which breeding program should be used to change allelic and phenotypic frequencies quickly and efficiently.

The middle section of this chapter describes various techniques that can be used to determine the frequencies of alleles in a population. One of the goals of broodstock management is to improve productivity. Genetically, that can be accomplished by using selection to change the frequencies of alleles, eliminating [frequency $(f) = 0\%$] those that depress productivity and profits, and fixing $(f = 100\%)$ those that improve productivity and profits. Consequently, it is important to be able to determine the frequencies of alleles during such programs in order to assess progress.

The final section of this chapter describes programs and techniques that can be used to eliminate undesirable qualitative phenotypes and to fix those that maximize productivity and profits, which is one of the ultimate goals of broodstock management. Different phenotypes have different economic values, and the value of a population depends on its ability to produce the desired phenotypes more often than the undesired phenotypes. The value of a population is maximized, for any phenotype, when that population will breed true and produce only the desired phenotype.

SINGLE AUTOSOMAL GENES

Genes can be located on either the autosomes or the sex chromosomes. When deciphering the genetics of a qualitative phenotype, it is important to determine whether the gene or genes that produce the phenotype are located on an autosome or a sex chromosome, because the inheritance of a phenotype controlled by an autosomal gene differs from that of one controlled by a sex-linked gene. Consequently, the management and control of these two kinds of phenotypes also differ.

In general, autosomal genes express themselves in either an additive or a nonadditive manner. In additive gene action, each allele produces an equal unidirectional phenotypic effect. In nonadditive gene action, one allele is expressed more strongly than the other and has a greater influence on the phenotype.

Complete Dominant Gene Action

Dominance occurs when one allele is expressed more strongly than the other. The allele that is expressed more strongly is called the dominant allele, while the other is called the recessive allele. The phenotype controlled by the dominant allele is called the dominant phenotype, while that controlled by the recessive allele is called the recessive phenotype.

When the mode of gene action is complete dominance, there are only two phenotypes, because the presence of the dominant allele masks the expression of the recessive allele in the heterozygous genotype. Thus, there are three genotypes but only two phenotypes. Examples of phenotypes controlled by autosomal genes with complete dominant gene action are shown in Table 3.1. This table is far from complete.

Albinism in channel catfish will be used to illustrate this type of

Table 3.1 Phenotypes Controlled by Single Autosomal Genes with Complete Dominant Gene Action

Species	Dominant allele	Recessive allele	Dominant phenotype	Recessive phenotype	Reference
Channel catfish	+	a	normal pigmentation	albino	Bondari (1984A)
Common carp	B	b	normal pigmentation	blue	Moav and Wohlfarth (1968); Wlodek (1968)
	G	g	normal pigmentation	gold	Moav and Wohlfarth (1968)
	Gr	gr	normal pigmentation	grey	Moav and Wohlfarth (1968)
	D	d	light yellow band on dorsal fin; yellow on head	normal pigmentation	Katasonov (1973)
Rainbow trout	A	a	normal pigmentation	albino	Bridges and von Limbach (1972)
	B	b	normal pigmentation	iridescent metallic blue	Kincaid (1975)
Tilapia nilotica	Bl	bl	normal pigmentation	blond	Scott et al. (1987)
	Er−	Er+	normal pigmentation	syrup	McAndrew et al. (1988)
	R	r	red	normal pigmentation	McAndrew et al. (1988)
	B	b	normal pigmentation	light-colored (pink)	Mires (1988)
	+	m	normal	caudal deformity syndrome	Mair (1992)
Goldfish	B	b	orange-red	blue	S.C. Chen (1934)
	D	d	normal eyes	telescope eyes	Matsui (1934)
	+	+ne	normal scales	nacreous-like scales	Yamamoto (1977)

	S_n	S_c			
Guppy			normal spine	lordosis (curvature of spine)	Rosenthal and Rosenthal (1950)
	B	b	grey	blond	Goodrich et al. (1944)
	G	g	grey	gold	Goodrich et al. (1944)
Medaka	+	+^{i}	orange-red	albino	Yamamoto (1969A)
	F	f	normal vertebrae	fused vertebrae	Aida (1930)
	W	w	normal spine	scoliosis (curvature of spine)	Aida (1930)
Platyfish	St	st	stippled	unstippled	Gordon (1927)
Montezuma swordtail	At	at	atromaculatus	unspotted	Kallman (1971)
Jewel tetra	Cam	cam	carbomaculatus	unspotted	Kallman (1971)
Convict cichlid	S	s	spotted	unspotted	Frankel (1982)
Swordtail (domestic stock)	P	p	grey	pink	Itzkovich et al. (1981)
	Mo	mo	Montezuma	unspotted	Gordon (1938)
Threespine stickleback	A	a	normal pigmentation	albino	Bakker et al. (1988)
Eye-spot rasbora	A	a	blue	silver	Frankel (1987)
Angelfish	Hg	hg	silver	Hong Kong gold	Norton (1982)
Guatopote del Mocorito (bisexual form)	+	tr	pigmented	transparent	Moore (1974)

gene action and to show how these phenotypes are inherited. Albinism in channel catfish is controlled by an autosomal recessive gene: the *a* gene (Bondari 1984A). The symbol + is used to designate the dominant allele that produces normal melanistic pigmentation (+ is often used to designate the allele that produces the common or wild-type phenotype), and the symbol *a* is used to designate the allele that produces albinism. Figure 3.1 shows these two phenotypes.

Because + is completely dominant over *a*, there are three possible genotypes at this locus, but these genotypes can produce only two phenotypes:

Genotype	Phenotype
++	normally pigmented
+*a*	normally pigmented
aa	albino

Thus, when a normally pigmented channel catfish with the ++ genotype is mated to an albino, the following progeny are produced:

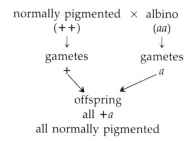

The recessive *a* allele is masked by the dominant + allele and is unable to express itself phenotypically, so only normally pigmented offspring are produced.

The only way an albino channel catfish can be produced is when a sperm that carries an *a* allele fertilizes an egg that also carries an *a* allele. Thus, both parents must either be albinos or carry the *a* allele in the heterozygous state. To illustrate this, the following occurs when two heterozygous (+*a*) normally pigmented channel catfish mate:

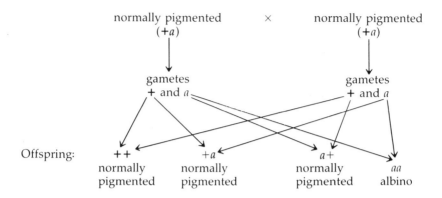

The preceding mating between two heterozygotes is the key mating that is used to unlock the mode of gene action. Because both parents are heterozygotes, they produce offspring which have all possible genotypes and all possible phenotypes. The phenotypic ratio of these progeny (the F_2 generation; the heterozygous parents are called the F_1 generation) differentiates the various types of gene action. F_2 phenotypic ratios that differentiate the major types of autosomal gene action are listed in Table 3.2. Throughout this chapter, I will use the mating of two heterozygotes and the F_2 phenotypic ratios that they produce to illustrate and to distinguish the various types of gene action.

Figure 3.1 Normally pigmented and albino channel catfish.
Source: Jack Turner

Genotypic and phenotypic ratios of the progeny can be easily determined by using a Punnett square. To set up a Punnett square, all possible gametes from the male are placed in columns along the top axis of a box and all possible gametes from the female are placed in rows along the left axis of a box. The combinations of the gametes from the two sexes form chambers or cells which correspond to the various possible zygotic combinations of the gametes (offspring genotypes). Once the genotypes are known, the corresponding phenotypes are also known.

The Punnett square for the mating of two heterozygous +a normally pigmented channel catfish is

normally pigmented ♀ × normally pigmented ♂
(+a)   (+a)

Male gametes

	+	a
+	++ normally pigmented	+a normally pigmented
a	a+ normally pigmented	aa albino

Female gametes

Phenotypic and genotypic ratios of the progeny are calculated by totaling the appropriate chambers:

Genotypic ratio: 1 ++ : 2 +a : 1 aa

Phenotypic ratio: 3 normally pigmented : 1 albino.

Punnett squares can be either complicated or simple, depending on the number of possible kinds of gametes that each sex can produce. For example, the mating of a heterozygous +a normally pigmented female channel catfish to an albino male produces this Punnett square:

normally pigmented ♀ × albino ♂
(+a)   (aa)

Male gametes

	a	a
+	+a normally pigmented	+a normally pigmented
a	aa albino	aa albino

Female gametes

Table 3.2 F₂ Phenotypic Ratios Produced by Mating Two Heterozygous F₁ Parents[a]

F₂ phenotypic ratio	Type of gene action
	Single autosomal gene
3:1	complete dominance
1:2:1	incomplete dominance; additive; codominance
	Two autosomal genes, each producing different phenotypes
9:3:3:1	two genes with complete dominance
3:6:3:1:2:1	two genes: one with complete dominance; the other with either additive, incomplete dominant, or codominant gene action
1:2:1:2:4:2:1:2:1	two genes: any combination of genes with additive, codominant, or incomplete dominant gene action
	Two autosomal genes producing the phenotypes through non-epistatic interaction
9:3:3:1	interaction of complementary genes
1:4:6:4:1	additive
	Two autosomal genes producing the phenotypes through epistatic interaction
12:3:1	dominant epistasis
9:3:4	recessive epistasis
9:6:1	duplicate genes with cumulative effects
15:1	duplicate dominant genes
9:7	duplicate recessive genes
13:3	dominant and recessive interaction

[a] All loci are autosomal genes with 2 alleles per locus. There are no lethal genotypes.

Because the male can produce only a gametes, the left and right columns are identical. Consequently, the right column can be omitted without altering the information or phenotypic and genotypic ratios:

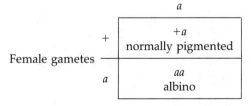

Male gametes

		a
Female gametes	$+$	$+a$ normally pigmented
	a	aa albino

In both cases

Genotypic ratio: 1 +a: 1 aa

Phenotypic ratio: 1 normally pigmented: 1 albino

Table 3.3 Possible Matings Among Normally Pigmented and Albino Channel Catfish, and the Genotypic and Phenotypic Ratios of the Progeny[a]

Mating	Progeny ratios	
	Genotypic ratio	Phenotypic ratio
normally pigmented (++) × normally pigmented (++)	all ++	all normally pigmented
normally pigmented (++) × normally pigmented (+a)	1 ++ : 1 +a	all normally pigmented
normally pigmented (++) × albino (aa)	all +a	all normally pigmented
normally pigmented (+a) × normally pigmented (+a)	1 ++ : 2 +a : 1 aa	3 normally pigmented : 1 albino
normally pigmented (+a) × albino (aa)	1 +a : 1 aa	1 normally pigmented : 1 albino
albino (aa) × albino (aa)	all aa	all albino

[a] Parental genotypes are given in parentheses after the phenotypes.

All possible matings among normally pigmented and albino channel catfish and the genotypic and phenotypic ratios of the progeny that are produced by the matings are shown in Table 3.3. The phenotypic and genotypic ratios of the progeny that are produced by the different matings in Table 3.3 are typical for qualitative phenotypes that are controlled by single autosomal genes with complete dominant gene action.

The information in Table 3.3 points out two important facts that are true for phenotypes controlled by single autosomal genes with complete dominant gene action: (1) Only homozygous fish breed true, i.e., they produce only one type of gamete. Heterozygotes do not breed true, because they produce two kinds of gametes, and produce them in equal numbers, i.e., in a 1 : 1 ratio. (2) Fish with the recessive phenotype breed true because they produce only one type of gamete. Fish with the dominant phenotype may not breed true, because it is impossible to distinguish homozygous from heterozygous dominant fish.

Thus, only albinos will breed true, because they produce only one type of gamete: *a*. Normally pigmented channel catfish, on the other hand, may produce both + and *a* gametes, because it is impossible to distinguish between homozygous and heterozygous normally pigmented channel catfish.

Incomplete Dominant Gene Action

Another form of dominance occurs when the dominant allele expresses itself more strongly than the recessive allele, but not strongly enough to make the heterozygous phenotype identical to the homozygous dominant phenotype. This type of dominance is called incomplete dominance. Genes that have incomplete dominant gene action produce three genotypes and three phenotypes, a separate phenotype for each genotype. Examples of phenotypes that are controlled by autosomal genes with incomplete dominant gene action are shown in Table 3.4.

The G gene in *Tilapia mossambica* will be used to illustrate this type of gene action and to show how these phenotypes are inherited. The G gene determines the number of melanophores and xanthophores, and those in turn affect body color (Tave *et al.* 1989A; Fig. 3.2). Because the dominant allele is not completely dominant, the heterozygous genotype produces a phenotype that approximates the dominant phenotype more than the recessive phenotype, but it is a unique phenotype. Consequently, there are three genotypes and three phenotypes. The geno-

Table 3.4 Phenotypes Controlled by Single Autosomal Genes with Incomplete Dominant Gene Action

Species	Dominant allele	Recessive allele	Dominant phenotype	Heterozygous phenotype	Recessive phenotype	Reference
Common carp	L	l	death	light colored saddleback (abnormal dorsal fin)	normal pigmentation	Katasonov (1973, 1976)
Tilapia aurea	S	$+$	death	normal	normal	Tave et al. (1983)
Tilapia mossambica	G	g	black (normal pigmentation)	bronze	gold	Tave et al. (1989A)
Goldfish	T'	T	transparent scales	calico	normal scales	S.C. Chen (1928)
Siamese fighting fish	V	v	steel blue	blue	green	Wallbrunn (1958)
Sailfin molly	M	m	melanistic	spotted	unspotted	Angus (1983)
Guppy	Pl	pl^+	death	palla (fused vertebrae)	normal vertebrae	Lodi (1978)
Angelfish	Lf	$+$	longfinned (full phenotype)	longfinned (almost full phenotype)	normal fins	Schröder (1976)
Sailfin molly	Sm	$sm+$	chocolate	smokey	silver	Norton (1982)
	G	g	normal pigmentation	increased no. xantho-phores; normally pigmented	gold	Angus (1991); Angus and Blanchard (1991)

types and phenotypes in *T. mossambica* are

Genotype	Phenotype
GG	black (normally pigmented)
Gg	bronze
gg	gold

Because each genotype produces a unique phenotype, a black, bronze, or gold *T. mossambica*'s genotype can be deciphered simply by an examination of the phenotype.

Phenotypic and genotypic ratios of the progeny from any mating among black, bronze, or gold *T. mossambica* can be generated with a Punnett square as described earlier. The Punnett square and phenotypic and genotypic ratios of the progeny for the mating of two bronze *T. mossambica* are

bronze × bronze
(*Gg*)   (*Gg*)

Male gametes

		G	g
	G	GG black	Gg bronze
Female gametes	g	gG bronze	gg gold

Genotypic ratio: 1 *GG* : 2 *Gg* : 1 *gg*

Phenotypic ratio: 1 black : 2 bronze : 1 gold

This example, like the previous one showing the mating of two heterozygous +*a* normally pigmented channel catfish, shows the 1 : 2 : 1 genotypic ratio that is generated when two heterozygotes mate. But because the *G* gene has incomplete dominant gene action, the phenotypic ratio is also 1 : 2 : 1 instead of the 3 : 1 ratio that was seen for phenotypes that are controlled by genes with complete dominant gene action. All possible matings among black, bronze, and gold *T. mossambica*, as well as the genotypic and phenotypic ratios of the progeny that are produced by the matings, are shown in Table 3.5. The genotypic and phenotypic ratios that are produced by the different matings in Table 3.5

Figure 3.2 Black (normally pigmented) (a), bronze (b), and gold (c) *Tilapia mossambica*.
Source: Tave et al. 1989A

Table 3.5 Possible Matings Among Black, Bronze, and Gold *Tilapia mossambica* and the Genotypic and Phenotypic Ratios of the Progeny[a]

| Mating | Progeny ratios | |
	Genotypic ratio	Phenotypic ratio
black (*GG*) × black (*GG*)	all *GG*	all black
black (*GG*) × bronze (*Gg*)	1 *GG* : 1 *Gg*	1 black : 1 bronze
black (*GG*) × gold (*gg*)	all *Gg*	all bronze
bronze (*Gg*) × bronze (*Gg*)	1 *GG* : 2 *Gg* : 1 *gg*	1 black : 2 bronze : 1 gold
bronze (*Gg*) × gold (*gg*)	1 *Gg* : 1 *gg*	1 bronze : 1 gold
gold (*gg*) × gold (*gg*)	all *gg*	all gold

[a] Parental genotypes are given in parentheses after the phenotypes.

are typical for qualitative phenotypes that are controlled by single autosomal genes with incomplete dominant gene action.

The information in Table 3.5 reconfirms the fact that only homozygous genotypes will breed true. However, when the mode of gene action is incomplete dominance, two phenotypes will breed true because the homozygous dominant phenotype and the heterozygous phenotype are distinguishable. Thus, both black and gold *T. mossambica* will breed true. The only phenotype which cannot breed true is bronze because it is heterozygous (*Gg*).

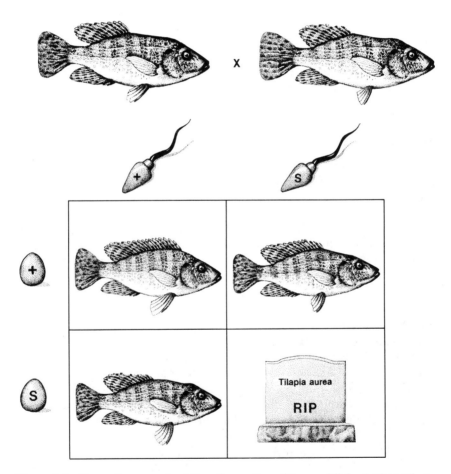

Figure 3.3 Punnett square depicting the mating of two saddleback (*S+*) *Tilapia aurea*. The *S* gene exhibits incomplete dominant gene action. Note that the *SS* genotype is lethal. Only two genotypes survive: homozygous recessive ++ (normal) and heterozygous *S+* (saddleback).

It is interesting to note that a number of the genes that exhibit incomplete dominance are dominant lethal genes, i.e., the homozygous dominant genotype causes death (Table 3.4). The *S* gene in *Tilapia aurea* is an example of a dominant lethal gene (Tave *et al.* 1983; Fig. 3.3). The genotypes and phenotypes (also see Fig. 3.9 on p. 73) for the *S* gene in *T. aurea* are

Genotype	Phenotype
SS	death
S+	saddleback (abnormal dorsal fin)
++	normal

The mating of two saddleback *T. aurea* produces the expected $1:2:1$ genotypic and phenotypic ratios, but because all homozygous dominant fish are aborted, what is seen are 2 *S+* : 1 *++* genotypic and 2 saddleback : 1 normal phenotypic ratios (Fig. 3.3).

Additive Gene Action

When neither allele is dominant, both contribute equally to the production of the phenotype in a unidirectional additive stepwise manner, and the heterozygous phenotype is intermediate between the two homozygous phenotypes.

The only way to distinguish autosomal genes that have additive gene action from those that have incomplete dominant gene action is to decide whether the heterozygous phenotype is intermediate between the two homozygous phenotypes or whether it approximates one homozygous phenotype more than the other. This is often difficult, because these phenotypes are qualitative and cannot be measured; consequently, the decision is judgmental. The exact classification is not crucial, because in both cases each genotype produces a unique phenotype. As a result, both types are managed and exploited using the same breeding techniques that will be shown later in Chapter 3.

The *G* gene in rainbow trout is an example of a gene with additive gene action (Wright 1972). The *G* gene produces the golden, palomino, and normally pigmented body color phenotypes:

Genotype	Phenotype
G'G'	golden
G'G	palomino
GG	normally pigmented

Table 3.6 Possible Matings among Normally Pigmented, Golden, and Palomino Rainbow Trout, and the Phenotypic and Genotypic Ratios of the Progeny[a]

Mating	Progeny ratios	
	Genotypic ratio	Phenotypic ratio
golden ($G'G'$) × golden ($G'G'$)	all $G'G'$	all golden
golden ($G'G'$) × palomino (GG')	1 $G'G'$:1 $G'G$	1 golden:1 palomino
golden ($G'G'$) × normally pigmented (GG)	all GG'	all palomino
palomino (GG') × palomino (GG')	1 $G'G'$:2 GG':1 GG	1 golden:2 palomino:1 normally pigmented
palomino (GG') × normally pigmented (GG)	1 GG':1 GG	1 palomino:1 normally pigmented
normally pigmented (GG) × normally pigmented (GG)	all GG	all normally pigmented

[a] Parental genotypes are given in parentheses following the phenotypes.

As with genes that exhibit incomplete dominance, genes that act in an additive manner produce three genotypes and three phenotypes, a unique phenotype for each genotype. Because each genotype has a distinct phenotype, an individual's phenotype tells its genotype.

Because each genotype produces a unique phenotype, the mating of two heterozygotes will produce progeny with $1:2:1$ genotypic and phenotypic ratios. The Punnett square and genotypic and phenotypic ratios of the progeny for the mating of two palomino rainbow trout are

palomino ♀ × palomino ♂
(GG')   (GG')

Male gametes

		G	G'
Female gametes	G	GG normally pigmented	GG' palomino
	G'	G'G palomino	G'G' golden

Genotypic ratio: 1 GG:2 GG':1 $G'G'$

Phenotypic ratio: 1 normally pigmented:2 palomino:1 golden

All possible matings among normally pigmented, golden, and palomino rainbow trout and the genotypic and phenotypic ratios of the progeny produced by the matings are shown in Table 3.6. The phenotypic and genotypic ratios of the progeny that are produced by the different matings in Table 3.6 are typical for qualitative phenotypes that are controlled by single autosomal genes with additive gene action.

An examination of Table 3.6 shows that genes with additive gene action produce two genotypes and two phenotypes that breed true—the two homozygotes. Only the heterozygotes cannot breed true.

DIHYBRID INHERITANCE

When two or more genes are inherited independently (i.e., they are not linked—linkage is discussed later in Chapter 3), and each gene controls a different phenotype, it is rather easy to work with the different phenotypes and to understand and to exploit the genetics either when the phenotypes are taken separately or when they are taken in combination. Because each gene is inherited independently, each phenotype is also inherited independently. As a result, the frequencies and probabilities for the simultaneous occurrence of specific combinations of phenotypes and genotypes is nothing more than the product of the frequencies and probabilities of the occurrences for each phenotype and genotype taken independently.

For example, gold body coloration in the guppy is controlled by an autosomal gene with complete dominant gene action: G (Goodrich *et al.* 1944). The dominant G allele produces grey guppies, and the recessive *g* allele produces gold guppies. Curvature of the spine is also controlled by an autosomal gene with complete dominant gene action: *Cu* (Rosenthal and Rosenthal 1950). The dominant *Cu* allele produces normal spines, while the recessive *cu* allele causes curvature of the spine. Gametes produced during meiosis will contain random combinations of the parents' alleles at both loci, since the two genes are inherited independently (Law of Independent Assortment). For example, the following gametes will be produced by a guppy that is heterozygous at both the G and *Cu* loci (a Punnett square can also be used to calculate gametic production):

Gamete production for a *Gg,Cucu* guppy

G locus

	G	*g*
Cu	G,*Cu*	*g*,*Cu*
cu	G,*cu*	*g*,*cu*

Cu locus

Gametes: 1 G,*Cu* : 1 *g*,*Cu* : 1 G,*cu* : 1 *g*,*cu*

The Punnett square and phenotypic and genotypic ratios of the progeny for the mating of two heterozygous grey guppies with normal spines (*Gg,Cucu*) are

grey and normal spined ♀ × grey and normal spined ♂
(*Gg,Cucu*)   (*Gg,Cucu*)

Male gametes

		G,Cu	G,cu	g,Cu	g,cu
Female gametes	G,Cu	GG,CuCu grey and normal spine	GG,Cucu grey and normal spine	Gg,CuCu grey and normal spine	Gg,Cucu grey and normal spine
	G,cu	GG,cuCu grey and normal spine	GG,cucu grey and curved spine	Gg,cuCu grey and normal spine	Gg,cucu grey and curved spine
	g,Cu	gG,CuCu grey and normal spine	gG,Cucu grey and normal spine	gg,CuCu gold and normal spine	gg,Cucu gold and normal spine
	g,cu	gG,cuCu grey and normal spine	gG,cucu grey and curved spine	gg,cuCu gold and normal spine	gg,cucu gold and curved spine

Genotypic ratio: 1 *GG,CuCu* : 2 *GG,Cucu* : 2 *Gg,CuCu* : 4 *Gg,Cucu* : 1 *GG,cucu* : 2 *Gg,cucu* : 1 *gg,CuCu* : 2 *gg,Cucu* : 1 *gg,cucu*

Phenotypic ratio: 9 grey and normal spine : 3 grey and curved spine : 3 gold and normal spine : 1 gold and curved spine

The 9 : 3 : 3 : 1 phenotypic ratio is always produced when two heterozygotes mate and the two phenotypes are each controlled by single autosomal genes with complete dominant gene action.

If each phenotype is considered separately, the mating of the two heterozygous guppies in the preceding example produces the classic 3 grey : 1 gold and 3 normal spine : 1 curved spine phenotypic ratios that were seen earlier for phenotypes that are controlled by single autosomal genes with complete dominant gene action. This can be verified by an examination of the Punnett square for the mating of two heterozygous *Gg,Cucu* guppies. If each phenotype is considered separately, a summation of the cells in the preceding Punnett square shows that 12 grey guppies will be produced for every 4 gold guppies and 12 normal spined guppies will be produced for every 4 with curved spines; 12:4 ratios are the same as 3:1 ratios.

Genotypic and phenotypic ratios for the simultaneous occurrence of many phenotypes are easy to calculate once the mode of gene action for each phenotype is known.

TWO OR MORE AUTOSOMAL GENES

Many phenotypes are controlled by a combination of two or more genes, and no single gene alone will produce these phenotypes. When two or more genes control a phenotype, one of two possible modes of gene interaction is in effect: non-epistatic or epistatic interaction.

Non-epistatic Interaction

Interaction of Complementary Genes The interaction of complementary genes occurs when two independent genes each produce two separate phenotypes, and the simultaneous occurrence of these traits produces "new" or different phenotypes that can be produced only by the simultaneous expression of the two traits. For example, body color in fish is often produced by the simultaneous expression or non-expression of different pigment cells, each of which is controlled by independent genes.

When this type of interaction occurs, the F_2 phenotypic ratio is the classic 9:3:3:1 ratio that is observed for the simultaneous expression of two independently segregating genes, each of which exhibits complete dominance and each of which controls a unique phenotype. However, in this case, the F_2 phenotypic categories are single phenotypes rather than the simultaneous expression of two phenotypes, as was illustrated in the previous section on dihybrid inheritance.

Gray, gold, ghost, and normally pigmented wild-type (olivaceous) body colors in the platyfish will be used to illustrate this type of inheritance. The St gene controls the production of micromelanophores, and the R gene controls the production of xanthophores (Kallman and Brunetti 1983). Both genes exhibit complete dominance: the dominant St allele produces micromelanophores, while the recessive st allele produces virtually no micromelanophores; the dominant R allele produces xanthophores, while the recessive r allele produces no xanthophores. The simultaneous presence or absence of these pigment cells produces normally pigmented (olivaceous), gray, gold, or ghost body colors:

Genotype	Phenotype
StSt,RR	normally pigmented
StSt,Rr	normally pigmented
StSt,rr	gray
Stst,RR	normally pigmented
Stst,Rr	normally pigmented
Stst,rr	gray
stst,RR	gold
stst,Rr	gold
stst,rr	ghost

The Punnett square and F_2 phenotypic ratio for the mating of two heterozygous (*Stst,Rr*) normally pigmented platyfish are

normally pigmented ♀ × normally pigmented ♂
(*Stst,Rr*)   (*Stst,Rr*)

Male gametes

		St,R	St,r	st,R	st,r
Female gametes	St,R	StSt,RR normally pigmented	StSt,Rr normally pigmented	Stst,RR normally pigmented	Stst,Rr normally pigmented
	St,r	StSt,rR normally pigmented	StSt,rr gray	Stst,rR normally pigmented	Stst,rr gray
	st,R	stst,RR normally pigmented	stSt,Rr normally pigmented	stst,RR gold	stst,Rr gold
	st,r	stSt,rR normally pigmented	stSt,rr gray	stst,rR gold	stst,rr ghost

Phenotypic ratio: 9 normally pigmented : 3 gray : 3 gold : 1 ghost

The F_2 genotypic ratio is $1:2:2:4:1:2:1:2:1$. The genotypic ratio is the same as that shown for the dihybrid mating in the previous section. F_2 genotypic ratios for two independently assorting genes always exhibit this ratio. F_2 phenotypic ratios are what differentiate different modes of inheritance.

Additive   Additive gene interaction with two or more loci is similar to that seen for single locus additive gene action. However, because more than one gene is involved, there are more possible phenotypes because there are more possible genotypes.

Melanistic body coloration in domesticated stocks of the molly is an example of a phenotype that is controlled by two genes with additive gene interaction (Schröder 1976). Body coloration ranges from a uniformly grey fish with light irises to a solid black fish with black irises. These phenotypes are controlled by the M and N genes. The phenotypes and the genotypes that control the phenotypes are shown in Table 3.7.

The melanistic phenotypes in the molly are determined by the number of color alleles. The only reason that a phenotypic distinction is made between the MM,nn and mm,NN and the Mm,Nn genotypes (two color alleles) is that body coloration is slightly different at birth. At sexual maturity the three genotypes produce phenotypes that are difficult to distinguish.

An examination of Table 3.7 shows that some of the molly's genotypes can be deciphered by an examination of the phenotypes, because some phenotypes are produced by only one genotype. Other phenotypes are controlled by one of two possible genotypes, so it is impossible to do more than narrow the genotype to one of two possibilities unless the fish are progeny tested (more about this later). Only two of the phenotypes will breed true: (1) color class IVb, the phenotype with 4 color alleles (MM,NN), and (2) color class I, the phenotype with no color alleles (mm,nn). All other phenotypes can produce either two or four different kinds of gametes and cannot breed true.

Variability in gametic production is determined by the number of heterozygous loci [Eq. (2.1)]:

$$\text{no. possible gametes} = 2^{(\text{no. heterozygous genes})}$$

For example, the MM,Nn and Mm,NN genotypes can produce $2^1 = 2$ different types of gametes:

Genotype	Gametes
MM,Nn	$M,n; M,N$
Mm,NN	$M,N; m,N$

while the Mm,Nn genotype can produce $2^2 = 4$ different types of gametes:

Genotype	Gametes
Mn,Nn	$M,N; M,n; m,N; m,n$

Table 3.7 Melanistic Coloration in Domesticated Stocks of the Molly

Genotype	Number of color alleles	Color class	Phenotype	
			Coloration at birth	Coloration at maturity
MM,NN	4	IVb	black; dark underside; dark iris	totally black; dark iris
MM,Nn; Mm,NN	3	IVa	black; lighter underside; light iris	totally black; dark iris
Mm,Nn	2	IIIb	slightly mottled; light iris	strongly mottled; light iris
MM,nn; mm,NN	2	IIIa	uniformly grey; unspotted; light iris	strongly mottled; light iris
Mm,nn; mm,Nn	1	II	uniformly grey; unspotted; light iris	slightly mottled; light iris
mm,nn	0	I	uniformly grey; unspotted; light iris	uniformly grey; unspotted; light iris

Source: After Schröder (1976).

Only one type of mating will produce all possible genotypes and phenotypes: color class IIIb ♀ × color class IIIb ♂. The Punnett square and genotypic and phenotypic (color class) ratios of the progeny for this mating are

IIIb ♀   ×   IIIb ♂
(*Mm,Nn*)   (*Mm,Nn*)

Male gametes

		M,N	*M,n*	*m,N*	*m,n*
	M,N	*MM,NN* IVb	*MM,Nn* IVa	*Mm,NN* IVa	*Mm,Nn* IIIb
	M,n	*MM,nN* IVa	*MM,nn* IIIa	*Mm,nN* IIIb	*Mm,nn* II
Female gametes	*m,N*	*mM,NN* IVa	*mM,Nn* IIIb	*mm,NN* IIIa	*mm,Nn* II
	m,n	*mM,nN* IIIb	*mM,nn* II	*mm,nN* II	*mm,nn* I

Phenotypic ratio: 1 IVb:4 IVa:4 IIIb:2 IIIa:4 II:1 I

This phenotypic ratio is a variation on the $1:4:6:4:1$ F_2 phenotypic ratio that is normally seen with this type of gene interaction. It is not the classic ratio, since phenotype III is split into IIIa and IIIb because they are slightly different at birth. By maturity, the two are difficult to distinguish, and the ratio becomes $1:4:6:4:1$.

Epistatic Interaction

Epistasis is a type of gene interaction where an allele at one locus modifies or suppresses the phenotypic expression of an allele at another locus. Epistatic interaction between two loci produces variations on the $9:3:3:1$ F_2 phenotypic ratio that occurs when there are two dominant genes which produce different phenotypes (Table 3.2). When there is epistasis, the number of F_2 phenotypes is reduced from four to either two or three, depending on the type of epistasis. Examples of phenotypes that are controlled by epistasis are listed in Table 3.8.

Table 3.8 Autosomal Phenotypes Controlled by Epistasis

Species	Genes	Phenotype	Reference
Common carp	S, N	scale pattern	Wohlfarth *et al.* (1963); Kirpichnikov (1981)
	P, R	pale red	Nagy *et al.* (1979)
Chinook salmon	A, B	flesh color	Withler (1986)
Goldfish	B_1, B_2	orange body color	Katasonov (1978)
	M, S	albino	Yamamoto (1973)
	Dp_1, Dp_2	depigmentation of melanophores	Kajishima (1977)
Guppy	Kal, Sup	veilfin	Schröder (1969)
Mexican cave characin	ab, bw	eye color	Sadoglu and McKee (1969)
Sumatran tiger barb	A, B	trunk stripe	Frankel (1985)
Half-banded barb	A, B	vertical banding; spotted	Frankel (1991)
Siamese fighting fish	C, B, V, Ri	body color	Wallbrunn (1958)

Body color in many tropical fish is controlled by epistatic interactions between or among two or more loci. Body and fin color in the Siamese fighting fish, examples of phenotypes that are controlled by epistatic interactions among four genes, are shown in Table 3.9.

Dominant Epistasis Dominant epistasis occurs when the dominant allele at one locus (the epistatic locus) produces a particular phenotype, regardless of the genotype at the second locus. In other words, the dominant allele at the epistatic locus prevents the second gene from producing either of its phenotypes. The second gene can express its phenotypes only when the epistatic locus is homozygous recessive. When that happens, the second gene can produce two additional phenotypes. Dominant epistasis produces a $12:3:1$ F_2 phenotypic ratio.

Albinism in goldfish is an example of a phenotype that is controlled by dominant epistasis. Albinism in goldfish is controlled by the M and the S genes (Yamamoto 1973). The M gene is the epistatic locus. A single dominant M allele produces dark goldfish, regardless of the genotype at the S locus.

When the M locus is homozygous recessive (mm), the S locus can produce either light (SS and Ss) or albino (ss) goldfish. Consequently, albinos can be produced only when a goldfish is homozygous recessive at both loci (mm,ss). The Punnett square for the mating of two heterozygous Mm,Ss dark goldfish and the F_2 phenotypic ratio are

dark ♀ × dark ♂
(Mm,Ss) (Mm,Ss)

Male gametes

		M,S	M,s	m,S	m,s
	M,S	MM,SS dark	MM,Ss dark	Mm,SS dark	Mm,Ss dark
	M,s	MM,sS dark	MM,ss dark	Mm,sS dark	Mm,ss dark
Female gametes	m,S	mM,SS dark	mM,Ss dark	mm,SS light	mm,Ss light
	m,s	mM,sS dark	mM,ss dark	mm,sS light	mm,ss albino

Phenotypic ratio: 12 dark : 3 light : 1 albino

Table 3.9 Body and Fin Color in the Siamese Fighting Fish Produced by Epistatic Interactions Among the *V*, *C*, *B*, and *Ri* Genes. The *riri*, *bb*, and *cc* Genotypes Alter the Phenotypes Produced by the *VV*, *Vv*, and *vv* Genotypes.

Genotype	*riri,VV* *riri,Vv* *riri,vv*	*RiRi,vv* *Riri,vv*	*RiRi,Vv* *Riri,Vv*	*RiRi,VV* *Riri,VV*
CC,BB *Cc,BB* *CC,Bb* *Cc,Bb*	Body: Red brown to black Fins: Dark red to black	Body: Green (over red brown to black) Fins: Green (and often red)	Body: Blue (over red brown to black) Fins: Blue (and often red)	Body: Steel blue (over red brown to black) Fins: Steel blue and often red
CC,bb *Cc,bb*	Body: Pale amber to bright red Fins: Red	Body: Red with overlying green Fins: Red and green	Body: Blue over red = purple Fins: Purple	Body: Steel blue over red = mauve Fins: Mauve
cc,BB *cc,Bb* *cc,bb*	Body: Light pink to red Fins: Colorless to red	Body: Light pink to red with overlying green Fins: Red and green to complete green	Body: Light pink to red with overlying blue Fins: Red and blue to complete blue	Body: Light pink to red with overlying silver Fins: Red and silver to complete silver

Source: After Wallbrunn (1958).

Albinism in goldfish illustrates an important concept: Do not assume that you know the genetics of a phenotype, even if your assumption works with similar phenotypes in other species. Albinism is one of the most universal abnormalities and has been described in dozens of species. When it has been studied, it is usually controlled by a single autosomal recessive allele, as is the case with channel catfish. But in goldfish, it is controlled by a specific combination of four alleles.

Assumptions that are made about a mode of inheritance, without statistically verifiable data gathered from a properly designed experiment, are of value only as working hypotheses. Similar phenotypes can be produced in different ways in different populations or different species. For example, Kallman (1970) found that a certain pigment pattern in the platyfish is controlled differently in two populations. The production of this pigment pattern by two different types of gene action means that the phenotype evolved independently and along different lines in the two populations.

In food fish culture, scale pattern in common carp (Fig. 3.4) is probably the most important phenotype controlled by epistasis. An understanding of the inheritance of scale pattern in common carp is of great importance because common carp with a reduced scale pattern can command a higher market price in Europe, whereas the common wild-type scale pattern (scaled) is often more desired in Asia. Since common carp are one of the world's most important cultured food fish, the ability to produce the desired scale phenotype can be of tremendous economic importance.

Scale pattern in common carp is controlled by the S and the N genes (Wohlfarth et al. 1963; Kirpichnikov 1981). They produce the phenotypes through a type of dominant epistasis where the N gene is the epistatic locus, but it is a dominant lethal epistatic gene (it exhibits incomplete dominance and is lethal in the homozygous state).

The S gene controls the scaliness, and the N gene modifies the pattern. The S allele is completely dominant over the s allele. The dominant phenotype controlled by the S allele is the common or wild-type scale pattern (scaled); the recessive phenotype controlled by the s allele is a reduced number of scales, and those that remain are greatly enlarged (mirror). A single N allele changes scaled common carp into line common carp (scales limited to the dorsal and ventral margins and the lateral line) and changes mirror common carp into leather common carp (no or virtually no scales). The N allele is lethal in the homozygous dominant state. The n allele has no effect on scale pattern (see frontispiece). All possible genotypes and phenotypes for scale pattern in com-

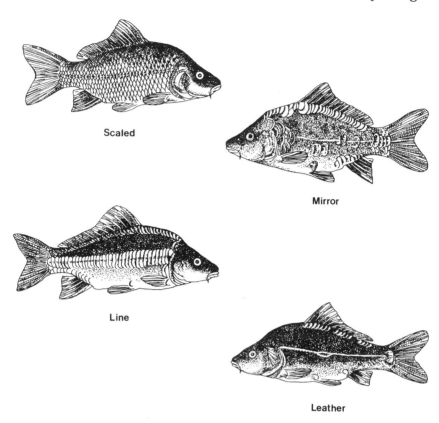

Figure 3.4 Scaled, mirror, line, and leather common carp.
Source: After Kirpichnikov (1981). Courtesy of V. S. Kirpichnikov and Nauka Publishers.

mon carp are

Genotype	Phenotype
SS,nn	scaled
Ss,nn	scaled
ss,nn	mirror
SS,Nn	line
Ss,Nn	line
ss,Nn	leather
SS,NN	death
Ss,NN	death
ss,NN	death

The Punnett square and the F_2 phenotypic ratio for the mating of two heterozygous (*Ss,Nn*) line common carp are

line ♀ × line ♂
(*Ss,Nn*)   (*Ss,Nn*)

Male gametes

		S,N	*S,n*	*s,N*	*s,n*
	S,N	SS,NN death	SS,Nn line	Ss,NN death	Ss,Nn line
	S,n	SS,nN line	SS,nn scaled	Ss,nN line	Ss,nn scaled
Female gametes	*s,N*	sS,NN death	sS,Nn line	ss,NN death	ss,Nn leather
	s,n	sS,nN line	sS,nn scaled	ss,nN leather	ss,nn mirror

Phenotypic ratio: 4 death : 6 line : 2 leather : 3 scaled : 1 mirror

Because the epistatic locus is a dominant lethal instead of a simple dominant epistatic gene, the F_2 phenotypic ratio deviates from the classic 12 : 3 : 1 F_2 phenotypic ratio that is usually seen for this type of mating.

An examination of the genotypes and phenotypes reveals that only mirror common carp will breed true. Line common carp and leather common carp cannot breed true because the *N* gene must be in the heterozygous state to produce both phenotypes. Scaled common carp may not breed true because this phenotype can be produced by a genotype with one gene in the heterozygous condition (*Ss,nn*). Because populations of scaled common carp can carry the *s* allele, they can produce both scaled common carp and mirror common carp unless broodstock have been certified to be free of the *s* allele by progeny testing (more about this later).

On the other hand, it is easy to establish a true-breeding population of mirror common carp because their genotype is the homozygous recessive genotype (*ss,nn*). Mirror common carp can produce only mirror common carp. The introduction of an *S* or *N* allele by mutation or accidental stocking can be remedied instantly because these alleles will be expressed phenotypically as scaled common carp (*Ss,nn*), leather common carp (*ss,Nn*), or line common carp (*Ss,Nn*). Thus, a single culling (removal) of the unwanted phenotypes before the common carp are allowed to reproduce will remove either or both dominant alleles, and the mirror common carp population will again breed true.

Because leather common carp have no or very few scales and are, therefore, easier to clean, they command a higher market price in certain areas. As a result, attempts have been made to produce populations of leather common carp for commercial production. But that is impossible, because leather common carp are heterozygous at the *N* locus. The *N* allele, which is needed in the heterozygous state to produce leather common carp, is lethal in the homozygous state. Thus, no parent can produce 100% *N* gametes. The maximum percentage of leather common carp that can be produced is 50%. Leather common carp are sometimes mated in the mistaken assumption that they will produce a population of leather common carp, but they cannot breed true. The mating of leather common carp will produce progeny with three phenotypes. The Punnett square and genotypic and phenotypic ratios of the progeny for the mating of two leather common carp are

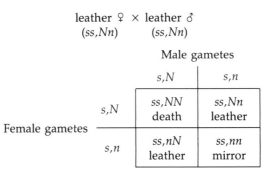

leather ♀ × leather ♂
(*ss,Nn*)   (*ss,Nn*)

Male gametes

		s,N	*s,n*
Female gametes	*s,N*	*ss,NN* death	*ss,Nn* leather
	s,n	*ss,nN* leather	*ss,nn* mirror

Genotypic ratio: 1 *ss,NN*:2 *ss,Nn*:1 *ss,nn*

Phenotypic ratio: 1 death:2 leather:1 mirror

Only one half the total progeny or two-thirds of the living progeny from this mating will be leather common carp.

From a geneticist's standpoint, the most desirable phenotype is the mirror common carp, because it is homozygous recessive at both loci and will breed true:

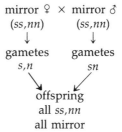

mirror ♀ × mirror ♂
(*ss,nn*)   (*ss,nn*)
↓   ↓
gametes   gametes
s,n   *sn*

offspring
all *ss,nn*
all mirror

In certain locales, line or leather common carp may have a greater market value, but the production of those phenotypes is, genetically, only 50% as efficient as the production of mirror common carp. Thus, it costs more to produce line and leather common carp.

Recessive Epistasis Recessive epistasis occurs when the recessive genotype at one locus (the epistatic locus) suppresses phenotypic expression by a second gene. The genotypes produced by the second locus can only be expressed when there is a dominant allele at the epistatic locus. Recessive epistasis produces a $9:3:4$ F_2 phenotypic ratio.

Eye color in Mexican cave characins is an example of a phenotype that is controlled by recessive epistasis. Black, brown, and pink eye color are controlled by the *ab* and *bw* genes (Sadoglu and McKee 1969). The *ab* locus is the epistatic locus in that the *abab* genotype produces pink eyes, regardless of the *bw* genotype. A single dominant *ab* allele (+) allows the *bw* gene to produce either black (+ + or +*bw*) or brown eyes (*bwbw*). The Punnett square and F_2 phenotypic ratio for the mating of two heterozygous *ab+,bw+* black-eyed Mexican cave characins are

black eyes ♀ × black eyes ♂
(*ab+,bw+*)   (*ab+,bw+*)

Male gametes

		+,+	+,*bw*	*ab*,+	*ab,bw*
	+,+	++,++ black eyes	++,+*bw* black eyes	+*ab*,++ black eyes	+*ab*,+*bw* black eyes
Female gametes	+,*bw*	++,*bw*+ black eyes	++,*bwbw* brown eyes	+*ab,bw*+ black eyes	+*ab,bwbw* brown eyes
	ab,+	*ab*+,++ black eyes	*ab*+,+*bw* black eyes	*abab*,++ pink eyes	*abab*,+*bw* pink eyes
	ab,bw	*ab*+,*bw*+ black eyes	*ab*+,*bwbw* brown eyes	*abab,bw*+ pink eyes	*abab,bwbw* pink eyes

Phenotypic ratio: 9 black eyes : 3 brown eyes : 4 pink eyes

Duplicate Genes with Cumulative Effects Duplicate genes with cumulative effects occur when two genes produce the same phenotype when either, but not both, has a genotype with a dominant allele, i.e., either locus, but not both, is either heterozygous or homozygous dominant. When there are dominant alleles at both loci, a second phenotype is produced, one that is a "cumulative" phenotype. A third phenotype is produced when both genes are homozygous recessive. Duplicate genes with cumulative effects produce a $9:6:1$ F_2 phenotypic ratio.

Trunk striping in the Sumatran tiger barb (Fig. 3.5) is an example of a phenotype that is controlled by duplicate genes with cumulative effects (Frankel 1985). Trunk striping is controlled by the A and B genes. The recessive genotype (*aa,bb*) produces the half-banded phenotype; when either, but not both, the A or B locus has a dominant allele (*Aa,bb; AA,bb; aa,Bb; aa,BB*), the incompletely banded phenotype is produced; when a dominant allele exists at both loci (*Aa,Bb; AA,Bb; Aa,BB; AA,BB*) the completely banded phenotype is produced. The Punnett square and F_2 phenotypic ratio for the mating of two heterozygous *Aa,Bb* completely banded Sumatran tiger barbs are

completely banded ♀ × completely banded ♂
(*Aa,Bb*)   (*Aa,Bb*)

Male gametes

		A,B	*A,b*	*a,B*	*a,b*
	A,B	*AA,BB* complete band	*AA,Bb* complete band	*Aa,BB* complete band	*Aa,Bb* complete band
	A,b	*AA,bB* complete band	*AA,bb* incomplete band	*Aa,bB* complete band	*Aa,bb* incomplete band
Female gametes	*a,B*	*aA,BB* complete band	*aA,Bb* complete band	*aa,BB* incomplete band	*aa,Bb* incomplete band
	a,b	*aA,bB* complete band	*aA,bb* incomplete band	*aa,bB* incomplete band	*aa,bb* half band

Phenotypic ratio: 9 completely banded : 6 incompletely banded : 1 half banded

Duplicate Dominant Gene Interaction   In duplicate dominant gene interaction, the dominant alleles at the two loci produce the same phenotype, but there is no cumulative effect. The only genotype that can produce a different phenotype is the homozygous recessive genotype. Duplicate dominant gene interaction produces a $15:1$ F_2 phenotypic ratio.

Transparent scales in the goldfish is an example of a phenotype that

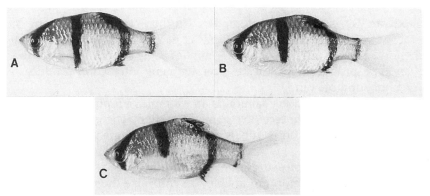

Figure 3.5 Trunk striping in the Sumatran tiger barb: (A) completely banded; (B) incompletely banded; (C) half-banded.
Source: Frankel (1985). Copyright 1985 by the American Genetic Association

is controlled by duplicate dominant gene interaction. Depigmentation of the melanophores in the scales is controlled by the Dp_1 and Dp_2 genes (Kajishima 1977). Normal pigmentation in the scales is produced only when both loci are homozygous recessive. All other genotypes produce transparent (unpigmented) scales. The Punnett square and F_2 phenotypic ratio for the mating of two heterozygous (Dp_1dp_1 , Dp_2dp_2) transparent-scaled goldfish are

transparent-scaled ♀ × transparent-scaled ♂
(Dp_1dp_1 , Dp_2dp_2)   (Dp_1dp_1 , Dp_2dp_2)

Male gametes

		Dp_1 , Dp_2	Dp_1 , dp_2	dp_1 , Dp_2	dp_1 , dp_2
	Dp_1 , Dp_2	Dp_1Dp_1 , Dp_2Dp_2 transparent scales	Dp_1Dp_1 , Dp_2dp_2 transparent scales	Dp_1dp_1 , Dp_2Dp_2 transparent scales	Dp_1dp_1 , Dp_2dp_2 transparent scales
	Dp_1 , dp_2	Dp_1Dp_1 , dp_2Dp_2 transparent scales	Dp_1Dp_1 , dp_2dp_2 transparent scales	Dp_1dp_1 , dp_2Dp_2 transparent scales	Dp_1dp_1 , dp_2dp_2 transparent scales
Female gametes	dp_1 , Dp_2	dp_1Dp_1 , Dp_2Dp_2 transparent scales	dp_1Dp_1 , Dp_2dp_2 transparent scales	dp_1dp_1 , Dp_2Dp_2 transparent scales	dp_1dp_1 , Dp_2dp_2 transparent scales
	dp_1 , dp_2	dp_1Dp_1 , dp_2Dp_2 transparent scales	dp_1Dp_1 , dp_2dp_2 transparent scales	dp_1dp_1 , dp_2Dp_2 transparent scales	dp_1dp_1 , dp_2dp_2 pigmented scales

Phenotypic ratio: 15 transparent scales : 1 pigmented scales

Duplicate Recessive Gene Interaction   In duplicate recessive gene interaction, the two recessive genotypes at each locus produce the same phenotype. Only when there is a dominant allele at both loci is another phenotype produced. Duplicate recessive gene interaction produces a $9:7$ F_2 phenotypic ratio.

Flesh color in chinook salmon is an example of a phenotype that is controlled by duplicate recessive gene interaction. The genetics of flesh color in chinook salmon is apparently different from that in any other species of salmon.

One of the great pleasures of eating salmon is its visual appeal; the flesh is beautiful. Salmon are unable to synthesize the carotenoid pigments which give their flesh the desired color. These pigments must be acquired in the diet. Salmon farmers realize this and also recognize the fact that visual appearance is more important than taste when it comes to consumer appeal, so they add carotenoid pigments to salmon feed in order to produce farm-raised salmon with the desired flesh color.

Unfortunately, some chinook salmon are unable to absorb the carotenoid pigments, so their flesh is white. While this has no effect on flavor, it creates a marketing problem. Consequently, the ability to produce a true-breeding red-fleshed chinook salmon is of great economic importance for the Canadian salmon farming industry.

Flesh color in chinook salmon is controlled by the A and B genes through duplicate recessive gene interaction (Withler 1986). When either the A or B gene or both are homozygous recessive, chinook salmon have white flesh. Red flesh is produced only when a chinook salmon has at least one dominant A and one dominant B allele. All possible genotypes and phenotypes for flesh color in chinook salmon are

Genotype	Phenotype
AA,BB	red flesh
AA,Bb	red flesh
AA,bb	white flesh
Aa,BB	red flesh
Aa,Bb	red flesh
Aa,bb	white flesh
aa,BB	white flesh
aa,Bb	white flesh
aa,bb	white flesh

The Punnett square and the F_2 phenotypic ratio for the mating of two heterozygous Aa,Bb red-fleshed chinook salmon are

red flesh ♀ × red flesh ♂
(*Aa,Bb*)   (*Aa,Bb*)

Male gametes

		A,B	*A,b*	*a,B*	*a,b*
Female gametes	*A,B*	*AA,BB* red flesh	*AA,Bb* red flesh	*Aa,BB* red flesh	*Aa,Bb* red flesh
	A,b	*AA,bB* red flesh	*AA,bb* white flesh	*Aa,bB* red flesh	*Aa,bb* white flesh
	a,B	*aA,BB* red flesh	*aA,Bb* red flesh	*aa,BB* white flesh	*aa,Bb* white flesh
	a,b	*aA,bB* red flesh	*aA,bb* white flesh	*aa,bB* white flesh	*aa,bb* white flesh

Phenotypic ratio: 9 red flesh : 7 white flesh

Only one genotype will breed true and produce 100% red-fleshed chinook salmon—*AA,BB*. Unfortunately, it is impossible to visually separate *AA,BB* red-fleshed chinook salmon from the other three red-fleshed genotypes (*AA,Bb; Aa,BB; Aa,Bb*). The only way to produce a true-breeding population of red-fleshed chinook salmon (provided the population is producing both flesh colors) is to progeny test the red-fleshed fish. Unfortunately, the ability to identify *AA,BB* red-fleshed chinook salmon by progeny testing depends on the ability to identify *aa,bb* white-fleshed chinook salmon, which is also impossible.

Dominant and Recessive Interaction   In dominant and recessive interaction, the dominant genotype at one locus (homozygous dominant and heterozygous) and the recessive genotype at a second locus produce the same phenotype. A second phenotype is produced when the first locus is homozygous recessive and the other has at least one dominant allele. Dominant and recessive interaction produces a 13 : 3 F_2 phenotypic ratio.

SEX-LINKED GENES

So far, the discussion about qualitative phenotypes has centered around those that are controlled by genes that are located on autosomal chromosomes. Qualitative phenotypes may also be controlled by genes located on one of the sex chromosomes. When this occurs, phenotypes are controlled by sex-linked genes. The inheritance of sex-linked phenotypes is different from that seen for autosomal phenotypes because, when a species has sex chromosomes, one sex is homogametic while the other sex is heterogametic. Sex-linked phenotypes have been discovered in relatively few species of fish, and most of the information about sex-linked phenotypes in fish comes from only two species: the guppy (Winge 1927; Dzwillo 1959; Yamamoto 1975A; Fernando and Phang 1989; Phang et al. 1989; Fujio et al. 1990) and the platyfish (Bellamy and Queal 1951; Kallman 1975). To date, all sex-linked phenotypes that are known in fish are controlled by genes that are located on the X and/or Y chromosome. No genes restricted to the W or Z chromosomes have been discovered.

Y-Linked Genes

Genes located on the Y chromosome are transmitted from father to son, and unless they cross over to the X chromosome, they will never exist in normal females (XX). Thus, Y-linked phenotypes are seen in only one sex—males. Examples of phenotypes that are controlled by sex-linked genes located on the Y chromosome in the guppy are listed in Table 3.10.

The maculatus gene in the guppy, which controls the maculatus pigment pattern (black spot on the dorsal fin and a red spot on the body), will be used to illustrate the inheritance of a Y-linked phenotype (Winge 1927). The symbol for the maculatus gene is Y_{Ma}. The symbol for the allele which produces the wild-type or unspotted phenotype is Y:

Genotype	Phenotype
XX	grey female
XY_{Ma}	maculatus male
XY	grey male

Only two types of matings are possible: grey ♀ × grey ♂ and grey ♀ × maculatus ♂. Only one mating will produce any maculatus progeny: grey ♀ × maculatus ♂. The Punnett square and phenotypic and genotypic ratios of the progeny for this mating are

grey ♀ × maculatus ♂
(XX)   (XY_{Ma})

Male gametes

		X	Y_{Ma}
Female gametes	X	XX grey female	XY_{Ma} maculatus male

Genotypic ratio: 1 XX : 1 XY_{Ma}

Phenotypic ratio: 1 grey female : 1 maculatus male

Table 3.10   Phenotypes Controlled by Sex-linked Genes in the Guppy[a]

Gene	Phenotype	Reference
Y-linked genes		
Y_{Ma}	maculatus pigmentation	Winge (1927)
Y_{Ir}	iridescens pigmentation	Winge (1927)
Y_{Ar}	armatus pigmentation	Winge (1927)
Y_{Sa}	sanguineus pigmentation	Winge (1927)
Y_{Pa}	pauper pigmentation	Winge (1927)
Y_{Oc}	oculatus pigmentation	Winge (1927)
Y_{Fe}	ferrugineus pigmentation	Winge (1927)
Y_{Va}	variabilis pigmentation	Winge (1927)
Y_{Ds}	double sword tail	Dzwillo (1959)
Y_{Fil}	filigran pigmentation	Dzwillo (1959)
Y_{Ssb}	snakeskin body	Phang et al. (1989)
Y_{Sst}	snakeskin tail	Phang et al. (1989)
X-linked genes		
X_{Ti}	tigrinus pigmentation	Winge (1927)
X_{Co}	coccineus pigmentation	Winge (1927)
X_{Vi}	vitellinus pigmentation	Winge (1927)
X_{Ci}	cinnamoneus pigmentation	Winge (1927)
X_{Lu}	luteus pigmentation	Winge (1927)
X_{El}	elongatus pigmentation; lengthening of the caudal fin	Winge (1927)
X_{NiII}	nigrocaudatus pigmentation, type II	Dzwillo (1959)
X_{Cp}	caudalis pigmentation	Dzwillo (1959)
X_{Grt}	golden yellow tail	Phang et al. (1989)
X_{Blt}	blue tail	Fernando and Phang (1989)
X_{Rdt}	red tail	Fernando and Phang (1989)
X_R	low-temperature resistant	Fujio et al. (1990)

[a] All phenotypes are sex-limited to the males except nigrocaudatus, caudalis, and temperature resistance. All genes exhibit dominant gene action. The dominant alleles and dominant phenotypes are listed in the table. The recessive body colors are the absence of the dominant phenotypes. The recessive X_r allele produces low-temperature-sensitive fish.

Thus, other than by mutation the only way to produce maculatus males is by spawning maculatus males.

This example illustrates an important point. If phenotypic ratio, without regard to sex, were taken in this mating, it would be 1:1 and would be totally indistinguishable from the 1:1 phenotypic ratio that is produced when a recessive fish is mated to a heterozygote in situations where the phenotype is controlled by a single autosomal gene with complete dominant gene action. Therefore, phenotypic ratios within both sexes are important bits of information that should be gathered when deciphering the inheritance of a phenotype. These data help differentiate between autosomal and sex-linked genes.

X-Linked Genes

Sex-linked genes can also be located on the X chromosome. Examples of phenotypes that are controlled by genes that are located on the X chromosome in the guppy are listed in Table 3.10. The mode of gene action for most X-linked genes is simple dominance.

Caudalis pigmentation (a darkly pigmented tail) and transparent tail in the guppy will be used to illustrate this type of gene action and to show how these phenotypes are inherited. These phenotypes are produced by the dominant X_{Cp} allele and the recessive X_{ch} allele (Dzwillo 1959). A single X_{Cp} allele produces caudalis pigmentation in either sex. The X_{ch} allele must be present in the homozygous state to produce the transparent phenotype in females ($X_{ch}X_{ch}$), but a single X_{ch} allele will produce the phenotype in males ($X_{ch}Y$):

Genotype	Phenotype
$X_{Cp}X_{Cp}$	caudalis female
$X_{Cp}X_{ch}$	caudalis female
$X_{ch}X_{ch}$	transparent-tailed female
$X_{Cp}Y$	caudalis male
$X_{ch}Y$	transparent-tailed male

The inheritance of X-linked phenotypes follows what is called a criss-cross pattern. The father determines his daughters' phenotypes, while the mother determines her sons' phenotypes. A father with the dominant phenotype can produce daughters with only the dominant phenotype. A father with the recessive phenotype can produce only recessive X gametes, so his daughters' phenotypes depend on his mate's genotype. On the other hand, if the mother has the recessive phenotype, all her sons have the recessive phenotype; if she has the dominant phenotype, either one half of her sons or all of her sons will have the dominant phenotype, depending on whether she is homozygous or heterozygous.

The only way transparent-tailed males are produced is when the female has the recessive X_{ch} allele. The Punnett square and phenotypic and genotypic ratios of the progeny for the mating of a heterozygous caudalis female with a caudalis male are

caudalis ♀ × caudalis ♂
$(X_{ch}X_{Cp})$ $(X_{Cp}Y)$

Male gametes

	X_{Cp}	Y
X_{Cp}	$X_{Cp}X_{Cp}$ caudalis female	$X_{Cp}Y$ caudalis male
X_{ch}	$X_{ch}X_{Cp}$ caudalis female	$X_{ch}Y$ transparent-tailed male

Female gametes

Genotypic ratio: 1 $X_{Cp}X_{Cp}$:1 $X_{Cp}X_{ch}$:1 $X_{Cp}Y$:1 $X_{ch}Y$

Phenotypic ratio: 2 caudalis females: 1 caudalis male:1 transparent-tailed male

A transparent-tailed (homozygous recessive) female can produce only one kind of son—transparent-tailed. The Punnett square and genotypic and phenotypic ratios of the progeny for the mating of a transparent-tailed female with a caudalis male are

transparent-tailed ♀ × caudalis ♂
$(X_{ch}X_{ch})$ $(X_{Cp}Y)$

Male gametes

	X_{Cp}	Y
X_{ch}	$X_{ch}X_{Cp}$ caudalis female	$X_{ch}Y$ transparent-tailed male

Female gametes

Genotypic ratio: 1 $X_{ch}X_{Cp}$:1 $X_{ch}Y$

Phenotypic ratio: 1 caudalis female:1 transparent-tailed male

This mating demonstrates the criss-cross inheritance pattern.

Transparent-tailed females can be produced only if the father has a transparent tail and the mother has at least one X_{ch} allele. The Punnett square and genotypic and phenotypic ratios of the progeny for the mating of a heterozygous caudalis female and a transparent-tailed male are

caudalis ♀ × transparent-tailed ♂
$(X_{ch}X_{Cp})$ $(X_{ch}Y)$

Male gametes

	X_{ch}	Y
X_{ch}	$X_{ch}X_{ch}$ transparent-tailed female	$X_{ch}Y$ transparent-tailed male
X_{Cp}	$X_{Cp}X_{ch}$ caudalis female	$X_{Cp}Y$ caudalis male

Female gametes

Genotypic ratio: 1 $X_{ch}X_{ch}$: 1 $X_{Cp}X_{ch}$: 1 $X_{ch}Y$: 1 $X_{Cp}Y$

Phenotypic ratio: 1 transparent-tailed female : 1 caudalis female : 1 transparent-tailed male : 1 caudalis male

Sex-Limited Phenotypes   X-linked alleles follow definite genetic patterns, but the phenotypes do not always follow the expected ratios. This is because many X-linked phenotypes (and some autosomal phenotypes too) are sex-limited in that the phenotype is expressed in only one sex.

For example, the tigrinus phenotype (stripes on the body) in the guppy is controlled by the X_{Ti} dominant sex-linked allele (Winge 1927). Under normal conditions, the tigrinus phenotype is not expressed in females, no matter the genotype:

Genotype	Phenotype
XX	grey female
XX_{Ti}	grey female
$X_{Ti}X_{Ti}$	grey female
XY	grey male
$X_{Ti}Y$	tigrinus male

Many sex-limited phenotypes need testosterone in order to be expressed (Hildemann 1954). Most of the X-linked genes in the guppy produce phenotypes that are sex-limited to the males (Table 3.10). The addition of methyltestosterone to the water or feed will allow the phenotypes to be expressed in females (Emmens 1970). If done properly, the masculinizing hormone will not damage the females, and this procedure allows a breeder to decipher (or partially decipher) the females' genotypes and thus improve the success of a breeding program.

GENES WITH MULTIPLE ALLELES

For the sake of simplicity, all concepts and examples that have been used up to this point were constructed using the premise that each gene has only two alleles. This is a false assumption because, in a population, the number of alleles for a given gene can range from one to over a dozen. Many genes have three or more alleles. The ideas that are discussed in this chapter also work when a gene has more than two alleles; all that is required is a little expansion of the concepts and formulas to include the different possible genotypes and phenotypes.

The B gene, which controls melanin formation in the medaka's melanophores, is an example of an autosomal gene with three alleles (Aida 1921). The B allele is dominant over the B' and b alleles, and the B' allele is dominant over the b allele; the b allele is recessive to the other two:

Genotypes	Phenotypes
BB, BB', Bb	full melanin production
$B'B'$, $B'b$	varigated or mottled pigmentation
bb	no or minimal melanin production

If there are more than two alleles at a given locus and if the mode of gene action for some of the alleles is either incomplete dominance, additive, or codominance, the number of phenotypes that can be produced by that locus can be quite large. For example, the D gene, which controls body color in angelfish, has four alleles—D, D^m, d^+, and D^{ng}—and they produce nine phenotypes (Norton 1982). The D allele exhibits incomplete dominance over the D^m, d^+, and d^{ng} alleles; the D^m allele exhibits incomplete dominance over the d^+ and d^{ng} alleles; the d^+ allele exhibits complete dominance over the d^{ng} allele; the d^{ng} allele is recessive to the other alleles. These alleles produce 10 genotypes and 9 phenotypes:

Genotype	Phenotype
DD	true black
DD^m	marble lace
Dd^+	black lace
Dd^{ng}	dark-new gold
D^mD^m	dark marble
D^md^+	light marble
D^md^{ng}	marble with jet black
d^+d^+	silver
d^+d^{ng}	silver
$d^{ng}d^{ng}$	new gold

The best example of a single gene which can produce many phenotypes is the P gene in the platyfish. The P gene controls spotting pattern, and there are nine known alleles at the P locus: P^+, P^M, P^{Mc}, P^T, P^{Co}, P^C, P^{Cc}, P^O, P^D (Gordon 1956; Kallman 1975). The P^+ allele produces the unspotted phenotype, and it is recessive to all other alleles. The remaining alleles are codominant, which means that the phenotypes that are produced by the alleles are always expressed, and heterozygous combinations of the alleles produce phenotypes which are combinations of the two separate phenotypes. The alleles at the P locus can, in theory, produce 37 different phenotypes, but a number of patterns overlap, which reduces the number of visible phenotypes to 27 (Kallman 1975). Figure 3.6 shows the phenotypes that are produced by the nine alleles at the P locus in the platyfish.

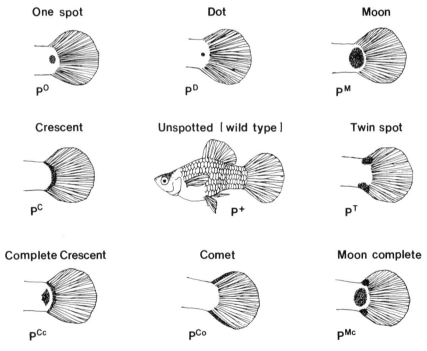

Figure 3.6 Spotting pattern in the platyfish controlled by the P gene. There are nine alleles: P^O, P^D, P^M, P^C, P^T, P^{Cc}, P^{Co}, P^{Mc}, and P^+. The P^+ allele is recessive to all other alleles, which are all codominant. The allele that controls the pigment pattern is listed below the phenotype.

Source: After Gordon (1956). By permission of New York Zoological Society

Sex-linked genes can also have multiple alleles. The R gene on the sex chromosomes in the medaka is an example of a sex-linked gene with three alleles (Aida 1921; Goodrich 1929; Yamamoto 1969A). The R alleles are responsible for carotenoids in the xanthophores, and this helps determine the medaka's color in an epistatic combination with the B and the I genes (Yamamoto 1969A).

GENES THAT EXHIBIT PLEIOTROPY, VARIABLE PENETRANCE, AND/OR VARIABLE EXPRESSIVITY

Pleiotropy

Another assumption that has been made throughout this discussion is the fact that a gene produces a specific phenotype in a one-to-one process and that a particular genotype always produces a given phenotype. This assumption is not always correct: A single gene can affect multiple phenotypes; phenotypes are not always expressed; and the expression of the phenotype is not always constant.

Genes affect biochemical pathways, and because biochemical pathways are interconnected, genes do not always influence phenotypes in a simple one-to-one process. The substitution of one allele for another can affect more than one biochemical pathway, and if more than one pathway is affected, more than one phenotype can be altered. Additional, secondary phenotypic effects are called pleiotropic effects. To a hatchery manager, pleiotropic effects may be minor and insignificant if no economically important characteristics are altered. However, when pleiotropic effects either increase or decrease viability, productivity, or market value, they become significant and can actually become more important than the phenotype.

Pleiotropy has been extensively studied in common carp because it is one of the most important cultured food fish in the world. Many color genes in common carp have detrimental pleiotropic effects. The L, D, B, and G color genes have many pleiotropic effects (Wohlfarth and Moav 1970; Katasonov 1974). For example, Wohlfarth and Moav (1970) found that both blue (bb) and gold (gg) common carp have lowered growth rates as a pleiotropic effect. Figure 3.7 shows the growth rates of normally pigmented, blue, and gold common carp in Israel. On the other hand, Wlodek (1968) found that Polish blue common carp grew better than normally pigmented common carp. The different pleiotropic effects

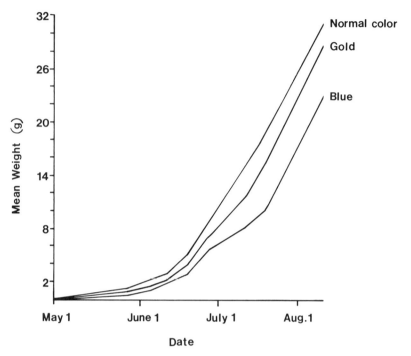

Figure 3.7 Growth curves of gold, blue, and normally pigmented common carp that were spawned on the same day. The growth curves show the negative pleiotropic effect that the *gg* (gold) and *bb* (blue) genotypes have on growth.
Source: After Wohlfarth and Moav (1970)

suggest either that Polish blue common carp and Israeli blue common carp may be different blue phenotypes, or that Polish blue common carp and Israeli blue common carp may be produced by slightly different biochemical pathways.

Mutant red and gold tilapia have captured the fancy of many tilapia farmers because these fish can command higher market prices in locales where consumers prefer light-skinned fish. Some of the pleiotropic effects have been investigated to help determine if the increased market price produces greater profits for the farmers. Gold *T. mossambica* have significantly whiter flesh than black (normally pigmented) ones, which is a marketing benefit (Tave *et al.* 1990B). Gold *T. mossambica*, as well as blond and light-colored (pink) *T. nilotica* have non-pigmented peritoneal membranes (Scott *et al.* 1987; Mires 1988; Tave *et al.* 1990B), which means that no black pigment has to be removed during processing. The bright body color of gold *T. mossambica* subjected them to significantly greater predation rates by largemouth bass, a predator that has color vision (Tave *et al.* 1991); on the other hand, gold body color did not adversely affect predation rates by dragonfly nymphs, a predator which does not have color vision (Tave *et al.* 1990C).

Viabilities of red *T. aurea* and *T. nilotica* hybrids were significantly lower than those of their normally pigmented sibs. El Gamal *et al.* (1988) found that hatchabilities of red embryos were significantly lower than those of the normally pigmented embryos and that post sex-reversal viabilities of red tilapia were significantly lower than those of the normally pigmented fish. Figure 3.8 shows the survival curves for red *T. aurea* and *T. nilotica* hybrids from fertilization until harvest.

It is important to quantify pleiotropic effects because color genes are often used as genetic markers to identify a particular group. Figures 3.7 and 3.8 show that such markers may have an adverse effect on the group that has been marked. Thus, the pleiotropic effects of the marker may confound the genetics of the groups that are being examined, and this will produce data that cannot be interpreted properly.

Scale patterns in common carp are important qualitative phenotypes when they are grown for food. Because of this, the pleiotropic effects of the *S* and *N* genes have been extensively studied in Russia, Eastern Europe, and Israel. The Russian and Eastern European literature was reviewed by Kirpichnikov (1981), and the pleiotropic effects that have been found in Israel were reviewed by Wohlfarth *et al.* (1963). Seventeen pleiotropic effects that have been detected in mirror, line, and leather common carp are listed in Table 3.11.

The genes which control the only qualitative phenotypes that have been deciphered in *T. aurea* and in channel catfish have some interesting pleiotropic effects. The *S* allele in *T. aurea* produces saddlebacks in the heterozygous state ($S+$). Pleiotropic effects of the $S+$ genotype are vertebral anomalies in vertebrae 1, 2, and 3; abnormal pelvic, pectoral, or anal fins; skeletal abnormalities in the pectoral and pelvic girdles; missing pterygiophores; abnormal caudal skeletons; lowered disease resistance; and reduced viability (Tave *et al.* 1983).

The *a* allele in channel catfish produces albinism in the homozygous state (*aa*). Pleiotropic effects of the *aa* genotype are that albinos spawn later; produce smaller egg masses; produce poorer quality eggs that have a poorer percent hatch; produce progeny that are less viable and have a poorer growth rate than normally pigmented channel catfish (Bondari 1984C); moreover, because of their bright color, albinos are subject to more predation than normally pigmented channel catfish (Prather 1961).

Pleiotropy does not always mean that an allele carries excess baggage. For example the *b* allele, which produces iridescent metallic blue

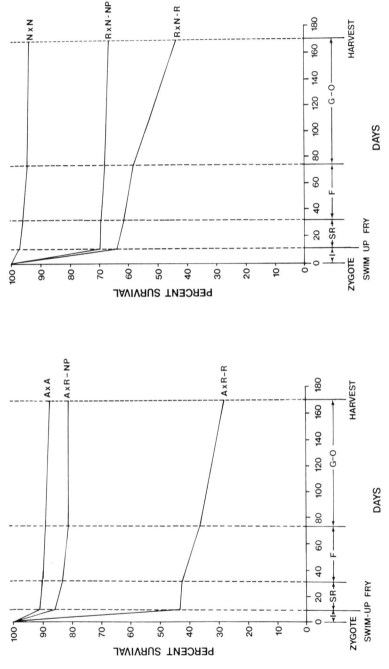

Figure 3.8 Viabilities of red *Tilapia aurea* (A) and red *T. nilotica* (B) hybrids. Viability of red *T. aurea* (A × R-R) is compared to their normally pigmented sibs (A × R-NP) and to a control *T. aurea* population (A × A) which had no red individuals. Viability of red *T. nilotica* (R × N-R) is compared to their normally pigmented sibs (R × N-NP) and to a control *T. nilotica* (N × N) population which had no red individuals. Viabilities over four stages of production are shown: I = artificial incubation of fertilized eggs; SR = sex reversal; F = fingerling production; G-O = grow-out. At harvest, fish averaged 250 g.
Source: after El Gamal (1987)

Table 3.11 Some Pleiotropic Effects of Scale Genotypes in Common Carp

Pleiotropic effects	Scaled (SS,nn; Ss,nn)	Mirror (ss,nn)	Line (SS,Nn; Ss,Nn)	Leather (ss,Nn)
			Scale phenotypes and genotypes	
Weight of 1-year-old fish under favorable conditions[a]	100	93–96	85–88	79–80
Weight of 1-year-old fish under unfavorable conditions[a]	100	83–94	42–70	37–72
Weight of 2-year-old fish[a]	100	94–96	86–91	83–84
Mean no. soft rays in dorsal fin	18.8	18.7	16.4	15.4
Mean no. soft rays in anal fin	4.96	5.00	3.82	3.56
Mean no. rays in pelvic fin	14.7	14.3	14.3	13.1
Mean no. gill lamelle	88.6	83.5	82.3	83.2
Mean no. pharyngeal teeth	9.22	9.58	7.63	7.44
Ability to regenerate fins[a]	100	76	39	19
Erythrocyte count (10⁶ cells/ml)	1.93	1.99	1.76	1.69
Hemoglobin (g/%)	9.02	8.87	8.18	8.28
Survival time in minutes under oxygen deficit	210	210	132	132
Immunologic reactivity	fast	fast	slow	slow
Resistance to dropsy	—	increased	—	decreased
Intensity of fat metabolism	low	low	high	very high
Total survival of 1-year-old fish under optimal conditions[a]	100	91–98	87–93	80–92
Total survival of 1-year-old fish under unfavorable conditions[a]	100	93–95	36–37	28–60

Source: After Kirpichnikov (1981). Courtesy of V. S. Kirpichnikov and Nauka Publishers.
[a] Expressed as percentage of the value of the scaled phenotype; scaled phenotype is 100%.

body color in rainbow trout, improved growth rate (Kincaid 1975). Iridescent metallic blue rainbow trout were 23% heavier after one year's
growth: normally pigmented rainbow trout averaged 196 g, while iridescent metallic blue rainbow trout averaged 236 g. The incorporation of
such a gene into a population is one way to increase productivity and
profits.

Penetrance and Expressivity

Gene action usually determines phenotypic expression, i.e., which of
two or more phenotypes is expressed, but this is not always true. There
are times when a particular genotype does not produce the expected
phenotype. When this occurs, the gene in question is said to exhibit
variable penetrance and/or variable expressivity.

Penetrance refers to the percentage of individuals that exhibit the
expected phenotype. If the phenotype is always expressed, penetrance
is 100%; if one half of the individuals exhibit the phenotype, penetrance
is 50%; if no individuals exhibit the phenotype, penetrance is 0%. Sex-
limited phenotypes have a penetrance of 0% in normal females or 0% in
normal males, depending on the particular phenotype and the sex in
which the phenotype is sex-limited.

Expressivity refers to the physical manifestation of a phenotype,
i.e., the range of phenotypic expression. Some phenotypes never vary;
others are expressed differently in different fish.

The *Sc* gene in the Montezuma swordtail (Kallman 1971) and the *M*
gene in the sailfin molly (Angus 1983) are examples of autosomal genes
that show variable penetrance and variable expressivity. The dominant
Sc allele produces the spotted caudal phenotype (irregular longitudinal
streaks or spots consisting of melanophores on the caudal fin) in Montezuma swordtails. The *Sc* allele has penetrance of 30–88% in the homozygous and heterozygous states, and the expression of the phenotype
ranges from a small elongated streak on the caudal fin to melanomas
(pigmented tumors) (Kallman 1971).

The *M* gene, which produces black spotting in the sailfin molly,
exhibits incomplete dominance as well as variable penetrance and variable expressivity (Angus 1983). The *M* allele is incompletely dominant
over the *m* allele:

Genotype	Phenotype
MM	highly spotted, mostly black
Mm	lightly spotted
mm	no spots

Figure 3.9 Normal (a) and saddleback (b-n) *Tilapia aurea*. Note the variable expressivity of the saddleback phenotype and some of the pleiotropic effects of the $S+$ genotype [missing or abnormal pectoral, pelvic, or anal fins (i-n)]. *Source: Tave et al. (1983)*

When water temperature is 20°C, penetrance is 100%. At warmer temperatures, penetrance is less than 100%, and at 26°–28°C penetrance is nearly zero in the heterozygotes. Expressivity (melanin production) is far less at higher temperatures.

Temperature also alters the penetrance of the Y_m gene in the eastern mosquitofish. The Y_m allele produces melanistic spotting in males (Angus 1989). When water temperature is 22°C, penetrance is 100%, but when water temperature is 26°–29°C, penetrance drops to 42%.

Temperature appears to be the ultimate agent that causes variable penetrance and expressivity for the M gene in sailfin mollies and the Y_m gene in the eastern mosquitofish, but the reason for variable penetrance and expressivity is usually unknown. The biological explanation is simple: The biochemical pathway that controls phenotypic expression is affected at different stages in different fish. The mechanism behind this explanation, however, is often difficult to determine. For example, the B gene in rainbow trout exhibits variable penetrance in that 10% of rainbow trout with the bb genotype do not express the iridescent metallic blue phenotype; penetrance of the bb genotype is 90%. The reason that penetrance is not 100% is not known, but the likely reason is light intensity (Kincaid 1987). The S allele in $T.$ $aurea$ is an example of an allele that produces variable expressivity for reasons that are unknown. In the heterozygous state, the S allele produces the saddleback phenotype, which is an abnormal dorsal fin. Expressivity is highly variable, ranging from saddlebacks that lack only the first spine to individuals that have no dorsal fin (Tave et $al.$ 1983); see Fig. 3.9.

Genes that exhibit variable penetrance and variable expressivity can cause problems during genetic analyses of phenotypes, because phenotypic categories can be erroneous, producing confusing phenotypic ratios. They can also cause trouble during selection programs if the allele to be culled is not always expressed.

LINKAGE

Two or more phenotypes, each controlled by single genes and with no epistatic interaction between the genes, are inherited independently only if the genes responsible for the phenotypes are located on different chromosomes. Obviously, this is not always true since there are hundreds to thousands of genes on each chromosome. When genes are located on the same chromosome, they are said to be linked. When the genes are linked, they are often inherited as pairs or groups, which means that the phenotypes are also inherited as pairs or groups. When the genetics of any animal or plant is studied in detail, linkage groups soon appear. In fact, it is inevitable once the number of known genes exceeds the number of chromosomes. Linkage groups in fish were first detected by Winge (1923).

When two genes are linked, gametes do not receive the alleles in an independent manner; rather, alleles are parceled out to the gametes as pairs or, in the case of three or more linked genes, as sets of alleles. For example, the X_{Vi} and X_{El} genes in the guppy are linked on the X chromosome (Winge and Ditlevsen 1947). Although both genes may cross over to the Y chromosome, this discussion assumes that they are located only on the X chromosome. Because the X_{Vi} and X_{El} genes are linked, a heterozygous female produces only two kinds of gametes. The gametes she produces depend on which of the following ways the alleles are paired:

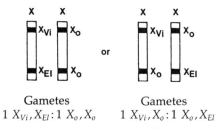

Gametes

$1\ X_{Vi}, X_{El} : 1\ X_o, X_o$ $1\ X_{Vi}, X_o : 1\ X_o, X_{El}$

Crossing over can occur between the two homologous chromosomes during the early stages of meiosis. If this occurs, a fish with a particular linkage group can produce gametes with all possible allelic combinations, but the production of the different possible gametes will not be equally likely. The production of the crossover gametes will depend on the crossing over frequency which, in turn, depends on the linear distance between the genes on a chromosome. The closer the genes, the less frequently crossing over occurs. As genes are separated by longer and longer distances, the cross over frequency increases until the genes are transmitted as if they were on separate chromosomes.

Distances between genes and the linear sequence of genes on chromosomes can be mapped by crossover frequencies. Linkage maps for some chromosomes have been developed in the guppy (Winge and Ditlevsen 1947), medaka (Yamamoto 1964A; 1975B), salmonids (May and Johnson 1990), *Xiphophorous* (Morizot 1990; Harless *et al.* 1991; Morizot *et al.* 1991), *Poeciliopsis* (Morizot 1990), *Fundulus* (Morizot 1990), and *Lepomis* (Morizot 1990). Winge and Ditlevsen's (1947) maps of the X and Y chromosomes in the guppy are shown in Fig. 3.10. One ultimate goal in animal or plant breeding is to produce detailed maps of each chromo-

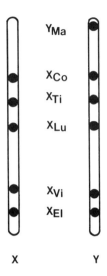

Figure 3.10   Chromosome maps of the X and Y chromosomes in the guppy showing the location of the Y_{Ma}, X_{Co}, X_{Ti}, X_{Lu}, X_{Vi}, and X_{El} genes.
Source: After Winge and Ditlevsen (1947)

some, because such maps enable breeders to manipulate many genes simultaneously and also allow them to predict the effect that selection for one gene will have on others.

POPULATION GENETICS

Understanding the genetics that controls the biological blueprint for a particular phenotype makes it possible to determine which breeding program can be used to change phenotypic, genotypic, and gene frequencies. Different phenotypes often have different economic values, and when this occurs it is desirable to change the frequencies of the alleles that produce the phenotypes, increasing the frequency of the allele that controls the desired phenotype while decreasing the frequency of the allele that controls the undesired phenotype. If the desired allele can be fixed [frequency $(f) = 100\%$] and the undesired allele can be eliminated $(f = 0\%)$, the population will breed true and will be more valuable as brood stock. During a selection program, it is often useful to know the frequencies of the alleles in the population at time zero, otherwise known as the parental (P_1) generation, and in subsequent $(F_1, F_2, \ldots, F_n)$ generations.

Autosomal Genes

Incomplete Dominance or Additive Gene Action   When a gene has two alleles that produce three genotypes and each genotype produces a unique phenotype (incomplete dominant and additive gene action), it is easy to determine allelic frequencies. The frequency of an allele is simply its proportion in a population:

$$f(\text{allele}) = \frac{2\begin{bmatrix} \text{no. fish with} \\ \text{homozygous phenotype} \\ \text{produced by the allele} \end{bmatrix} + \begin{bmatrix} \text{no. fish with} \\ \text{heterozygous} \\ \text{phenotype} \end{bmatrix}}{2(\text{no. fish in the population})} \quad (3.1)$$

The number of fish with the homozygous phenotype is multiplied by two because each fish carries two copies of the allele; heterozygotes have only one copy. The number of fish in the population is multiplied by two because each fish is a diploid and has two alleles per locus.

For example, if a trout farmer finds some golden and palomino rainbow trout at his farm and wants to know the frequencies of the alleles responsible for the production of these phenotypes in his population of rainbow trout, all he has to do is count the phenotypes, add the alleles, and calculate a simple percentage. This is because these phenotypes are controlled by a single gene with additive gene action, so each genotype produces a unique phenotype: GG = normally pigmented; GG' = palomino; $G'G'$ = golden. A census of his population shows the following:

Phenotype	Genotype	Number
normally pigmented	GG	360
golden	G'G'	160
palomino	GG'	480
	Total	1000

The frequency of the G [$f(G)$] and G' [$f(G')$] alleles can be determined by calculating the number of G and G' alleles that exist in the population, and by then determining their proportions:

To determine the frequency of the G allele you need to determine the number of G alleles and then the total number of alleles at the G locus ($G + G'$). To determine the number of G alleles you add the number of palomino rainbow trout to twice the number of normally pigmented rainbow trout. Normally pigmented rainbow trout are counted twice because each carries two G alleles. Golden rainbow trout have no G alleles. To determine the total number of alleles at the G locus, you multiply the total number of rainbow trout in the population (golden + palomino + normally pigmented) by two. To determine the frequency of the G allele, you

divide the number of G alleles by the total number of alleles at the G locus
[Eq. (3.1)]:

$$f(G) = \frac{\text{no. } G \text{ alleles}}{\text{total no. alleles at } G \text{ locus}}$$

$$f(G) = \frac{2(360 \text{ normally pigmented}) + 480 \text{ palomino}}{2(1000 \text{ rainbow trout})} \qquad (3.1)$$

$$f(G) = \frac{1200 \ G \text{ alleles}}{2000 \text{ total alleles}}$$

$$f(G) = 0.6$$

In a similar manner, the frequency of the G' allele can be determined by
summing the G' alleles that exist in the population and by then determining its proportion. The number of G' alleles is determined by adding the
number of palomino rainbow trout to twice the number of golden rainbow
trout. Golden rainbow trout are counted twice because each golden rainbow trout has two G' alleles. Normally pigmented rainbow trout have no
G' alleles. The frequency of the G' allele is determined by using Eq. (3.1):

$$f(G') = \frac{\text{no. } G' \text{ alleles}}{\text{total no. alleles at } G \text{ locus}}$$

$$f(G') = \frac{2(160 \text{ golden}) + 480 \text{ palomino}}{2(1000 \text{ rainbow trout})}$$

$$f(G') = \frac{800 \ G' \text{ alleles}}{2000 \text{ total alleles}}$$

$$f(G') = 0.4$$

Note that

$$f(G) + f(G') = 0.6 + 0.4 = 1.0.$$

The sum of the frequencies for the alleles at a locus must equal 1.0. If it
does not, an error has been made.

Another way to calculate the frequency of an allele is to take the
square root of the frequency of its homozygous phenotype:

$$f(\text{allele}) = \sqrt{f(\text{homozygous phenotype produced by allele})} \qquad (3.2)$$

From the preceding example:
To calculate $f(G)$:

$$f(G) = \sqrt{\frac{\text{no. } GG \text{ phenotypes (normally pigmented rainbow trout)}}{\text{total population}}}$$

$$f(G) = \sqrt{\dfrac{360 \text{ normally pigmented rainbow trout}}{1000 \text{ total rainbow trout}}}$$

$$f(G) = \sqrt{0.36}$$

$$f(G) = 0.6$$

To calculate $f(G')$:

$$f(G') = \sqrt{\dfrac{\text{no. } G'G' \text{ phenotypes (golden rainbow trout)}}{\text{total population}}}$$

$$f(G') = \sqrt{\dfrac{160 \text{ golden rainbow trout}}{1000 \text{ total rainbow trout}}}$$

$$f(G') = \sqrt{0.16}$$

$$f(G') = 0.4$$

As before, $f(G) + f(G') = 0.6 + 0.4 = 1.0$.

This method works because the frequency of a homozygous genotype is the square of the frequency of the allele in question. For example, the GG genotype has two G alleles, so the probability (and frequency) of producing a GG genotype is $f(G) \times f(G)$ or $f(G)^2$; therefore, $f(G) = \sqrt{f(GG)}$.

An alternate way to demonstrate that the square root of the frequency of the homozygous phenotype will give the frequency of that allele is to set up a Punnett square. This time, gametic frequencies are included in the Punnett square because it is a population problem, not an individual mating:

Genotypic frequencies at the G locus that are produced in a population where $f(G) = 0.6$ and $f(G') = 0.4$ are

		Male gametes	
		G ($f = 0.6$)	G' ($f = 0.4$)
Female gametes	G ($f = 0.6$)	GG normally pigmented $f(GG) = (0.6)(0.6) = 0.36$	GG' palomino $f(GG') = (0.6)(0.4) = 0.24$
	G' ($f = 0.4$)	$G'G$ palomino $f(G'G) = (0.4)(0.6) = 0.24$	$G'G'$ golden $f(G'G') = (0.4)(0.4) = 0.16$

Phenotype	Genotype	Frequency
normally pigmented	GG	0.36
palomino	GG'	0.24 ⎫
palomino	G'G	0.24 ⎭ 0.48 total
golden	G'G'	0.16
	Total	1.00

The frequencies of the homozygous genotypes are the squares of the allelic frequencies responsible for the genotypes; conversely, the frequencies of the alleles are the square roots of the frequencies of the homozygous genotypes [e.g., $f(GG) = f(G)^2 = (0.6)^2 = 0.36$; $f(G) = \sqrt{f(GG)} = \sqrt{0.36} = 0.6$].

Complete Dominance When the mode of gene action is complete dominance, the allelic frequencies must be calculated by the square-root technique. This is because this type of gene action produces three genotypes but only two phenotypes. Because the heterozygous phenotype is indistinguishable from the homozygous dominant phenotype, allelic frequencies cannot be determined by counting the alleles, since there is no way to know which individuals with the dominant phenotype have two dominant alleles and which have only one.

When the mode of gene action is complete dominance, the recessive genotype is the only genotype that can be deciphered by an examination of the phenotypes. Because the frequency of the recessive phenotype is the square of the frequency of the recessive allele, the square root of the frequency of the recessive phenotype will give the frequency of the recessive allele. The frequency of the dominant allele is then calculated by subtracting the recessive allele's frequency from 1.0. (Remember, the sum of the frequencies of the alleles at each locus must equal 1.0.)

For example, say a carp farmer finds some blue common carp at his hatchery; upon discovering this, he wants to know the frequencies of the alleles responsible for the production of blue and normally pigmented common carp. Blue is controlled by the recessive *b* allele, and normal pigmentation is controlled by the dominant *B* allele. To determine the frequencies of the *B* and *b* alleles, he must use the square root technique. A census of his population shows the following:

Phenotype	Genotype	Number
blue	*bb*	90
normally pigmented	*BB* and *Bb*	910
	Total	1000

To calculate $f(b)$ use Eq. (3.2):

$$f(b) = \sqrt{f(bb)} = \sqrt{f(\text{blue common carp})}$$

$$f(b) = \sqrt{\frac{90 \text{ blue common carp}}{1000 \text{ total common carp}}}$$

$$f(b) = \sqrt{0.09}$$

$$f(b) = 0.3$$

To calculate $f(B)$, subtract $f(b)$ from 1.0:

$$f(B) = 1.0 - 0.3$$

$$f(B) = 0.7$$

The preceding Punnett square (see page 79), which incorporated allelic frequencies, shows how to calculate genotypic frequencies from allelic frequencies even for genes that have complete dominance:

$$f(\text{homozygous genotype for allele one}) = f(\text{allele one})^2 \qquad (3.3)$$

$$f(\text{heterozygous genotype}) = 2[f(\text{allele one})][f(\text{allele two})] \quad (3.4)$$

$$f(\text{homozygous genotype for allele two}) = f(\text{allele two})^2 \qquad (3.5)$$

When determining the frequency of the heterozygotes [Eq. (3.4)], the product of the two alleleic frequencies must be multiplied by two because there are two ways to produce a heterozygote:

allele one (mother) allele two (father);   allele one (father) allele two (mother)

Using formulas (3.3), (3.4), and (3.5) to calculate genotypic frequencies in the common carp population used in the previous example shows the following:

1. Calculate the frequency of homozygous (*BB*) normally pigmented common carp using Eq. (3.3):

$$f(BB) = f(B)^2$$

$$f(BB) = (0.7)^2$$

$$f(BB) = 0.49$$

2. Calculate the frequency of heterozygous (*Bb*) normally pig-
    mented common carp using Eq. (3.4):

$$f(Bb) = 2[f(B)][f(b)]$$
$$f(Bb) = 2(0.7)(0.3)$$
$$f(Bb) = 0.42$$

3. Calculate the frequency of blue (*bb*) common carp using Eq.
    (3.5):

$$f(bb) = f(b)^2$$
$$f(bb) = (0.3)^2$$
$$f(bb) = 0.09$$

Note that

$$f(BB) + f(Bb) + f(bb) = 0.49 + 0.42 + 0.09 = 1.0.$$

The sum of the frequencies must equal 1.0; if it does not, an error has
been made.

Since there are 1000 common carp in this population, these frequencies mean that there are 490 *BB* normally pigmented common carp, 420
Bb normally pigmented common carp, and 90 blue (*bb*) common carp.

The use of Eqs. (3.2), (3.3), (3.4), and (3.5) to provide either gene or
genotypic frequencies should be done with a little caution because they
may provide incorrect frequencies. The use of these equations depends
on one precondition: The population must be in Hardy-Weinberg equilibrium for the gene in question. Several conditions must be met before a
population can be in Hardy-Weinberg equilibrium: (1) the population
must be infinitely large; (2) mating must be random; (3) there is no
mutation; (4) there is no selection; (5) there is no migration (movement
of fish from one population to another). These five preconditions combine to produce one effect: Gene frequencies and genotypic frequencies
will not change from generation to generation. Thus, genotypic frequencies can be used to calculate gene frequencies, and gene frequencies can
be used to calculate genotypic frequencies.

As can be imagined, few populations are ever in Hardy-Weinberg
equilibrium for all genes, but many are in Hardy-Weinberg equilibrium
at one or more loci. Despite the fact that few populations are in Hardy-
Weinberg equilibrium for all genes, formulas (3.2), (3.3), (3.4), and (3.5)

are routinely used to determine gene and genotypic frequencies. If possible, Eq. (3.1) should be used to determine gene frequencies, because that formula will always provide the correct frequencies. If formula (3.1) can be used to determine gene frequencies, genotypic frequencies can be determined by censusing the population.

When the mode of gene action is complete dominance, Eq. (3.2) must be used to determine gene frequencies. When Eq. (3.2) is used to determine gene frequencies, the frequencies that it, as well as Eqs. (3.3), (3.4), and (3.5), generates are usually either correct or are good approximations to the correct frequencies.

Two or More Genes That Produce Separate Phenotypes $\quad$ Calculation of allelic frequencies at two or more loci when each locus produces a separate phenotype is a simple extension of the methods used for a single locus. The simultaneous occurrence of two separate events can be separated into individual events. Because each locus produces a different phenotype, each phenotype and the alleles that produce it can be separated, and allelic frequencies can then be calculated independently for each phenotype. For example, say a tropical fish farmer finds the following guppies in a pond on his farm:

Phenotype	Genotype	Number
grey and normal spine	GG,CuCu; Gg,CuCu; Gg,Cucu; GG,Cucu	8316
grey and curved spine	GG,cucu; Gg,cucu	84
gold and normal spine	gg,CuCu; gg,Cucu	1584
gold and curved spine	gg,cucu	16
	Total	10,000

and he wants to know the frequencies of the G, g, Cu, and cu alleles. To calculate the frequencies, he must calculate the frequencies at the G locus and the Cu locus as if each were the only one that existed.

Step 1. $\quad$ Calculate the allelic frequencies at the G locus. Regroup the phenotypes as if the gold and grey phenotypes were the only ones that existed:

$$\text{grey } (GG + Gg) = 8316 + 84 = \quad 8400$$
$$\text{gold } (gg) = 1584 + 16 = \underline{\quad 1600}$$
$$\text{Total} = 10,000$$

To calculate $f(g)$ use Eq. (3.2):

$$f(g) = \sqrt{f(gg)} = \sqrt{f(\text{gold guppies})}$$

$$f(g) = \sqrt{\frac{1600 \text{ gold guppies}}{10,000 \text{ total guppies}}}$$

$$f(g) = \sqrt{0.16}$$

$$f(g) = 0.4$$

To calculate $f(G)$, subtract $f(g)$ from 1.0:

$$f(G) = 1.0 - f(g)$$

$$f(G) = 1.0 - 0.4$$

$$f(G) = 0.6$$

Step 2.   Calculate the allelic frequencies at the Cu locus. Regroup the phenotypes as if the curved spine and normal spine phenotypes were the only ones that existed:

$$\text{normal spine } (CuCu + Cucu) = 8316 + 1584 = \quad 9900$$

$$\text{curved spine } (cucu) = 84 \quad + 16 \quad = \quad \underline{100}$$

$$\text{Total} = 10,000$$

To calculate $f(cu)$ use Eq. (3.2):

$$f(cu) = \sqrt{f(cucu)} = \sqrt{f(\text{curved spine guppies})}$$

$$f(cu) = \sqrt{\frac{100 \text{ curved spine guppies}}{10,000 \text{ total guppies}}}$$

$$f(cu) = \sqrt{0.01}$$

$$f(cu) = 0.1$$

To calculate $f(Cu)$, subtract $f(cu)$ from 1.0:

$$f(Cu) = 1.0 - f(cu)$$

$$f(Cu) = 1.0 - 0.1$$

$$f(Cu) = 0.9$$

Genes That Produce Phenotypes through Epistasis   A simple extension of the ideas that have been discussed in this section makes it easy to calculate gene frequencies for phenotypes that are controlled by two or more genes. As with phenotypes that are controlled by one gene, the key is this: If the genotypes can be deciphered from an examination of the phenotypes, it is easy to determine gene frequencies.

For example, say a carp farmer wants to know the frequencies of the alleles responsible for scale pattern in his population of common carp. A census of his common carp reveals the following:

Phenotype	Genotype	Number
scaled	$SS,nn; Ss,nn$	1370
mirror	ss,nn	250
line	$SS,Nn; Ss,Nn$	310
leather	ss,Nn	70
	Total	2000

To calculate the frequencies of the S, s, N, and n alleles in this population, the phenotypes must be regrouped so that the genotypes can be deciphered:

Step 1.   Calculate the allelic frequencies at the S locus. To do this, regroup the four phenotypes so that the new genotypic categories are (1) ss (recessive); (2) $SS + Ss$ (dominant):

$$ss = \text{mirror} + \text{leather} = 250 + 70 = \ 320$$

$$SS + Ss = \text{scaled} + \text{line} = 1370 + 310 = \underline{1680}$$

$$\text{Total} \quad 2000$$

To calculate $f(s)$ use Eq. (3.2):

$$f(s) = \sqrt{f(ss)} = \sqrt{\frac{\text{mirror common carp} + \text{leather common carp}}{\text{total population}}}$$

$$f(s) = \sqrt{\frac{320 \text{ mirror and leather common carp}}{2000 \text{ total common carp}}}$$

$$f(s) = \sqrt{0.16}$$

$$f(s) = 0.4$$

To calculate $f(S)$, subtract $f(s)$ from 1.0:

$$f(S) = 1.0 - f(s)$$

$$f(S) = 1.0 - 0.4$$

$$f(S) = 0.6$$

Step 2.   Calculate the allelic frequencies at the N locus. To do this, regroup the four phenotypes so that the new genotypic categories are nn (recessive) and Nn (heterozygous):

$$nn = \text{scaled} + \text{mirror} = 1370 + 250 = 1620$$

$$Nn = \text{line} + \text{leather} = 310 + 70 = \underline{380}$$

$$\text{Total} \quad 2000$$

Each of the new categories is produced by a singular genotype (nn or Nn), so the frequencies of the n and N alleles can be determined by adding the alleles and by then determining their proportions:

To calculate $f(n)$ use Eq. (3.1):

$$f(n) = \frac{2(nn) + Nn}{2(\text{total no. of common carp})}$$

$$f(n) = \frac{2(1620) + 380}{2(2000)}$$

$$f(n) = \frac{3620 \; n \text{ alleles}}{4000 \text{ total alleles}}$$

$$f(n) = 0.905$$

To calculate f(N) use Eq. (3.1). *Note:* There is no NN category because that genotype is lethal, so the number of N alleles is simply the number of common carp with the Nn genotype.

$$f(N) = \frac{Nn}{2(\text{total no. of common carp})}$$

$$f(N) = \frac{380}{2(2000)}$$

$$f(N) = \frac{380 \; N \text{ alleles}}{4000 \text{ total alleles}}$$

$$f(N) = 0.095$$

Unfortunately, it is sometimes impossible to calculate gene frequencies when a phenotype is controlled by two or more genes. When a combination of two or more genes is needed in order to produce a particular set of qualitative phenotypes, genotypic categories often overlap. When that occurs, it is impossible to calculate gene frequencies.

For example, transparent and pigmented scales in the goldfish are produced by the epistatic interaction between the Dp_1 and Dp_2 genes. The genes produce the phenotypes through duplicate dominant interaction, and any genotype with at least one dominant allele will produce transparent scales. The homozygous recessive genotype is the only one that will produce pigmented scales. There are nine genotypes, but they produce only two phenotypes:

Transparent scales	Pigmented scales
Dp_1Dp_1, Dp_2Dp_2	dp_1dp_1, dp_2dp_2
Dp_1Dp_1, Dp_2dp_2	
Dp_1Dp_1, dp_2dp_2	
Dp_1dp_1, Dp_2Dp_2	
Dp_1dp_1, Dp_2dp_2	
Dp_1dp_1, dp_2dp_2	
dp_1dp_1, Dp_2Dp_2	
dp_1dp_1, Dp_2dp_2	

It is not possible to regroup the phenotypes and genotypes in any other manner. Consequently, allelic frequencies at the Dp_1 and Dp_2 loci cannot be determined.

Sex-Linked Genes

As was the case with phenotypes that are controlled by autosomal genes, it is often desirable to calculate allelic frequencies of sex-linked genes as part of a breeding program that will manipulate the frequencies of sex-linked phenotypes.

Y-Linked Genes   It is easy to calculate the frequency of a Y-linked gene in a population. The Y chromosome is found only in normal males, so the frequency of any Y-linked gene is simply the percentage of males that exhibit the phenotype. With a Y-linked gene there are two genotypes and two phenotypes:

Genotype	Phenotype
XY	wild-type males
XY_{allele}	Y-linked phenotype males

and

$$f(\text{Y-linked allele}) = \frac{\text{no. } \male \text{ with Y-linked phenotype}}{\text{total no. } \male} \qquad (3.6)$$

For example, say a guppy farmer drained a pond and found the following:

Phenotype	Genotype	Number	
grey females	XX	1650	
maculatus males	XY_{Ma}	340 ⎫	850 total males
grey males	XY	510 ⎭	

The frequency of the Y_{Ma} gene in this population is

$$f(Y_{Ma}) = \frac{\text{maculatus } \male}{\text{total } \male}$$

$$f(Y_{Ma}) = \frac{340}{850}$$

$$f(Y_{Ma}) = 0.4$$

X-Linked Genes   To calculate allelic frequencies for X-linked genes, three separate steps are needed: (1) calculate allelic frequencies in the males; (2) calculate allelic frequencies in the females; (3) calculate overall allelic frequencies based on the proportional numbers of alleles in the two sexes.

The frequencies of the X-linked alleles in males are determined by calculating the percentage of males that exhibit the phenotype. The procedure is identical to that described for Y-linked genes. In normal males (XY), there is only one X chromosome, so there are two genotypes and two phenotypes:

Genotype	Phenotype
$X_{dominant\ allele}Y$	male with dominant X-linked phenotype
$X_{recessive\ allele}Y$	male with recessive X-linked phenotype

Consequently

$$f(\text{X-linked recessive allele} - \male) = \frac{\text{no. } \male \text{ with recessive X-linked phenotype}}{\text{total no. } \male} \quad (3.7)$$

$$f(\text{X-linked dominant allele} - \male) = \frac{\text{no. } \male \text{ with dominant X-linked phenotype}}{\text{total no. } \male} \quad (3.8)$$

The frequencies of the X-linked alleles in females are determined by using Eq. (3.9), a modified form of Eq. (3.2). Equation (3.9) must be used because an X-linked gene expresses itself in females in a manner similar to that seen for autosomal genes with complete dominance. In females X-linked genes produce three genotypes but only two phenotypes:

Genotype	Phenotype
$X_{dominant\ allele}X_{dominant\ allele}$	female with dominant X-linked phenotype
$X_{dominant\ allele}X_{recessive\ allele}$	female with dominant X-linked phenotype
$X_{recessive\ allele}X_{recessive\ allele}$	female with recessive X-linked phenotype

Consequently

$$f(\text{X-linked recessive allele}-♀) = \sqrt{\frac{\text{no. ♀ with recessive X-linked phenotype}}{\text{total no. ♀}}} \qquad (3.9)$$

The frequency of the dominant X-linked allele in females can be determined by subtracting the frequency of the recessive X-linked allele in females from 1.0:

$$f(\text{X-linked dominant allele}-♀) = 1.0 - f(\text{X-linked recessive allele}-♀)$$

The overall frequency of either allele is determined from the frequency in both sexes and the proportional number of alleles in the two sexes:

$$f(\text{X-linked}-\text{overall}) = \frac{2(\text{no. }♀)[f(\text{allele }♀)] + (\text{no. }♂)[f(\text{allele }♂)]}{2(\text{no. }♀) + (\text{no. }♂)} \qquad (3.10)$$

The number of females is multiplied by two because each female has two X chromosomes.

For example, say a guppy farmer drains a pond and finds that he has both nigrocaudatus (black body) and grey guppies; upon discovering this, he wants to know the frequencies of the alleles responsible for these phenotypes. The nigrocaudatus phenotype is controlled by the dominant X_{Nill} allele, and the grey phenotype is controlled by the recessive X allele (Dzwillo 1959). A census of the population shows the following:

Phenotype	Genotype	Number	
nigrocaudatus males	$X_{Nill}Y$	200 ⎱	
grey males	XY	300 ⎰	500 males
nigrocaudatus females	$X_{Nill}X$ and $X_{Nill}X_{Nill}$	546 ⎱	
grey females	XX	54 ⎰	600 females

What are the frequencies of the X_{Nill} and the X alleles?

Step 1.   Calculate the frequencies in the males using Eqs. (3.7) and (3.8):

$$f(X_{Nill} - \male) = \frac{\text{nigrocaudatus } \male}{\text{total } \male}$$

$$f(X_{Nill} - \male) = \frac{200}{500}$$

$$f(X_{Nill} - \male) = 0.4$$

and

$$f(X - \male) = \frac{\text{grey } \male}{\text{total } \male}$$

$$f(X - \male) = \frac{300}{500}$$

$$f(X - \male) = 0.6$$

Since the sum of the allelic frequencies must equal 1.0, $f(X - \male)$ can also be calculated by subtracting $f(X_{Nill} - \male)$ from 1.0:

$$f(X - \male) = 1.0 - f(X_{Nill} - \male)$$
$$f(X - \male) = 1.0 - 0.4$$
$$f(X - \male) = 0.6$$

Step 2.   Calculate the frequencies in the females using Eq. (3.9):

$$f(X - \female) = \sqrt{\frac{\text{grey } \female}{\text{total } \female}}$$

$$f(X - \female) = \sqrt{\frac{54}{600}}$$

$$f(X - \female) = \sqrt{0.09}$$

$$f(X - \female) = 0.3$$

To calculate $f(X_{Nill} - \female)$, subtract $f(X - \female)$ from 1.0:

$$f(X_{Nill} - \female) = 1.0 - f(X - \female)$$
$$f(X_{Nill} - \female) = 1.0 - 0.3$$
$$f(X_{Nill} - \female) = 0.7$$

Step 3. Calculate the overall frequencies of $f(X_{Nill}$—overall) and $f(X$—overall) in the population by using Eq. (3.10):

$$f(X_{Nill}-\text{overall}) = \frac{2(600\ ♀)(0.7) + (500\ ♂)(0.4)}{2(600\ ♀) + 500\ ♂}$$

$$f(X_{Nill}-\text{overall}) = \frac{1040}{1700}$$

$$f(X_{Nill}-\text{overall}) = 0.6118$$

and

$$f(X-\text{overall}) = \frac{2(600\ ♀)(0.3) + (500\ ♂)(0.6)}{2(600\ ♀) + 500\ ♂}$$

$$f(X-\text{overall}) = \frac{660}{1700}$$

$$f(X-\text{overall}) = 0.3882$$

Because the sum of the frequencies must be equal to 1.0, $f(X$—overall) can also be calculated by subtracting $f(X_{Nill}$—overall) from 1.0:

$$f(X-\text{overall}) = 1.0 - f(X_{Nill}-\text{overall})$$

$$f(X-\text{overall}) = 1.0 - 0.6118$$

$$f(X-\text{overall}) = 0.3882$$

Frequencies of the X_{Nill} and X alleles in the population

	Males	Females	Overall
$f(X_{Nill})$	0.4	0.7	0.6118
$f(X)$	0.6	0.3	0.3882
Total	1.0	1.0	1.0

Sex-limited phenotypes The allelic frequencies of sex-limited X-linked genes are impossible to determine in females until you reveal the "unexpressed" phenotypes by the addition of methyltestosterone to either the feed or water (Emmens 1970). Once you do this, it is done as described in the preceding section.

SELECTION

As I mentioned earlier, different phenotypes often have different economic values, and the value of a population depends on its ability to produce the desired phenotypes more often than the undesired phenotypes. The value of a population is maximized, for any given phenotype, when that population will breed true and produce only the desired phenotype. Thus, the elimination of alleles responsible for the production of undesired phenotypes and the fixation of those responsible for the production of desired traits by selection are important aspects of fish stock management.

Autosomal Genes

Selection for qualitative autosomal phenotypes can be either simple or difficult depending on the mode of gene action. If the desired phenotype is produced only by a homozygous genotype, it is easy to produce a true-breeding population by culling (selecting against) the undesired phenotype(s). If the desired phenotype is produced by more than one genotype and some individuals cannot breed true because they are heterozygotes, it is impossible to produce a true-breeding population by culling the undesired phenotype(s). If the desired phenotype is produced by a heterozygous genotype, it is impossible to produce a true-breeding population by selection. Selection cannot fix a heterozygous genotype, because heterozygotes cannot breed true. The only way to produce 100% heterozygotes is to mate the two homozygotes.

Complete Dominant Gene Action   Autosomal genes that exhibit complete dominant gene action produce two phenotypes: a dominant and a recessive phenotype. Selection can easily fix the recessive phenotype and produce a true-breeding population, but selection cannot fix the dominant phenotype and produce a true-breeding population.

   Selection for the recessive phenotype   Selection can fix the recessive phenotype and eliminate the undesired dominant phenotype in a single generation. To do this, all individuals that express the dominant phenotype are culled. When this is done, the only fish that remain are those that have the recessive phenotype, and because recessive individuals have only the desired recessive allele (they are homozygous recessive), the dominant allele will not exist in the new population. The fixation of a recessive allele and the elimination of a dominant allele takes only one generation (it can actually be done in a single day). A single generation of selection against the dominant phenotype will produce a population that breeds true.

For example, if a catfish farmer wants to produce a true-breeding population of albino channel catfish, he can accomplish his goal in a single generation by culling all normally pigmented channel catfish. A single culling of all normally pigmented channel catfish will cull all + alleles. Albinos will be the only fish that remain in the population, and since albinos are homozygous recessive (*aa*), the *a* allele will be fixed, and the farmer will have a true-breeding population of albino channel catfish:

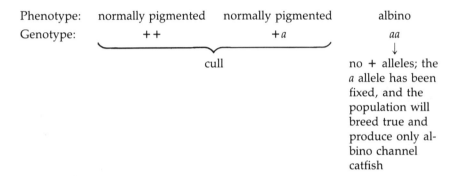

Phenotype: normally pigmented   normally pigmented   albino

Genotype: + +   + *a*   *aa*

                              cull   no + alleles; the
                                                            a allele has been
                                                            fixed, and the
                                                            population will
                                                            breed true and
                                                            produce only al-
                                                            bino channel
                                                            catfish

Selection for the dominant phenotype Selection cannot fix the dominant phenotype and eliminate the recessive phenotype. Consequently, it is impossible to eliminate undesired recessive alleles via selection. This is unfortunate, because many undesired phenotypes are controlled by recessive autosomal alleles.

Genes with complete dominant gene action produce two types of dominant phenotypes: the homozygous and the heterozygous. Phenotypically, they are identical, so it is impossible to separate homozygous dominant fish from heterozygous dominant fish. Consequently, recessive alleles are hidden and cannot be culled during selection.

For example, if a catfish farmer has a population which contains albino channel catfish and he wants to produce a true-breeding population of normally pigmented channel catfish, he will find that selection against the albino phenotype will not eliminate the *a* allele:

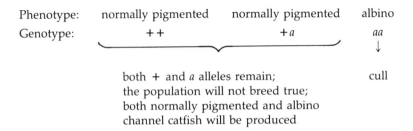

Phenotype:	normally pigmented	normally pigmented	albino
Genotype:	++	+a	aa

both + and *a* alleles remain;   cull
the population will not breed true;
both normally pigmented and albino
channel catfish will be produced

Because the *a* allele is carried by some normally pigmented channel catfish and is hidden from his selection program, the farmer will discover that he cannot eliminate the *a* allele by culling albinos, and his population will continue to produce albinos.

The effect of trying to eliminate a recessive allele by culling the recessive phenotype can be determined by using the following formula:

$$q_n = \frac{q_0}{1 + (Nq_0)} \tag{3.11}$$

where q_0 is the initial frequency of the recessive allele, q_n is the frequency of the recessive allele after n generations, and N is the number of generations.

The effect of such a program is shown in Fig. 3.11. If initial allelic frequency is high, the first few generations of selection lower the frequency precipitously. But when the frequency is low, the rate of change is very small, and allelic frequency approaches zero asymptotically, because as the frequency of a recessive allele decreases, the percentage of the alleles that are hidden by the heterozygous dominant phenotype increases. For example

Allelic frequencies		Genotypic frequencies			
$f(+)$	$f(a)$	$f(++)$	$f(+a)$	$f(aa)$	Percent *a* in heterozygotes
0.1	0.9	0.01	0.18	0.81	10%
0.5	0.5	0.25	0.50	0.25	50%
0.9	0.1	0.81	0.18	0.01	90%

When $f(a) = 0.9$, 90% of all *a* alleles can be culled, but when $f(a) = 0.1$, only 10% of the *a* alleles can be culled.

Since selection cannot eliminate a recessive allele, the question becomes: How many generations will it take to reduce the frequency of a recessive allele to an acceptable level? For example, how many genera-

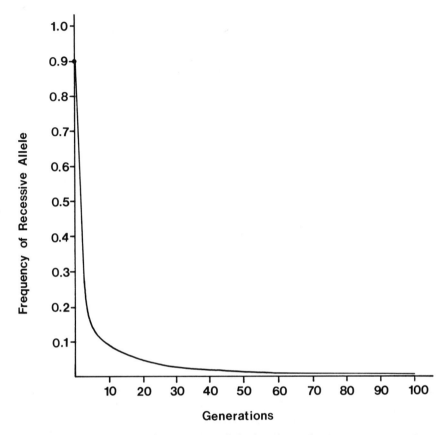

Figure 3.11 Frequency of a recessive allele during a selection program against the recessive phenotype. Initial allelic frequency was 0.9.

tions of selection against albino channel catfish are necessary to reduce the frequency of the a allele to the point where the population will breed 99.99% true? This question can be answered by using the following formula:

$$N = \frac{1}{q_n} - \frac{1}{q_0} \tag{3.12}$$

where N is the number of generations, q_n is the desired frequency of the recessive allele, and q_0 is the frequency when the question is asked.

For example, say the farmer's population of channel catfish is 9% albinos, and he wants to reduce the frequency of albinos to 0.01% (a population of normally pigmented channel catfish that breeds 99.99% true). How many generations of selection against albinos will be necessary to achieve his goal?

Step 1. q_0, the initial frequency of the a allele, is determined by using Eq. (3.2):

$$q_0 = \sqrt{f(\text{albinos})}$$

$$q_0 = \sqrt{0.09}$$

$$q_0 = 0.3$$

Step 2. q_n, the desired frequency of the a allele, is also determined by using Eq. (3.2):

$$q_n = \sqrt{f(\text{desired frequency of albinos})}$$

$$q_n = \sqrt{0.0001}$$

$$q_n = 0.01$$

Step 3.   The number of generations needed to reduce $f(a)$ from 0.3 to 0.01 is determined by using Eq. (3.12):

$$N = \frac{1}{0.01} - \frac{1}{0.3}$$

$$N = 96.67$$

It will take 97 generations (96.67 generations is rounded up to 97 generations) or about 388 years (in channel catfish, 1 generation is 4 years) to reduce $f(a)$ from 0.3 to 0.1. Obviously, this is not an efficient way to cull a recessive allele, since the frequency will not reach 0.01 until the end of the 24th century, and even then, the population will not breed true. But despite the obvious inefficiency inherent in this type of selection, it is commonly practiced in the erroneous assumption that it will succeed.

Progeny testing   Since selection against the recessive phenotype will not eliminate the recessive allele and produce a population that will breed true, the question now becomes: Is it possible to eliminate an undesired recessive allele, fix the desired dominant allele, and produce a population that will breed true and produce only the dominant phenotype? The answer is yes, and it is accomplished by using a technique called progeny testing, which is used to identify and to cull the heterozygotes. This approach eliminates hidden recessive alleles. Progeny testing is a program in which a fish's genotype is deciphered by mating it to a test fish (a fish whose genotype is known) and by then determining the phenotypic ratios of its progeny. Progeny testing is used to answer the following question: Is the fish a homozygous dominant individual which will be saved, or is it a heterozygous dominant individual which will be culled?

To answer this question, fish with the dominant phenotype are mated to fish with the recessive phenotype. Individuals with recessive phenotypes are usually used as test animals because they always breed true and will produce only recessive gametes, and also because their genotype is known (homozygous recessive). Thus, the fish with the dominant phenotype will be the one which determines the progeny's phenotypes.

Continuing with albinism in channel catfish as our example, progeny testing sets up the following problem:

normally pigmented channel catfish × albino channel catfish (test fish)
$(+?)$ (aa)

You want to know: Is ? (the unknown allele) a + or an a allele?

To answer this question, you have to assume that the normally pigmented channel catfish is a heterozygote. This assumption must be made because if you assume that it is homozygous, you do not need to progeny test the fish. Thus, you use the null or working hypothesis: The normally pigmented channel catfish is a heterozygote. You then examine the progeny that are produced to determine whether the fish is homozygous or heterozygous:

Only normally pigmented progeny (1:0 ratio) means that $? = +$.

Normally pigmented and albino progeny (1:1 ratio) means that $? = a$.

The reason we can declare the normally pigmented channel catfish to be either homozygous $(++)$ or heterozygous $(+a)$ from this progeny test is due to the fact that each parent contributes one allele to each offspring. We know the albino test fish contributes an a allele to all offspring, so the offsprings' phenotypes tell us the normally pigmented parent's genotype:

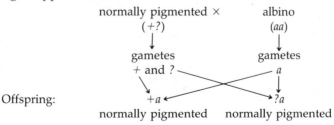

The only way a normally pigmented fish can be produced is if it receives a + allele from either parent. The only parent that can contribute a + allele is the normally pigmented one. In the preceding mating, only normally pigmented offspring were produced; therefore, the ? allele must be a + allele, and the normally pigmented parent must be homozygous ++.

Conversely:

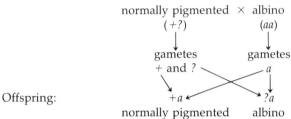

The only way an albino can be produced is if it receives an *a* allele from both parents. The production of an albino offspring means that the ? allele must be an *a* allele and that the normally pigmented parent must be heterozygous *+a*. Once you know the phenotypic ratio of the progeny, you either accept the null hypothesis and cull the normally pigmented brooder, or you reject the null hypothesis and keep it for breeding purposes.

How do you decide which ratio has been produced? As soon as you find one albino offspring you know that the normally pigmented channel catfish is a heterozygote and should be culled. You do not need to calculate whether the phenotypic ratio is statistically identical to a 1:1 ratio.

But what happens if you find no albinos? How many progeny do you need to examine before you decide that the normally pigmented channel catfish is homozygous and not heterozygous? Remember, the probability of producing an albino offspring from this mating is 0.5, because one half of the fish should be albinos, but that does not mean that one out of every two fish must be an albino. The same probability exists for the production of daughters and sons, but there are many aquaculturists who have four daughters and no sons.

Because phenotypic ratios are determined by probabilities, the number of offspring that must be examined when only the dominant phenotype is found, in order to reject the null hypothesis and declare the fish to be homozygous, is determined by the confidence level that you establish, e.g., 95, 99, or 99.9999%; i.e., you reject the null hypothesis at $P =$

0.05, 0.01, or 0.000001. The greater your confidence level, the more valuable the fish, because you can then give a better guarantee that the normally pigmented channel catfish will not carry the a allele and will breed true.

The probability that you use either to accept or to reject the null hypothesis establishes the odds of making the wrong decision when you decide that a fish is homozygous; $P = 0.05$, 0.01, and 0.000001 mean that the odds of incorrectly deciding that a fish is homozygous are 1 in 20, 1 in 100, and 1 in 1,000,000, respectively. Thus, if you use $P = 0.05$ to reject the null hypothesis and decide that 20 fish are homozygotes, the odds are that you will make one mistake, and one of the 20 fish will actually be a heterozygote. If you use $P = 0.01$, you will make such a mistake only one time out of every 100 decisions, and if you use $P = 0.000001$ that mistake will occur only once every 1,000,000 decisions.

Because one half of the offspring should be normally pigmented, the probability of producing a normally pigmented offspring is 0.5. The probability (P) of producing more than one normally pigmented off-spring without producing an albino is

$$P = (0.5)^N \qquad\qquad (3.13)$$

where 0.5 is the probability of producing a normally pigmented off-spring and N is the number of normally pigmented offspring. This for-mula will calculate the number of normally pigmented progeny that need to be observed without seeing an albino in order to accept or to reject the null hypothesis that the normally pigmented parent is a het-erozygote at a given probability (P).

For example, if a catfish farmer wants to progeny test a normally pigmented channel catfish, and he wants a 95% confidence level that he will correctly determine the fish's genotype ($P = 0.05$), he will mate the fish to an albino test fish. If he sees one albino offspring, he knows that the normally pigmented fish is a heterozygote and should be culled. If he sees no albino offspring, he can use Eq. (3.13) to determine how many offspring he must observe without seeing an albino in order to reject the null hypothesis and declare that the normally pigmented par-ent is homozygous at $P = 0.05$ (a 95% guarantee of being correct):

$$0.05 = (0.5)^N$$

To calculate N, you must convert the formula to logarithms (the logarithm of numbers can be found in many textbooks and can also be determined by inexpensive pocket calculators which are programmed to calculate logarithms at the touch of a button):

$$0.5 = (0.5)^N$$

$$\log 0.05 = \log (0.5)^N$$

$$\log 0.5 = (N)(\log 0.5)$$

$$\frac{\log 0.05}{\log 0.5} = N$$

$$N = \frac{-1.30103}{-0.30103}$$

$$N = 4.32$$

N is rounded up to 5, because you cannot have 0.32 of a fish. Consequently, a sample of five normally pigmented and no albino progeny must be observed in order to reject the null hypothesis at $P = 0.05$ and declare the normally pigmented parent to be homozygous. Actually, because N was rounded up, if five progeny are examined and no albino is in the sample, there is a probability of 96.88% that the normally pigmented channel catfish is a homozygote and a probability of 3.12% that it is a heterozygote. A probability of 96.88% may be impressive, but in a breeding program it is not impressive if you are trying to produce a true-breeding population. A probability of 96.88% means that 3.12% of the progeny-tested channel catfish that have been certified to be $++$ will actually be $+a$.

Fish are so prolific that there is no reason not to reduce the risk of incorrectly determining a fish's genotype to 0.000001. Only 20 consecutive normally pigmented progeny need to be observed to establish this probability. With this probability, only one in a million progeny-tested channel catfish that has been certified to be $++$ will actually be $+a$. If you increase the number to 50 consecutive normally pigmented progeny, you will reduce the probability to $P = 8.9 \times 10^{-16}$. This probability means that you will incorrectly determine a genotype less than once in a thousand trillion progeny-tested channel catfish that have been certified to be $++$. Obviously, that population will breed true. Because fish are so prolific, it is easy to progeny test both males and females quickly and reliably, something which is difficult in most forms of animal husbandry.

Progeny testing is so simple with fish that it is easy to progeny test for lethal genes or genes that cause sterility. When a recessive allele causes death or sterility in the recessive phenotype, it is impossible to use that phenotype as the test animal. In this case, the test animal becomes a known heterozygote (an animal that has been shown to produce both phenotypes). As before, the null hypothesis is that the fish whose genotype is being deciphered is a heterozygote. If recessive progeny are produced, the null hypothesis is accepted, the fish is declared a heterozygote and is culled. As before, if only normal offspring are produced you need to decide what confidence level you desire in order to reject the null hypothesis and to declare the fish to be homozygous. The probability of producing a normal offspring from this mating is 0.75, because the mating of two heterozygotes produces a 3 dominant progeny : 1 recessive progeny phenotypic ratio (75% dominant progeny). The probability of producing more than one normal offspring without producing a dead or sterile offspring is

$$P = (0.75)^N \tag{3.14}$$

where P is the probability of producing N normal offspring and 0.75 is the probability of producing a normal offspring.

Stubby (vertebral anomaly) in Guatopote culiche is an example of a recessive phenotype that is almost functionally sterile in males because the phenotype is also characterized by an abnormal gonopore which makes many stubby males unable to copulate. The St allele produces normal vertebrae, and the st allele produces the recessive phenotype (Schultz 1963).

If a tropical fish farmer finds stubby fish in his population of *Guatopote culiche* and wants to cull all heterozygous females, the only way to progeny test females is by using known heterozygous males as the test fish:

normal ♀ × heterozygous normal test ♂

($St?$)   ($Stst$)

The farmer wants to know: Is ? (the unknown allele) an St or an st allele?

As before, the null hypothesis is that the female is a heterozygote. Once this assumption is made, you can answer the question. To determine the female's genotype, you examine the progeny that are produced:

Normal and stubby progeny (3:1 ratio) means that ? = st.

Only normal progeny (1:0 ratio) means that ? = St.

As soon as the farmer finds one stubby offspring, he can accept the null hypothesis, declare the female to be a heterozygote, and cull her. If only normal progeny are seen, the farmer must use a probability either to accept or to reject the null hypothesis. For example, if the farmer wants to have a 95% confidence level ($P = 0.05$) for his decision, he can use Eq. (3.14) to calculate the number of offspring that must be observed without seeing a stubby fish in order to reject the null hypothesis and to declare the female to be a homozygote:

$$0.05 = (0.75)^N$$

As before, N must be solved by using logarithms:

$$0.05 = (0.75)^N$$
$$\log 0.05 = \log (0.75)^N$$
$$\log 0.05 = (N)(\log 0.75)$$
$$\frac{\log 0.05}{\log 0.75} = N$$
$$N = \frac{-1.30103}{-0.1249387}$$
$$N = 10.41$$

N is rounded up to 11, so a sample of 11 normal and no stubby progeny must be observed in order to reject the null hypothesis and to declare the fish to be homozygous. Actually, because N was rounded up, if 11 progeny are examined and no stubby fish are in the sample, there is a probability of 95.78% that the female is a homozygote and a probability of 4.22% that she is a heterozygote; this also means that 4.22% of the progeny-tested females that have been certified to be *StSt* will actually be *Stst*. As before, the probability can be reduced, and the guarantee can be increased without much additional effort. The numbers of offspring needed to establish various probability levels for rejecting the null hypothesis that a fish is heterozygous when only dominant progeny are detected are listed in Table 3.12.

Incomplete Dominance or Additive Gene Action   Autosomal genes that exhibit incomplete dominance or additive gene action produce three genotypes and three phenotypes. Because each genotype produces a unique phenotype, selection can fix either the dominant or recessive phenotype and either of the alleles in only one generation. A single culling of the undesired phenotypes will cull the undesired allele. The only phenotype that cannot be fixed by selection is the heterozygote.

Table 3.12 Number of Progeny with Dominant Phenotype, When Only That Phenotype Is Produced, That Must Be Observed to Reject the Null Hypothesis at Various Probabilities (P) That the Fish Being Progeny-Tested Is a Heterozygote

| | Number of progeny[a] | | |
| | Autosomal gene | | |
P[b]	Homozygous test parent	Heterozygous test parent	X-linked gene
0.05	5	11	5
0.01	7	17	7
0.001	10	25	10
0.0001	14	33	14
0.00001	17	41	17
0.000001	20	48	20
0.0000006	21	50	21
3.2×10^{-13}	42	100	42
8.9×10^{-16}	50	121	50
7.9×10^{-31}	100	241	100

[a] The number of fish are those that must be examined to produce probabilities equal to or just less than those that are listed (some of the numbers were rounded to the next higher whole number).

[b] Probabilities are the probabilities of incorrectly deciding that the fish is a homozygote. The guarantee of making a correct decision is $1.0 - P$.

Incomplete dominance A single generation of selection will produce a true-breeding population, regardless of whether the desired phenotype is the dominant or the recessive phenotype.

Selection for the recessive phenotype Because the desired phenotype is homozygous recessive, a single culling of the other phenotypes will cull the undesired dominant allele and produce a true-breeding population.

An example of a selection program to fix a recessive phenotype was the one done at Auburn University to fix the normal phenotype and the + allele and to eliminate all saddlebacks and the S allele in the Auburn strain of T. aurea (Tave et al. 1983). The dominant S allele is extremely deliterious. A single S allele (S+) produces saddlebacks, which are characterized by gross fin and skeletal abnormalities. A single S allele also reduces viability by 67% during the first 3 months of life and makes the saddlebacks far more sensitive to stress and disease. Two S alleles (SS) cause death prior to swim-up. Obviously, elimination of the S allele was necessary to improve the breeding value of the Auburn strain of T. *aurea*.

All S alleles were destroyed in a single generation by culling all saddlebacks, and the population now breeds true and produces only normal fish. The homozygous dominant fish culled themselves (died), and the heterozygotes were culled by a selective breeding program:

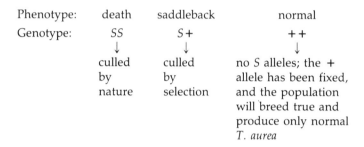

It would be equally effective to produce a true-breeding population of gold (gg) *T. mossambica*. If a tilapia farmer wanted to produce a true-breeding population, all he would have to do is cull all bronze (Gg) and black (GG) *T. mossambica*. Once he does, only gold tilapia will remain, and they will breed true because they are homozygous gg:

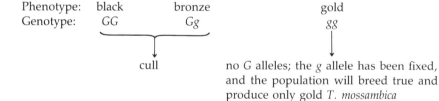

Selection for the dominant phenotype   Selection is equally effective if you want to fix the dominant phenotype. With incomplete dominant gene action, the homozygous dominant phenotype and the heterozygous phenotype are two separate phenotypes. Consequently, a single culling of the heterozygous and the homozygous recessive phenotypes will fix the dominant allele and produce a true-breeding population.

For example, if a tilapia farmer decides that he does not like gold tilapia and that he wants to produce a true-breeding population of black (GG) *T. mossambica*, all he would have to do is cull all bronze (Gg) and gold (*gg*) fish. Once he does, the only fish that remain are black, and they breed true because they are homozygous *GG:*

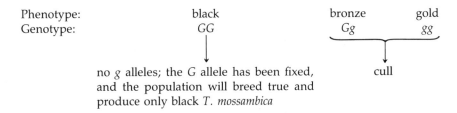

Phenotype: black   bronze   gold
Genotype: *GG*   *Gg*   *gg*

no *g* alleles; the *G* allele has been fixed,   cull
and the population will breed true and
produce only black *T. mossambica*

Selection for the heterozygous phenotype Selection cannot produce a true-breeding population of heterozygotes because they cannot breed true.

For example, if a tilapia farmer wants to produce a true-breeding population of bronze (*Gg*) *T. mossambica*, he will discover that he cannot achieve his goal by selection:

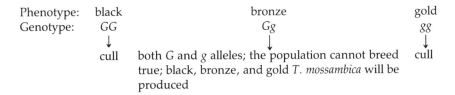

Phenotype: black   bronze   gold
Genotype: *GG*   *Gg*   *gg*

cull   both *G* and *g* alleles; the population cannot breed   cull
true; black, bronze, and gold *T. mossambica* will be
produced

If the farmer wants to produce 100% bronze *T. mossambica*, he must mate the two homozygotes (black × gold):

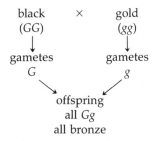

black   ×   gold
(*GG*)   (*gg*)

gametes   gametes
G   g

offspring
all *Gg*
all bronze

Additive gene action Selection for or against phenotypes controlled by genes with additive gene action is identical to that described for phenotypes that are controlled by genes with incomplete dominant gene action.

Selection for the homozygous phenotypes   Selection for the two homozygous phenotypes is equally effective, and the procedure to produce a true-breeding population is identical to that seen for phenotypes controlled by autosomal genes with incomplete dominant gene action.

For example, if a hatchery manager wants to produce both a true-breeding population of golden rainbow trout and a true-breeding population of normally pigmented rainbow trout, he can accomplish both goals in a single generation. To produce the true-breeding population of golden rainbow trout he must cull all normally pigmented and palomino rainbow trout. This will eliminate all G alleles. Once this is done, the only fish that remain will be golden rainbow trout, and they breed true because they are homozygous G'G':

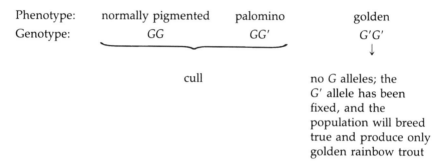

Phenotype: normally pigmented   palomino   golden
Genotype: GG   GG'   G'G'

                          cull   no G alleles; the
                                                         G' allele has been
                                                         fixed, and the
                                                         population will breed
                                                         true and produce only
                                                         golden rainbow trout

Conversely, to produce the true-breeding population of normally pigmented rainbow trout, all the manager needs to do is cull all golden and palomino rainbow trout. A single culling of these phenotypes will eliminate the G' allele. Once this is done, the only fish that remain will be normally pigmented rainbow trout, and they breed true because they are homozygous GG:

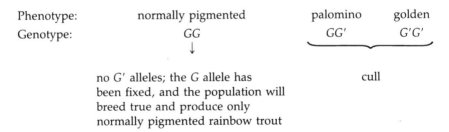

Phenotype: normally pigmented   palomino   golden
Genotype: GG   GG'   G'G'

              no G' alleles; the G allele has   cull
              been fixed, and the population will
              breed true and produce only
              normally pigmented rainbow trout

Selection for the heterozygous phenotype   As was seen earlier, selection cannot produce a true-breeding population of heterozygotes, because they cannot breed true.

For example, a trout farmer who wants to produce 100% palomino rainbow trout will discover that selection cannot achieve this goal:

Phenotype:	normally pigmented	palomino	golden
Genotype:	GG	GG'	$G'G'$
	↓	↓	↓
	cull	both G and G' alleles; the population cannot breed true; normally pigmented, palomino, and golden rainbow trout will be produced	cull

If the farmer wants to produce 100% palominos, he must cross normally pigmented rainbow trout with golden rainbow trout:

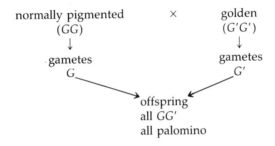

Two or More Autosomal Genes Selection for qualitative phenotypes that are controlled by two or more autosomal genes is similar to that which has been described for qualitative phenotypes that are controlled by single autosomal genes. If the desired phenotype is homozygous recessive at all loci, selection will fix the desired alleles in a single generation and produce a true-breeding population. Otherwise, selection will be a difficult way to produce a true-breeding population, and you should use progeny testing to cull recessive alleles. If the desired phenotype is produced by a genotype where one or more loci must be in the heterozygous state (e.g., you want to produce a true-breeding population of leather common carp), you will never be able to create a true-breeding population by selection.

Sex-Linked Genes

As was the case with phenotypes that are controlled by autosomal genes, sex-linked phenotypes often have different economic values, and the value of a population depends on its ability to produce the desired sex-linked phenotype more often than the undesired one. Thus, the elimination of alleles responsible for the production of undesired sex-linked phenotypes and the fixation of those responsible for the production of desired sex-linked phenotypes are important aspects of brood-stock management. Selection for or against sex-linked alleles can be either easy or difficult depending on the mode of gene expression and whether the gene is on the X or Y chromosome.

Y-Linked Genes   The elimination or fixation of a Y-linked allele is easy because there are two genotypes and two phenotypes. Every male with the desired allele expresses the desired phenotype, and every male with the undesired allele expresses the undesired phenotype. Consequently, a single culling of the males with the undesired phenotype will fix the desired allele and produce a true-breeding population.

For example, if a guppy breeder wants to eliminate the maculatus phenotype from one pond but fix it in another, he can accomplish both goals in a single generation:

Elimination of maculatus phenotype (XY_{Ma}):

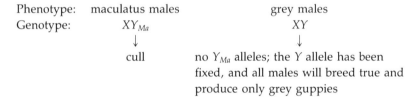

Fixation of maculatus phenotype in males (it cannot be fixed in females since normal females do not have a Y sex chromosome):

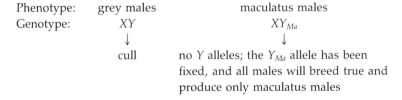

X-Linked Genes   The efficiency of fixing X-linked phenotypes by selection is similar to that seen for autosomal alleles. It is easy to fix a recessive X-linked phenotype by selecting against the dominant X-linked phenotype, but it is impossible to fix a dominant X-linked phenotype by selecting against the recessive X-linked phenotype.

Selection for recessive X-linked phenotypes Recessive X-linked phenotypes can be fixed in a single generation by culling all dominant X-linked phenotypes. Because every fish that has the undesired dominant X-linked allele expresses the undesired phenotype, all undesired X-linked alleles can be culled in a single generation.

For example, if a hatchery manager wants to produce a true-breeding population of transparent-tailed guppies, he can accomplish his goal in a single generation by culling all guppies with the caudalis phenotype:

Phenotype:	caudalis female	caudalis female	caudalis male	transparent-tailed female	transparent-tailed male
Genotype:	$X_{Cp}X_{Cp}$	$X_{Cp}X_{ch}$	$X_{Cp}Y$	$X_{ch}X_{ch}$	$X_{ch}Y$

cull

no X_{Cp} alleles; the X_{ch} allele has been fixed, and the population will breed true and produce only transparent-tailed guppies

Selection for dominant X-linked phenotypes As was the case with autosomal alleles, it is impossible to fix a dominant X-linked phenotype by selection, because it is impossible to eliminate a recessive X-linked allele by selection. This is because the dominant X-linked allele masks the recessive X-linked allele in heterozygous females. Selection against recessive X-linked alleles is more efficient than selection against autosomal recessive alleles because every male with a recessive allele can be culled, but the frequency of the allele will still approach zero asymptotically.

For example, if a guppy breeder wants to fix the caudalis phenotype, he will find that selection against the transparent-tailed phenotype will not eliminate the X_{ch} allele:

Phenotype:	caudalis female	caudalis female	caudalis male	transparent-tailed female	transparent-tailed male
Genotype:	$X_{Cp}X_{Cp}$	$X_{Cp}X_{ch}$	$X_{Cp}Y$	$X_{ch}X_{ch}$	$X_{ch}Y$

both the X_{Cp} and the X_{ch} alleles remain; the population will not breed true; both caudalis and transparent-tailed male guppies will be produced; all females, however, will have the caudalis pigmentation

cull

Because the X_{ch} allele is carried by heterozygous caudalis females, it is hidden, and the farmer will never be able to eliminate it by selection.

The effect of selection against a recessive X-linked phenotype in an effort to eliminate a recessive X-linked allele can be determined by using a series of four equations [(3.15), (3.16), (3.17), and (3.18)]. The effect of the first generation of selection against the recessive X-linked phenotype can be calculated by using Eqs. (3.15), (3.16), and (3.17).

In the first generation of selection, the effect of selection against a recessive X-linked phenotype in females is identical to that seen for selection against recessive autosomal phenotypes. Females with the recessive phenotype are culled, but the heterozygotes remain in the population. On the other hand, selection against a recessive X-linked phenotype in males is effective because all males that carry the undesired recessive X-linked allele express the undesired X-linked phenotype and can be culled. Consequently, the frequency of a recessive X-linked allele in males will drop to zero. The overall frequency is a function of the frequencies in each sex and the relative abundance of each sex:

1. Frequency in females:

$$f(X_{q1}-\female) = \frac{X_{q0}}{1 + X_{q0}}$$
(3.15)

where $X_{q1}-\female$ is the frequency of the recessive X-linked allele after one generation of selection, and X_{q0} is the initial frequency of the recessive X-linked allele in the population. Equation (3.15) is a modified form of Eq. (3.11) where $N = 1$.

2. Frequency in males:

$$f(X_{q1}-\male) = 0$$
(3.16)

3. Overall frequency:

$$f(X_{q1}-\text{overall}) = \frac{2(\female)(X_{q1}-\female)}{2(\female) + (\male)}$$
(3.17)

where $\female$ is the number of females and $\male$ is the number of males. Equation (3.17) is a modified form of Eq. (3.10) where the allele does not exist in males.

From this point on, selection will reduce the frequency of a recessive X-linked allele by one half every generation. This is because heterozygous females are the only broodfish that carry the recessive X-linked allele, and when they reproduce, it is split 50:50 between males and females. Males with the allele express the recessive X-linked phenotype and can be culled, but the allele will be hidden in females because all females with the allele will be heterozygotes. Thus, one half of the alleles can be culled each generation. After the first generation of selection, the effect of selection against a recessive X-linked phenotype can be determined by using the following formula:

$$X_{qn} = X_{q1}(0.5^N) \tag{3.18}$$

where X_{qn} is the frequency of the recessive X-linked allele after n generations of selection, X_{q1} is the frequency of the recessive X-linked allele after the first generation of selection, and N is the number of generations of selection after the first generation of selection.

For example, if a guppy farmer finds caudalis and transparent-tailed guppies in one of his ponds and wants to produce a population that will produce 100% caudalis females and 99.99% caudalis males [remember, no transparent-tailed females ($X_{ch}X_{ch}$) can be produced because all $X_{ch}Y$ males will be culled before they are allowed to reproduce], the number of generations required to achieve his goal by selection against transparent-tailed guppies can be determined as follows:

Phenotype	Genotype	Number	
transparent-tailed males	$X_{ch}Y$	450	500 males
caudalis males	$X_{Cp}Y$	50	
transparent-tailed females	$X_{ch}X_{ch}$	405	500 females
caudalis females	$X_{Cp}X_{Cp}$ and $X_{Cp}X_{ch}$	95	

Step 1. Calculate $f(X_{ch\text{-}0})$, the frequency of the allele before selection is initiated:

To calculate $f(X_{ch\text{-}0})$ in the males, use Eq. (3.7):

$$f(X_{ch\text{-}0}—\male) = \frac{\text{transparent-tailed } \male}{\text{total no. } \male}$$

$$f(X_{ch\text{-}0}—\male) = \frac{450}{500}$$

$$f(X_{ch\text{-}0}—\male) = 0.9$$

To calculate $f(X_{ch\text{-}0})$ in the females, use Eq. (3.9):

$$f(X_{ch\text{-}0}—\female) = \sqrt{\frac{\text{transparent-tailed } \female}{\text{total no. } \female}}$$

$$f(X_{ch\text{-}0}—\female) = \sqrt{\frac{405}{500}}$$

$$f(X_{ch\text{-}0}\text{---}\female) = \sqrt{0.81}$$

$$f(X_{ch\text{-}0}\text{---}\female) = 0.9$$

To calculate overall $f(X_{ch\text{-}0})$, use Eq. (3.10):

$$f(X_{ch\text{-}0}\text{---overall}) = \frac{2(500\female)(0.9) + (500\male)(0.9)}{2(500\female) + 500\male}$$

$$f(X_{ch\text{-}0}\text{---overall}) = \frac{1350}{1500}$$

$$f(X_{ch\text{-}0}\text{---overall}) = 0.9$$

Step 2. Calculate $f(X_{ch\text{-}1})$, the frequency of the allele after the first generation of selection:

To calculate $f(X_{ch\text{-}1})$ in females after one generation of selection, use Eq. (3.15):

$$f(X_{ch\text{-}1}\text{---}\female) = \frac{f(X_{ch\text{-}0})}{1 + f(X_{ch\text{-}0})}$$

$$f(X_{ch\text{-}1}\text{---}\female) = \frac{0.9}{1 + 0.9}$$

$$f(X_{ch\text{-}1}\text{---}\female) = 0.4736$$

To calculate $f(X_{ch\text{-}1})$ in males after one generation of selection, use Eq. (3.16):

$$f(X_{ch\text{-}1}\text{---}\male) = 0$$

To calculate overall $f(X_{ch\text{-}1})$ after one generation of selection use Eq. (3.17):

$$f(X_{ch\text{-}1}\text{---overall}) = \frac{2(50\female)(0.4736)}{2(50\female) + 50\male}$$

$$f(X_{ch\text{-}1}\text{---overall}) = \frac{47.36}{150}$$

$$f(X_{ch\text{-}1}\text{---overall}) = 0.3157$$

Note: The 50s used in the preceding equation represent an assumed 50:50 sex ratio.

Step 3. Calculate $f(X_{ch-n})$, the desired frequency of the allele:

The desired population will produce 0% transparent-tailed females and 0.01% transparent-tailed males.

Calculate $f(X_{ch-n}$—♂ $)$ in males:

After the first generation of selection, $f(X_{ch}) = 0$ in males that are used as broodstock. Some transparent-tailed males will be produced each generation, but they will be produced because of heterozygous females. Consequently $f(X_{ch-n}$—♂ $) = 0$.

After the first generation of selection, the X_{ch} alleles in the females will be split 50:50 between the daughters and the sons. Since all the sons with the allele will express the phenotype, the frequency of the X_{ch} allele in the females can be determined by multiplying the desired phenotypic frequency in the males by two:

$$f(X_{ch-n}\text{—♀}) = 2(0.0001)$$

$$f(X_{ch-n}\text{—♀}) = 0.0002$$

To calculate overall $f(X_{ch-n})$, use Eq. (3.17):

$$f(X_{ch-n}\text{—overall}) = \frac{2(50♀)(0.0002)}{2(50♀) + 50♂}$$

$$f(X_{ch-n}\text{—overall}) = \frac{0.02}{150}$$

$$f(X_{ch-n}\text{—overall}) = 0.0001333$$

Note: The 50s used in the preceding equation represent an assumed 50:50 sex ratio.

Step 4. Calculate the number of generations of selection after the first generation of selection that is required to reduce X_{ch} from 0.3157 to 0.0001333.

To do this, use Eq. (3.18):

$$f(X_{ch-n}) = f(X_{ch-1})(0.5^N)$$

$$0.0001333 = (0.3157)(0.5^N)$$

The formula must be converted to logarithms to solve for N:

$$0.0001333 = (0.3157)(0.5^N)$$

$$\frac{0.0001333}{0.3157} = (0.5)^N$$

$$0.0004222 = (0.5)^N$$

$$\log 0.0004222 = \log (0.5)^N$$

$$\log 0.0004222 = (N)(\log 0.5)$$

$$\frac{\log 0.0004222}{\log 0.5} = N$$

$$N = \frac{-3.3744444}{-0.30103}$$

$$N = 11.21$$

N is rounded up to 12. It will take a total of 13 generations to reduce $f(X_{ch})$ from 0.9 to 0.0001333. While selection against the recessive phenotype is far more effective in culling recessive X-linked alleles than recessive autosomal alleles, it is not possible, in either case, to produce a population that will breed true by selection alone.

Progeny testing The only way to eliminate recessive X-linked alleles is to progeny test the females and to cull the heterozygotes. The phenotype of the test male is unimportant because it is the female which contributes the X chromosome to her sons, and it is the sons' phenotypes which are examined in an X chromosome sex-linked progeny test. As was the case when progeny testing for autosomal genes, the female must be considered a heterozygote, and the null hypothesis is that the female is a heterozygote. Once this assumption is made, the progeny test sets up the following mating:

dominant ♀ × either phenotype ♂
$(X_{Dom}X_?)$ $(X_{does\ not\ matter}Y)$

You want to know: Is $X_?$ (the unknown allele) a dominant or a recessive X-linked allele? To answer this question, you examine the female's sons. You need not examine her daughters:

Only dominant sons (1:0 ratio) means that $X_? =$ the dominant X-linked allele.
Dominant and recessive sons (1:1 ratio) means that $X_? =$ the recessive X-linked allele.

If one son with the recessive X-linked phenotype is observed, the female can be considered a heterozygote and should be culled. When no sons with the recessive X-linked phenotype are observed, you must establish a desired confidence level in order to establish a probability that will enable you either to accept or to reject the null hypothesis that the female is a heterozygote.

The probability of producing a son with the dominant X-linked phenotype is 0.5, and the probability (P) of producing N consecutive dominant X-linked sons without producing a recessive X-linked son from this mating can be determined by using Eq. (3.13):

$$P = (0.5)^N$$

where 0.5 is the probability of producing a dominant son and N is the number of dominant sons. The number of sons that need to be sampled without observing a recessive X-linked son in order to produce various probabilities are listed in Table 3.12.

Sex-Limited Phenotypes Selection for sex-limited sex-linked phenotypes can be very difficult because one sex will not express the phenotype, regardless of genotype, and this will hide many undesired alleles. Selection for or against phenotypes that are sex-limited to males in some species of fish can be made efficient by the administration of small amounts of methyltestosterone to the feed or water to reveal the females' "unexpressed" phenotypes and thus decipher the genotypes (Emmens 1970). If this is done correctly, the females will not be harmed. Selection can then proceed as described for non-sex-limited sex-linked phenotypes.

CHAPTER 4

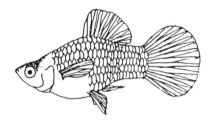

Genetics of Quantitative Phenotypes

4

Genetics of Quantitative Phenotypes

Quantitative phenotypes are those characteristics that are measured rather than described. These phenotypes do not segregate individuals into discrete phenotypic categories such as black or albino; instead, each quantitative characteristic exhibits a continuous distribution within a population. Phenotypic differences between and among fish are a matter of degree rather than of kind. Quantitative phenotypes are the important production traits such as length, weight, dressing percentage, viability, fat content, and fecundity, which are measured in millimeters, grams, inches, pounds, percentage, and eggs/kg female. For example, weight is a single phenotype and differences are measured as 0.1 g or 1 g or 1 oz or whatever unit is desired. Thus, phenotypic differences among individuals are matters of degrees: 0.1 g, 1 g, 1 oz, etc. Because these phenotypes are measured, they are also called metric traits.

Because each phenotype comprises only one distinct category and because the phenotypes are measured, variations within quantitative phenotypes are not given descriptive identities such as albino, palomino, and saddleback. Additionally, because quantitative phenotypes do not exhibit discrete variation, the phenotypes in a population cannot be described by a specific ratio. In a population, quantitative phenotypes are described by their central tendencies and the distributions around those tendencies: mean, variance, standard deviation, coefficient of variation, and range (see Appendix B for a detailed discussion about these values and how they are calculated). In a population, most quantitative phenotypes produce what is called a normal distribution. A theoretical normal distribution is illustrated in Fig. 4.1, and an actual frequency distribution for a quantitative phenotype is shown in Fig. 4.2.

Genetically, quantitative phenotypes are very complicated. Unlike qualitative phenotypes, which can be controlled by a single gene, a quantitative phenotype can be controlled by 20 or 50 or 100 or perhaps even 1000+ genes. The number of genes that control these phenotypes are unknown, and the modes of gene expression for the genes are also unknown.

This is unfortunate because of the importance of quantitative phenotypes in food fish, game fish, and bait fish farming. Qualitative phenotypes may be more important for ornamental fish farmers, but for other types of fish farming, the genetics of quantitative traits such as egg production, growth, survival, and dressing percentage are of paramount importance. Because these important phenotypes cannot be analyzed and understood by simple Mendelian genetics, it is impossible to

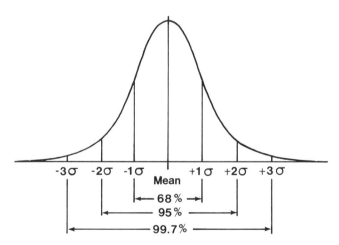

Figure 4.1 A theoretical normal distribution (a bell-shaped curve). Sixty-eight percent of the population is ±1σ (1 standard deviation) from the mean; 95% of the population is ±2σ from the mean; and 99.7% of the population is ±3σ from the mean.

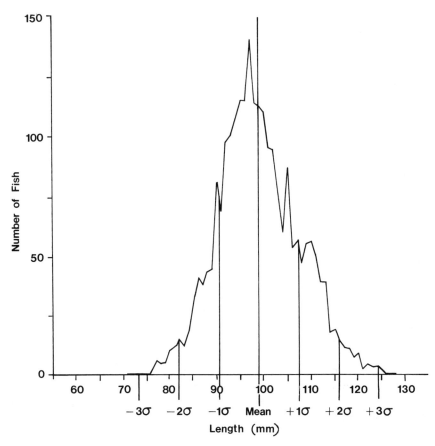

Figure 4.2 Frequency distribution of length at 3 months in a population of *Tilapia nilotica*. The distribution approximates a normal bell-shaped curve.

produce a population that will breed true for a given quantitative phenotype.

This chapter is divided into six sections. The first section describes the variance components of quantitative phenotypic variance and explains which ones can be exploited by breeding programs. The second section explores additive genetic variance, the most important genetic component of phenotypic variance, and shows how it can be exploited by selective breeding programs. The third section discusses dominance genetic variance and explains how it can be exploited by crossbreeding. The fourth section examines inbreeding and its relationship to productivity. A large portion of this section is devoted to the consequences of small breeding populations on productivity, through the accumulation of inbreeding and loss of genetic variance as a result of genetic drift. The fifth section describes genetic–environmental interaction variance and explains how to exploit this through yield trials. The final section examines the influence of environmental variance on quantitative phenotypes and discusses techniques that can be used to minimize its influence during breeding programs.

QUANTITATIVE PHENOTYPIC VARIATION AND ITS CONSTITUENT PARTS

Each gene that helps produce a quantitative phenotype exhibits discrete variance—a gamete either has one allele or it has another, and that is as discrete as possible. So why do quantitative phenotypes exhibit continuous variation? The answer is twofold: (1) It is true that each gene follows Mendelian genetics and that the two alleles at each locus segregate during meiosis so that a gamete will receive only one of these alleles. However, because so many loci are involved in the production of a quantitative phenotype and because each locus is undergoing segregation simultaneously and independently of all others (except for linked alleles), the end result is that the genetic potential of gametes, and ultimately offspring, vary to some degree, and in a population, the phenotypes produce a normal distribution (Fig. 4.3). (2) All phenotypes are controlled to some extent by the environment (a phenotype cannot be expressed unless the chemical precursors—proteins, carbohydrates, fats, vitamins, minerals—are provided in the diet), but the environment plays a major role in the expression of quantitative phenotypes. Furthermore, the environmental influence on the production of individual phe-

---

Figure 4.3 Theoretical distributions of phenotypes in the F_2 generation. The distributions were generated using the following assumptions: (1) heritability = 1.0; (2) a 12-unit difference between the parents is controlled by various numbers of genes with equal phenotypic effects; (3) there is no linkage; and (4) dominance is unidirectional. The graphs show that phenotypic variance becomes less discrete and more continuous as the number of genes which controls a phenotype increases.
Source: Principles of Plant Breeding by R. W. Allard. Copyright © 1960. Reprinted by permission of John Wiley & Sons, Inc.

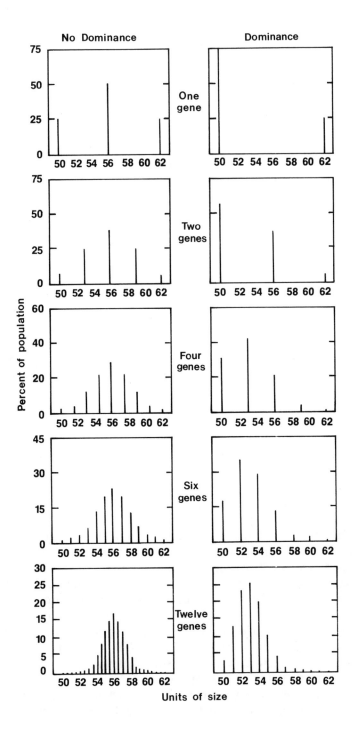

notypes also varies. Consequently, the environment is a major factor that contributes to the production of continuous distributions that are observed for quantitative phenotypes in a population (Fig. 4.4). Thus, the combined action of the environment and the simultaneous segregation of many genes produces continuous distributions.

Because quantitative phenotypes exhibit continuous variation, the only way to study them is to analyze the variance that exists in a population and to dissect it into its component parts. Once this has been done, the genetics of a trait can be understood and exploited. The phenotypic variance (V_P) that is observed for a quantitative trait is the sum of the genetic variance (V_G), the environmental variance (V_E), and the interaction that exists between the genetic and environmental variance (V_{G-E}):

$$V_P = V_G + V_E + V_{G-E}$$

Genetic variance is the component which is of greatest interest, because the object of any breeding program is to exploit or to change the genetics of a population in order to improve productivity and profits. But in order to work with V_G, it must be subdivided into its subcomponent parts. Genetic variance is the sum of the additive genetic variance (V_A), the dominance genetic variance (V_D), and the epistatic genetic variance (V_I):

$$V_G = V_A + V_D + V_I$$

The subcomponents of V_G do not refer to additive, dominant, and epistatic gene action. The redundant use of these terms is unfortunate because they have different meanings for each type of genetics. Additive genetic variance, V_D, and V_I refer to components of phenotypic variance, not the mode of gene action for a particular gene.

The differences among V_A, V_D, and V_I; how they are inherited; and their proportionate amounts are among the most important pieces of information in a breeding program. Each is different, and each is inherited in different ways, so different breeding programs are needed to exploit each type of genetic variance and thus improve productivity.

Dominance genetic variance is the variance that is due to the interaction of the alleles at each locus. Because of this, V_D cannot be inherited; it is, however, created anew each generation. The fact that some form of genetics cannot be inherited usually astonishes most people and causes confusion, but this is really a simple concept.

Dominance genetic variance is the interaction between alleles at each locus; as such, V_D is a function of the diploid state because that is

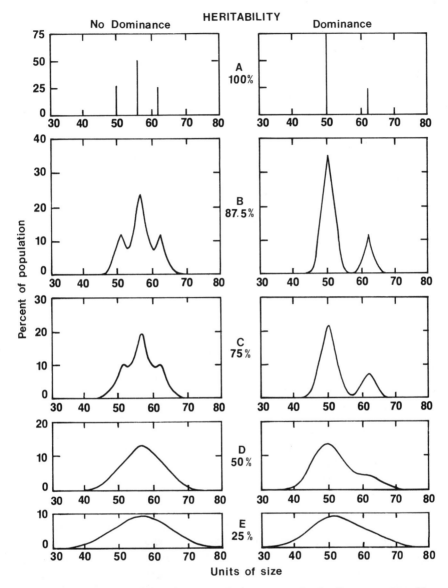

Figure 4.4 Theoretical distributions of phenotypes in the F$_2$ generation. The distributions were generated using the following assumptions: (1) the phenotype is controlled by a single gene; (2) there is a 12-unit difference between the parents; and (3) the effect of the environment ranges from 0% (heritability = 100%) to 75% (heritability = 25%). The graphs show that phenotypic variance becomes less discrete and more continuous as the environmental influence on a phenotype increases.

Source: *Principles of Plant Breeding by R. W. Allard. Copyright © 1960. Reprinted by permission of John Wiley & Sons, Inc.*

when alleles occur in pairs. During meiosis, homologous chromosomes and allelic pairs are separated during reduction division (Fig. 2.3, page 13), and the chromosome complement is reduced from the diploid to the haploid state. Gametes are haploid, so they contain no V_D, and that is why V_D cannot be inherited. The alleles that contribute to V_D are transmitted from a parent to its offspring, but the interaction (V_D) is destroyed during reduction division in meiosis. A fish cannot inherit V_D from a parent unless a gross error occurs during meiosis. When nondisjunction (the transmission of an allelic pair or a chromosomal pair) occurs, the end product is a genetic defect such as Down's syndrome in humans. Dominance genetic variance is created anew and in different combinations each generation.

Epistatic genetic variance is the variance that is due to the interaction of alleles between two or more loci. Epistatic genetic variance is the interaction that occurs across loci; V_D is within locus interaction.

Independent assortment and segregation during meiosis disrupt most V_I, so most V_I is not transmitted from parents to offspring but must be created anew each generation. Some V_I is transmitted, but only a small random sample of a parent's V_I is inherited by its progeny.

Additive genetic variance is the component that is due to the additive effect of the genes. Additive genetic variance is the sum of the effect of all alleles across all loci, taken independently, i.e., additive genetic variance is the sum of the effect of each allele that helps produce the phenotype. Additive genetic variance does not depend on specific interactions or combinations of alleles. Because V_A does not depend on interactions between or among alleles, it is not disrupted during meiosis. As a result, additive effects are transmitted from a parent to its offspring.

All of this theory is of value only if it is used to manipulate the population in order to improve productivity. Quantification of V_A, V_D, and V_I dictates the type of breeding program that must be used to exploit the phenotype's genetics and thus improve productivity. Once you decide to work with a quantitative phenotype, you must have some idea about the magnitude of V_P and the proportionate amounts of V_A, V_D, and V_I in order to choose the breeding program which will produce the best results.

Most breeders assume that $V_I \approx 0$. This is not true, but it is a practical assumption for two reasons: First, it is difficult to measure V_I. The breeding program needed to quantify V_I is very complicated, and because of that, few studies have attempted to quantify V_I for any phenotype. Second, it is virtually impossible to maximize combinations of alleles when you do not know what allelic combinations are desirable. It is often difficult to fix qualitative phenotypes that are controlled by epistatic gene interaction between just two genes. When you try to maximize interactions among dozens of genes without knowing what combination you want, it is extremely difficult to design a program that will produce the correct combination and improve productivity. It is often said that trying to exploit V_I is like trying to build a sand castle using dry sand. You can build the sand castle only so high before it collapses, because there is nothing to keep the sand grains in the right structural combination. You can exploit some V_I, but the improvement plateaus quickly.

Because V_I is difficult to exploit, the important genetic components of a quantitative phenotype are V_D and V_A. If you consider the definitions of V_A and V_D, you will get the impression that they are diametrically opposed components of genetic variance: V_D cannot be inherited, but V_A is inherited; V_D is created anew each generation, while V_A is never disrupted; V_D depends on allelic interactions, whereas V_A is additive. Consequently, these two forms of genetic variance require different breeding programs in order to manipulate and to exploit the benefits that they produce: V_A is exploited by selection; V_D is exploited by hybridization.

ADDITIVE GENETIC VARIANCE AND SELECTION

Selection is a breeding program in which individuals or families are chosen in an effort to change the population mean in the next generation. Selection is based on minimal performance levels: Fish that exceed minimal performance levels will be selected (saved) and used as broodstock; those that fall below minimal performance levels will be culled (removed). Selection is used in the hope that the select fish will produce offspring with a mean and range similar to that in the select population instead of the original mean and range.

Selection for quantitative phenotypes is more difficult than for qualitative phenotypes because of the number of genes involved and because of environmental influences. In order for selection to change the population mean of a quantitative phenotype, the genetic variance responsible for the production of the phenotype must be transmitted from parents to offspring in a predictable and reliable manner.

Of the three types of genetic variance, selection is able to exploit only one—V_A. The reason that selection can exploit only V_A is because V_A is a function of the alleles, whereas V_D is a function of the genotype—interactions between alleles. Parents do not transmit their genotypes to their offspring; instead they transmit haploid gametes, which are random samples of their alleles. Segregation and independent assortment during meiosis reduce genotypes from the diploid state to the haploid state and disrupt V_D. But because V_A is a function of the alleles rather than the genotype, it cannot be disrupted by meiosis. Consequently, V_D is created anew and in different combinations each generation, while V_A is transmitted in a predictable and reliable manner.

Since parents contribute alleles, not genotypes, to the next generation, individuals who are superior because of V_D will not contribute to progress in a selection program, because the underlying cause of their superiority cannot be transmitted. On the other hand, fish which are superior because of V_A will be able to transmit their additive effects to their progeny and will contribute to progress in a selection program.

Because V_A is transmitted in a predictable and reliable manner, and since fish that are superior because of V_A are able to transmit their additive effects and contribute to progress in a selection program, if you know the V_A for a quantitative phenotype, you should be able to predict the phenotypic mean in the next generation, based on the mean of the selected broodstock. Consequently, V_A is also called the variance of breeding values.

Heritability

It is important to know the proportionate amount of V_A so that you can predict whether selection will be effective. While this is not mandatory, the ability to predict the success of a selection program will enable you to decide whether the predicted success and effort needed to achieve it is worthwhile.

The proportionate amount of V_A for each quantitative phenotype is so important that it is among the most valuable pieces of information that can be distilled from a population. The proportionate amount of V_P that is controlled by V_A is called heritability (h^2):

$$h^2 = \frac{V_A}{V_P} \tag{4.1}$$

Heritability describes the percentage of V_P that is inherited in a predictable and reliable manner, because h^2 describes the genetic component that is not disrupted by meiosis.

Once you know the h^2 for a quantitative phenotype, you can predict the response to selection by using the following formula:

$$R = Sh^2 \tag{4.2}$$

where R is response (gain or loss) per generation, S is the selection differential or reach (the superiority or inferiority of the selected brood-stock over the population average), and h^2 is the proportionate amount of V_A. Equation (4.2) shows that h^2 can be considered the governor in any selection program: if $h^2 = 1.0$, $R = S$; if $h^2 = 0$, $R = 0$. Heritabilities between 0 and 1.0 allow you to gain or lose a proportionate amount of the reach. For example, say a catfish farmer decides to initiate a selection program for increased growth rate in the Marion strain of channel catfish, which currently averages 454 g at 18 months of age at his farm. To implement his program, the farmer selects 50 females that average 604 g and 40 males that average 692 g. What is the predicted average weight in the next generation? To answer this, you must (1) obtain a h^2; (2) calculate the selection differential or reach; (3) calculate the predicted response to selection:

Step 1. h^2 for growth in this strain is 0.50 (Dunham and Smitherman 1983).

Step 2. The selection differential (S) is calculated as follows:

$$S = \frac{\text{mean wt selected } ♀ + \text{mean wt selected } ♂}{2} - \text{population mean} \quad (4.3)$$

$$S = \frac{604 \text{ g} + 692 \text{ g}}{2} - 454 \text{ g}$$

$$S = 194 \text{ g}$$

Step 3. Calculate the predicted response to selection using Eq. (4.2):

$$R = Sh^2$$

$$R = (194 \text{ g})(0.5)$$

$$R = 97 \text{ g}$$

The next generation of Marion channel catfish at the hatchery (F_1) should average 97 g more than the parental generation (P_1) and will average

$$F_1 = P_1 \text{ mean weight} + \text{response to selection}$$

$$F_1 = 454 \text{ g} + 97 \text{ g}$$

$$F_1 = 551 \text{ g}$$

Figure 4.5 illustrates how h^2 determines response to selection.

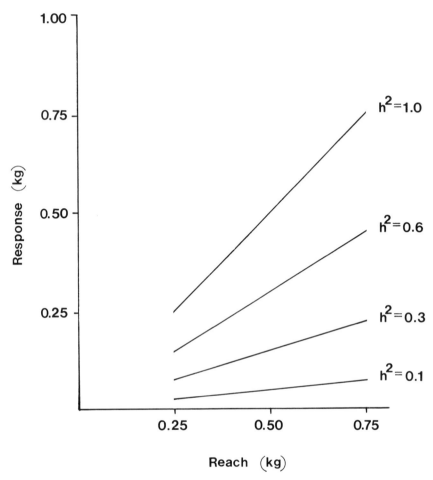

Figure 4.5 Response to selection in four different populations that have the same mean (1.0 kg) but different h^2s.

In the preceding example, the F_1 generation is predicted to average 97 g more than the P_1 generation. However, the actual gain can be greater or less than 97 g. The average weight of the F_1 population can even be less than the P_1 generation. Deviations from the predicted response are due to V_D and V_I effects, changes in the environment (V_E or V_{G-E}), and can also be due to sampling error as a result of working with a small population.

Phenotypes with h^2s $\gtrsim 0.25$ can be changed effectively by selection. Phenotypes with h^2s $\lesssim 0.15$ are not easy to change by selection. There is no magic number above which selection is effective, but the larger the h^2, the easier it will be to change a population mean by selection. The reason that it is difficult to change a phenotype with a small h^2 by selection is that nature may have already done it. Heritabilities are not immutable. The value can change for a number of reasons, the most important of which is changes in gene frequencies. Phenotypes with low h^2s are generally crucial in nature. Evolution has already exploited the heritable variance for that trait, and the phenotype has been changed to that which is most efficient in nature.

Reproductive phenotypes are examples of traits which have low h^2s, because of their importance in evolution. In evolution, reproduction is the name of the game: Produce more viable offspring than anyone else and you will have the highest fitness value. Consequently, natural selection has exploited most of the heritable genetic variance (V_A) of reproduction. Heritabilities for reproductive traits in livestock are small, and studies with fish suggest that this pattern also holds true in these animals (Table 4.1).

Meristic traits, on the other hand, are examples of phenotypes that generally have large h^2s (i.e., $h^2 \gtrsim 0.3$). The few studies that have been done on meristic phenotypes suggest that the number of spines in a dorsal fin or the number of vertebrae, while important, is not as crucial as reproductive performance. Heritabilities for meristic phenotypes have generally been greater than 0.3.

Heritabilities for growth rate and other production criteria generally fall between the values for reproductive and meristic phenotypes. More than 450 h^2s are listed in Table 4.1.

Heritabilities can also change because of changes in the environment. Heritability is the proportionate amount of V_A, so any factor that changes the denominator will change h^2:

$$h^2 = \frac{V_A}{V_A + V_D + V_I + V_E + V_{G-E}}$$

Several studies have shown that different environments change h^2. Hagen (1973) found that the h^2 for lateral plates in the three-spine stickleback was 0.50 at 21°C and 0.83 at 16°C. McIntyre and Blanc (1973) found that the h^2 for hatching time for steelhead trout eggs was 0 in incubators and 0.23 in troughs. Tave (1984A) found that the h^2 for dorsal fin ray number in the guppy was 0.41 at 19°C and 0.77 at 25°C. Beacham (1988) found that h^2s for hatching time, survival, weight, and length for both pink salmon and chum salmon were affected by temperature. Oldorf et al. (1989) found that h^2s for 136-day weight for the Manzala

Table 4.1   Heritabilities (h^2) in Fish

Species Phenotype	$h^2 \pm SE^{a,b}$	Reference
	Production Phenotypes	
Channel catfish		
30-day wt	0.92 ± 1.08	Reagan (1979)
60-day wt	0.32 ± 0.61	Reagan (1979)
90-day wt	0.18 ± 0.73	Reagan (1979)
120-day wt	0.92 ± 0.95	Reagan (1979)
150-day wt	0.98 ± 0.85	Reagan (1979)
5-month wt	0.61 ± 0.35	Reagan *et al.* (1976)
40-week wt	0.58 ± 0.29	Bondari (1983A)
40-week increased wt (realized)	0.10	Bondari (1983A)
40-week decreased wt (realized)	0.14	Bondari (1983A)
48-week wt, ♀	0.52 ± 0.42	El-Ibiary and Joyce (1978)
48-week wt, ♂	0.27 ± 0.37	El-Ibiary and Joyce (1978)
15-month wt	0.75 ± 0.53	Reagan *et al.* (1976)
18-month wt	0.41 ± 0.69	Reagan (1979)
18-month increased wt (realized) strains		
Rio Grande	0.24 ± 0.06	Dunham and Smitherman (1983)
Marion	0.50 ± 0.13	Dunham and Smitherman (1983)
Kansas	0.33 ± 0.10	Dunham and Smitherman (1983)
30-day length	1.22 ± 1.11	Reagan (1979)
60-day length	0.20 ± 0.64	Reagan (1979)
90-day length	0.61 ± 0.68	Reagan (1979)
120-day length	0.88 ± 0.91	Reagan (1979)
150-day length	1.13 ± 0.92	Reagan (1979)
5-month length	0.12 ± 0.30	Reagan *et al.* (1976)
40-week length	0.67 ± 0.33	Bondari (1983A)
40-week increased length (realized)	0.35	Bondari (1983A)
40-week decreased length (realized)	0.16	Bondari (1983A)
48-week length, ♀	0.81 ± 0.46	El-Ibiary and Joyce (1978)
48-week length, ♂	0.39 ± 0.39	El-Ibiary and Joyce (1978)
15-month length	0.67 ± 0.57	Reagan *et al.* (1976)
18-month length	0.40 ± 0.72	Reagan (1979)
feed conversion at 5 months, Tifton strain	0.0 to 0.38	Burch (1986)

[a] When the standard error (SE) was determined, it is listed with the h^2.
[b] h^2s followed by an asterisk were determined by a method which includes V_D.

Table 4.1 Heritabilities (h^2) in Fish (*continued*)

Species Phenotype	$h^2 \pm SE^{a,b}$	Reference
Production Phenotypes (continued)		
body depth, ♀	0.16 ± 0.32	El-Ibiary and Joyce (1978)
body depth, ♂	0.19 ± 0.36	El-Ibiary and Joyce (1978)
girth, ♀	0.39 ± 0.36	El-Ibiary and Joyce (1978)
girth, ♂	0.19 ± 0.37	El-Ibiary and Joyce (1978)
head wt, ♀	0.57 ± 0.46	El-Ibiary and Joyce (1978)
head wt, ♂	0.28 ± 0.38	El-Ibiary and Joyce (1978)
dressing wt, ♀	0.48 ± 0.40	El-Ibiary and Joyce (1978)
dressing wt, ♂	0.25 ± 0.36	El-Ibiary and Joyce (1978)
dressing wt	0.43 ± 0.85	Reagan (1979)
dressing percentage, ♀	0.0 ± 0.20	El-Ibiary and Joyce (1978)
dressing percentage, ♂	0.0 ± 0.27	El-Ibiary and Joyce (1978)
percent fat	0.61 ± 0.78	Reagan (1979)
fry mortality at 1.1 ppm O_2	0.9 ± 0.3* to 1.7 ± 0.1*	Durborow *et al.* (1985)
Rainbow trout/Steelhead trout		
4-month wt	0.52 ± 0.15	Gall and Huang (1988A)
147-day wt (realized)	0.26	Kincaid *et al.* (1977)
150-day wt	0.09 ± 0.10	Aulstad *et al.* (1972)
150-day wt	0.50 ± 0.07	von Limbach (1970)
9-month wt	0.2	Langholz and Hörstgen-Schwark (1987)
280-day wt	0.29 ± 0.20	Aulstad *et al.* (1972)
first-year wt	0.06 to 0.19	Refstie (1980)
334-day wt	0.82 ± 0.38	Klupp (1979)
1-year wt	0.38 ± 0.25	Linder *et al.* (1983)
1-year wt	0.20 ± 0.11	Gall and Huang (1988A)
16-month wt	0.2	Langholz and Hörstgen-Schwark (1987)
2-year wt	0.32 ± 0.11	Gjerde and Gjedrem (1984)
2-year wt	−0.01 to 0.34	Gunnes and Gjedrem (1981)
25-month wt	0.18 ± 0.12	Gall and Huang (1988A)
2.5-year wt	0.21	Gjerde and Schaeffer (1989)
2.5-year wt	0.38 ± 0.22	McKay *et al.* (1986)
4-year wt	0.27 ± 0.20	McKay *et al.* (1986)
mature wt	0.20 ±0.10	Gall and Huang (1988A)
wt at slaughter (~2.8 kg)	0.43	Pohar (1992)
150-day length	0.16 ± 0.14	Aulstad *et al.* (1972)
280-day length	0.37 ± 0.23	Aulstad *et al.* (1972)
first year length	0.20 to 0.24	Refstie (1980)

Table 4.1   Heritabilities (h^2) in Fish (*continued*)

Species Phenotype	$h^2 \pm SE^{a,b}$	Reference
Production Phenotypes (continued)		
2-year length	0.26 ± 0.11	Gjerde and Gjedrem (1984)
2-year length	-0.03 to 0.32	Gunnes and Gjedrem (1981)
2.5-year length	0.18	Gjerde and Schaeffer (1989)
2.5-year fork length	0.66 ± 0.27	McKay *et al.* (1986)
4-year fork length	0.48 ± 0.24	McKay *et al.* (1986)
K factor	-0.02 to 0.06	Gunnes and Gjedrem (1981)
2.5-year K factor	0.19	Gjerde and Schaeffer (1989)
28-month body circumference		
at pelvic fin	0.22	Gjerde and Schaeffer (1989)
at dorsal fin	0.28	Gjerde and Schaeffer (1989)
at anal fin	0.30	Gjerde and Schaeffer (1989)
28-month body height		
at pelvic fin	0.31	Gjerde and Schaeffer (1989)
at dorsal fin	0.34	Gjerde and Schaeffer (1989)
28-month body width		
at pelvic fin	0.26	Gjerde and Schaeffer (1989)
at dorsal fin	0.19	Gjerde and Schaeffer (1989)
28-month body thickness		
at pelvic fin	0.23	Gjerde and Schaeffer (1989)
at dorsal fin	0.21	Gjerde and Schaeffer (1989)
2.5-year belly thickness	0.32	Gjerde and Schaeffer (1989)
head length, 334 days	0.76 ± 0.40	Klupp (1979)
2-year gutted wt	0.19 ± 0.12	Gjerde and Gjedrem (1984)
2.5-year gutted wt	0.21 to 0.34	Gjerde and Schaeffer (1989)
viscera wt	0.33	Gjerde and Schaeffer (1989)
dressing percentage	0.01 ± 0.05	Gjerde and Gjedrem (1984)
dressing percentage	0.36	Gjerde and Schaeffer (1989)
meatiness score	0.14 ± 0.06	Gjerde and Gjedrem (1984)
meat color score	0.06 ± 0.08	Gjerde and Gjedrem (1984)
meat color score	0.27	Gjerde and Schaeffer (1989)
abdominal fat	0.25	Gjerde and Schaeffer (1989)
flesh composition		
percent water	0.33	Gjerde and Schaeffer (1989)
percent fat	0.47	Gjerde and Schaeffer (1989)
percent protein	0.03	Gjerde and Schaeffer (1989)
age at sexual maturity	0.21 ± 0.14	McKay *et al.* (1986)
sexual precosity	0.3	Burger and Chevassus (1987)
precocious spawning	0.1	Tofteberg and Hansen (1987)
age of spawning, ♀	0.38 ± 0.0	Gall *et al.* (1988)
age of spawning, ♀ (realized)	0.53 ± 0.05 to 0.55 ± 0.07	Siitonen and Gall (1989)

Table 4.1 Heritabilities (h^2) in Fish (*continued*)

Species Phenotype	$h^2 \pm SE^{a,b}$	Reference
Production Phenotypes (*continued*)		
age of spawning, ♀ (realized)	0.9	Sadler *et al.* (1992)
maturity at 2 years	−0.05 ± 0.12 to 0.28 ± 0.12	Gjerde and Gjedrem (1984)
age of maturation, 2nd year	0 to 0.09	Møller *et al.* (1979)
age of maturation, 3rd year	0.47 ± 0.74	Møller *et al.* (1979)
postspawning weight, ♀	0.37 ± 0.12* to 0.74 ± 0.15*	Gall and Gross (1978B)
postspawning wt, ♀	0.15 ± 0.14	Gall and Huang (1988B)
postspawning weight, ♂	0.14 ± 0.13* to 0.63 ± 0.18*	Gall and Gross (1978B)
gonad wt	0.19 ± 0.23	Gjerde and Schaeffer (1989)
egg size	0.20 ± 0.05*	Gall (1975)
egg size	0.05 ± 0.08* to 0.50 ± 0.14*	Gall and Gross (1978B)
egg size	0.28 ± 0.16	Gall and Huang (1988B)
egg no.	0.19 ± 0.06*	Gall (1975)
egg no.	0.16 ± 0.10* to 0.67 ± 0.15*	Gall and Gross (1978B)
egg no.	0.32 ± 0.14	Gall and Huang (1988B)
egg no.	0.33 ± 0.20	Haus (1984)
egg volume	0.20 ± 0.05*	Gall (1975)
egg volume	0.16 ± 0.10* to 0.76 ± 0.15*	Gall and Gross (1978B)
egg volume	0.44 ± 0.20	Haus (1984)
egg volume	0.30 ± 0.15	Gall and Huang (1988B)
volume of eggs retained in ovary after spawning	0.34 ± 0.22	H. Zhang *et al.* (1990)
no. eggs/kg	0.20 ± 0.05*	Gall (1975)
hatching time		
incubators	0.0	McIntyre and Blanc (1973)
troughs	0.23	McIntyre and Blanc (1973)
no. eyed eggs	0.09 ± 0.11* to 0.40 ± 0.13*	Gall and Gross (1978B)
percent dead eyed eggs	0.15 ± 0.06	Kanis *et al.* (1976)
percent dead alevins	0.14 ± 0.03	Kanis *et al.* (1976)
food conversion	0.41 ± 0.13*	Kinghorn (1983)
O_2 consumption	0.51 ± 0.12*	Kinghorn (1983)
percent fat	0.47 ± 0.34*	Kinghorn (1983)
N digestibility	−0.07 ± 0.24*	Kinghorn (1983)
formalin tolerance	0.41 ± 0.07	von Limbach (1970)
DDT tolerance (13.3 μg/l)	0.06 ± 0.03	von Limbach (1970)

Table 4.1   Heritabilities (h^2) in Fish (*continued*)

Species Phenotype	$h^2 \pm SE^{a,b}$	Reference
Production Phenotypes (continued)		
DDT tolerance (40.0 μg/l)	0.05 ± 0.03	von Limbach (1970)
resistance to 8 ppb mercury	0.49 ± 0.14	Blanc (1973)
tolerance to high temperature (realized)	0.48	Ihssen (1986)
tolerance to low temperature (realized)	0.03	Ihssen (1986)
total haemolytic activity	0.34 ± 0.40 to 0.96 ± 0.55	Røed *et al.* (1990)
non-specific haemolytic activity	0.08 ± 0.33 to 0.33 ± 0.27	Røed *et al.* (1990)
resistance to viral haemorrhagic septicaemia (VHS)	−0.10 ± 0.12 to 0.30 ± 0.09	Kaastrup *et al.* (1991)
Brown trout		
percent dead uneyed eggs	0.03 ± 0.06	Kanis *et al.* (1976)
percent dead eyed eggs	−0.01 ± 0.01	Kanis *et al.* (1976)
percent dead alevins	0.02 ± 0.01	Kanis *et al.* (1976)
survival in acidic water		
dead eggs	0.33	Edwards and Gjedrem (1979)
dead eyed eggs	0.27	Edwards and Gjedrem (1979)
percent live fingerlings	0.10	Edwards and Gjedrem (1979)
Brook trout		
144-day wt	0.08 ± 0.13	Robison and Luempert (1984)
243-day wt	0.60 ± 0.27	Robison and Luempert (1984)
survival to eyed embryo	0.09 ± 0.05 to 0.16 ± 0.0	Robison and Luempert (1984)
survival to hatch	0.08 ± 0.05 to 0.16 ± 0.0	Robison and Luempert (1984)
survival to 144 days	−0.04 ± 0.04 to 0.11 ± 0.0	Robison and Luempert (1984)
survival to 243 days	−0.02 ± 0.04 to 0.10 ± 0.0	Robison and Luempert (1984)
Atlantic salmon		
12-week wt	0.89 ± 0.32	Bailey and Loudenslager (1986)
6-month wt	0.40 ± 0.26	Bailey and Loudenslager (1986)
10-month wt	0.15 to 0.33	Refstie and Steine (1978)
15-month wt	0.67 ± 0.32	Bailey and Loudenslager (1986)

Table 4.1   Heritabilities (h^2) in Fish (*continued*)

Species Phenotype	$h^2 \pm SE^{a,b}$	Reference
	Production Phenotypes (*continued*)	
2-year wt	0.12 ± 0.05 to 0.38 ± 0.15	Standal and Gjerde (1987)
3-year wt	0.38 ± 0.10	Gjerde and Gjedrem (1984)
3-year wt	0.03 to 0.36	Gunnes and Gjedrem (1978)
3.5-year wt	0.34	Naevdal *et al.* (1976)
12-week length	0.79 ± 0.31	Bailey and Loudenslager (1986)
6-month length	0.15 to 1.0	Naevdal *et al.* (1975)
6-month length	0.57 ± 0.28	Bailey and Loudenslager (1986)
10-month length	0.07 to 0.38	Refstie and Steine (1978)
1-year length	0.03 to 0.64	Naevdal *et al.* (1975)
15-month length	0.73 ± 0.32	Bailey and Loudenslager (1986)
length after 1 year in net-pens	0.28* to 0.67*	Bailey and Friars (1989)
18-month length	0.03 to 0.57	Naevdal *et al.* (1975)
2-year length	0.07 to 0.36	Naevdal *et al.* (1975)
2-year length	0.08 ± 0.05 to 0.42 ± 0.16	Standal and Gjerde (1987)
3-year length	0.33 ± 0.10	Gjerde and Gjedrem (1984)
3-year length	0.15 ± 0.33*	Gunnes and Gjedrem (1978)
3½-year length	0.76 to 0.84	Naevdal *et al.* (1976)
fork length at grilse (realized)	0.27	Friars *et al.* (1990)
2-year K-factor	0.05 ± 0.03 to 0.34 ± 0.12	Standal and Gjerde (1987)
3-year K factor	−0.12 to 0.04	Gunnes and Gjedrem (1978)
3.5-year K factor	0.71 to 0.81	Naevdal *et al.* (1976)
percent smolting at year 1	−0.04 ± 0.09 to 0.16 ± 0.05	Refstie *et al.* (1977)
survival after 2 years in net-pens	0.03 ± 0.07 to 0.83 ± 0.13	Standal and Gjerde (1987)
age at sexual maturity	0.48 ± 0.20 to 0.66 ± 0.08	Gjerde (1984A)
maturity at 3 years	0.06 ± 0.05 to 0.39 ± 0.12	Gjerde and Gjedrem (1984)
egg size	0.44 ± 0.18	Halseth (1984)
egg volume	0.13 ± 0.15	Halseth (1984)
egg no.	0.30 ± 0.16	Halseth (1984)
percent dead uneyed eggs	0.32 ± 0.06	Kanis *et al.* (1976)
percent dead eyed eggs	0.05 ± 0.04	Kanis *et al.* (1976)
percent dead alevins	0.04 ± 0.01	Kanis *et al.* (1976)

Table 4.1 **Heritabilities (h^2) in Fish** (*continued*)

Species Phenotype	$h^2 \pm SE^{a,b}$	Reference
Production Phenotypes (continued)		
percent dead fry	-0.02 ± 0.01	Kanis *et al.* (1976)
percent smolt	$0.85 \pm 0.34^*$	Bailey and Friars (1990)
3-year gutted wt	0.44 ± 0.11	Gjerde and Gjedrem (1984)
dressing percentage	0.03 ± 0.02	Gjerde and Gjedrem (1984)
meat color score	0.01 ± 0.03	Gjerde and Gjedrem (1984)
survival in acid water		
1 month	0.54 ± 0.30	Schom (1986)
4 months	0.62 ± 0.26	Schom (1986)
5 months	0.57 ± 0.30	Schom (1986)
length in acid water		
4 months	0.72 ± 0.33	Schom (1986)
5 months	0.29 ± 0.18	Schom (1986)
6 months	0.39 ± 0.78	Schom (1986)
erythrocyte cell membrane fragility (strength)	0.60 ± 0.20	Gjedrem *et al.* (1991B)
susceptibility to *Aeromonas salmonicida*	0.48 ± 0.17	Gjedrem *et al.* (1991A)
tolerance to vibrio, *Vibrio anguillarum*	0.11 ± 0.06	Gjedrem and Aulstad (1974)
Coho salmon		
57-day post swim-up wt	0.61 ± 0.31	Iwamoto *et al.* (1982)
84-day post swim-up wt	0.38 ± 0.25	Iwamoto *et al.* (1982)
141-day post swim-up wt	0.25 ± 0.22	Iwamoto *et al.* (1982)
wt at transfer to salt water	0.25 ± 0.22 to 0.63 ± 0.19	Hershberger *et al.* (1990A)
wt after 8 months in salt water	0.19 ± 0.11 to 0.33 ± 0.10	Hershberger *et al.* (1990A)
wt after 8 months in salt water (realized)		
odd-year line	1.22 ± 0.32	Hershberger *et al.* (1990A)
even-year line	0.81 ± 0.30	Hershberger *et al.* (1990A)
wt 8 months after transfer to salt water	$0.29 \pm 0.13^*$ to $0.40 \pm 0.15^*$	Iwamoto *et al.* (1990)
84-day post swim-up length	0.30 ± 0.24	Iwamoto *et al.* (1982)
141-day post swim-up length	0.22 ± 0.21	Iwamoto *et al.* (1982)
length 8 months after transfer to salt water	$0.31 \pm 0.01^*$	Iwamoto *et al.* (1990)
length after 8 months in salt water	0.18 ± 0.11 to 0.30 ± 0.10	Hershberger *et al.* (1990A)
hatching time	0.26	Sato (1980)

Table 4.1 Heritabilities (h^2) in Fish (*continued*)

Species Phenotype	$h^2 \pm SE^{a,b}$	Reference
Production Phenotypes (*continued*)		
percent smolts	0.23 ± 0.07 to 0.25 ± 0.14	Saxton *et al.* (1984)
percent survival after transfer to salt water	0.07 ± 0.02 to 0.08 ± 0.07	Saxton *et al.* (1984)
sexual precocity		
normal males	0.03 to 0.30	Iwamoto *et al.* (1984)
jack males	0.10 to 0.29	Iwamoto *et al.* (1984)
flesh carotenoid level	0.30 ± 0.14* to 0.50 ± 0.16*	Iwamoto *et al.* (1990)
dressed wt	0.22 ± 0.43* to 0.36 ± 0.02*	Iwamoto *et al.* (1990)
percent moisture	0.14 ± 0.13* to 0.25 ± 0.19*	Iwamoto *et al.* (1990)
percent lipid	0.18 ± 0.13* to 0.19 ± 0.23*	Iwamoto *et al.* (1990)
Chinook Salmon		
smolt weight		
Big Qualicum strain	0.0 ± 0.59	Withler *et al.* (1987)
Harrison strain	0.19 ± 0.29	Withler *et al.* (1987)
Robertson strain	0.88 ± 0.72	Withler *et al.* (1987)
survival of uneyed fry		
Big Qualicum strain	0.07 ± 0.11	Withler *et al.* (1987)
Harrison strain	0.21 ± 0.21	Withler *et al.* (1987)
Robertson strain	0.0 ± 0.11	Withler *et al.* (1987)
fry survival		
Big Qualicum strain	0.0 ± 0.0	Withler et al. (1987)
Harrison strain	0.0 ± 0.02	Withler et al. (1987)
Robertson strain	0.0 ± 0.01	Withler *et al.* (1987)
alevin survival		
Big Qualicum strain	0.0 ± 0.01	Withler *et al.* (1987)
Harrison strain	0.03 ± 0.02	Withler *et al.* (1987)
Robertson strain	0.11 ± 0.09	Withler *et al.* (1987)
smolt survival		
Big Qualicum strain	0.03 ± 0.04	Withler *et al.* (1987)
Harrison strain	0.01 ± 0.16	Withler *et al.* (1987)
Robertson strain	0.0 ± 0.07	Withler *et al.* (1987)
Chum salmon		
alevin wt		
3°C	0.04 ± 0.02	Beacham (1988)
8°C	0.0 ± 0.01	Beacham (1988)
15°C	0.05 ± 0.02	Beacham (1988)

Table 4.1   **Heritabilities (h^2) in Fish** (*continued*)

Species Phenotype	$h^2 \pm SE^{a,b}$	Reference
Production Phenotypes (*continued*)		
alevin yolk wt		
3°C	0.02 ± 0.02	Beacham (1988)
8°C	0.01 ± 0.03	Beacham (1988)
16°C	0.01 ± 0.01	Beacham (1988)
alevin tissue wt		
3°C	0.06 ± 0.05	Beacham (1988)
8°C	0.14 ± 0.07	Beacham (1988)
16°C	0.45 ± 0.22	Beacham (1988)
fry wt		
3°C	0.0 ± 0.02	Beacham (1988)
8°C	0.08 ± 0.05	Beacham (1988)
16°C	0.15 ± 0.09	Beacham (1988)
fry yolk wt		
3°C	0.20 ± 0.10	Beacham (1988)
8°C	0.0 ± 0.04	Beacham (1988)
16°C	0.01 ± 0.04	Beacham (1988)
fry tissue wt		
3°C	0.0 ± 0.02	Beacham (1988)
8°C	0.13 ± 0.09	Beacham (1988)
16°C	0.22 ± 0.12	Beacham (1988)
alevin length		
3°C	0.29 ± 0.20	Beacham (1988)
8°C	0.12 ± 0.06	Beacham (1988)
15°C	0.67 ± 0.24	Beacham (1988)
fry length		
3°C	0.10 ± 0.11	Beacham (1988)
8°C	0.15 ± 0.11	Beacham (1988)
16°C	0.52 ± 0.25	Beacham (1988)
hatching time		
3°C	0.52 ± 0.04	Beacham (1988)
8°C	0.30 ± 0.18	Beacham (1988)
15°C	0.37 ± 0.22	Beacham (1988)
button-up time		
3°C	0.13 ± 0.11	Beacham (1988)
8°C	0.06 ± 0.04	Beacham (1988)
15°C	0.01 ± 0.05	Beacham (1988)
embryo survival		
3°C	0.0 ± 0.08	Beacham (1988)
8°C	0.0 ± 0.0	Beacham (1988)
15°C	0.0 ± 0.06	Beacham (1988)

Table 4.1 **Heritabilities (h^2) in Fish** (*continued*)

Species Phenotype	$h^2 \pm SE^{a,b}$	Reference
Production Phenotypes (*continued*)		
alevin survival		
3°C	0.0 ± 0.01	Beacham (1988)
8°C	0.0 ± 0.01	Beacham (1988)
15°C	0.03 ± 0.03	Beacham (1988)
no. days between spawning and emergence		
Kilches River strain	0.0 ± 1.1	Smoker (1981)
Whiskey River strain	0.8 ± 1.1	Smoker (1981)
tolerance to vibrio, *V. anguillarum*	0.5 ± 0.6	Smoker (1981)
Pink salmon		
alevin wt		
4°C	0.01 ± 0.02	Beacham (1988)
8°C	0.01 ± 0.01	Beacham (1988)
16°C	0.03 ± 0.03	Beacham (1988)
alevin yolk wt		
4°C	0.02 ± 0.02	Beacham (1988)
8°C	0.0 ± 0.0	Beacham (1988)
16°C	0.0 ± 0.02	Beacham (1988)
alevin tissue wt		
4°C	0.03 ± 0.05	Beacham (1988)
8°C	0.13 ± 0.03	Beacham (1988)
16°C	0.02 ± 0.03	Beacham (1988)
fry wt	0.22 ± 0.74	Beacham and Murray (1988B)
fry wt		
4°C	0.02 ± 0.04	Beacham (1988)
8°C	0.0 ± 0.02	Beacham (1988)
16°C	0.37 ± 0.12	Beacham (1988)
fry yolk wt		
4°C	0.0 ± 0.03	Beacham (1988)
8°C	0.0 ± 0.05	Beacham (1988)
16°C	0.12 ± 0.11	Beacham (1988)
fry tissue wt		
4°C	0.0 ± 0.04	Beacham (1988)
8°C	0.04 ± 0.01	Beacham (1988)
16°C	0.23 ± 0.09	Beacham (1988)
60-day wt	1.05 ± 0.66	Beacham and Murray (1988B)
75-day wt	0.03 ± 0.29	Beacham (1989)
75-day wt	0.44 ± 0.42	Beacham and Murray (1988A)
150-day wt	0.98 ± 0.46	Beacham and Murray (1988B)

Table 4.1   Heritabilities (h^2) in Fish (*continued*)

Species Phenotype	$h^2 \pm SE^{a,b}$	Reference
Production Phenotypes (continued)		
215-day wt	0.90 ± 0.43	Beacham and Murray (1988B)
255-day wt	0.58 ± 0.40	Beacham and Murray (1988A)
315-day wt	0.94 ± 0.46	Beacham and Murray (1988B)
345-day wt	0.69 ± 0.48	Beacham and Murray (1988A)
410-day wt	0.75 ± 0.39	Beacham and Murray (1988B)
500-day wt	0.94 ± 0.43	Beacham and Murray (1988B)
GSI, ♂	0.56 ± 0.31	Beacham and Murray (1988B)
GSI, ♀	0.07 ± 0.24	Beacham and Murray (1988B)
Sockeye salmon	0.27 ± 2.6 to	
resistance to IHN virus	0.38 ± 5.1	McIntyre and Amend (1978)
Amago salmon		
percent eyed eggs	0.10	Sato and Morikawa (1982)
hatchability	0.08	Sato and Morikawa (1982)
hatching time	0.22	Sato and Morikawa (1982)
duration of hatching	0.28	Sato and Morikawa (1982)
Arctic charr		
2-year wt	0.31 ± 0.21	Nilsson (1990)
3-year wt	0.59 ± 0.11	Nilsson (1990)
2-year length	0.31 ± 0.21	Nilsson (1990)
3-year length	0.55 ± 0.13	Nilsson (1990)
2-year K-factor	0.23 ± 0.20	Nilsson (1990)
3-year K-factor	0.53 ± 0.18	Nilsson (1990)
Common carp		
4-month wt	0.48*	Nagy *et al.* (1980)
1-year wt	0 ± 0.07 to	Nenashev (1966)
	0.34 ± 0.50	
1-year wt	0.49	Smíšek (1979A)
2-year wt	0.50 ± 0.08	Nenashev (1966)
2-year wt	0.15	Smíšek (1979A)
3-year wt	0.24	Smíšek (1979A)
4-year wt	0.21	Smíšek (1979A)
wt gain (realized)	0	Moav and Wohlfarth (1976)
decreased growth rate (realized)	0.3	Moav and Wohlfarth (1976)
wt gain	0.25 to 0.38	Smíšek (1979B)
wt gain	0.47	Brody *et al.* (1981)
growth rate	0.25	Smíšek (1979B)
1-year length	0.04 ± 0.07 to	Nenashev (1966)
	0.34 ± 0.05	
2-year length	0.55 ± 0.08	Nenashev (1966)
1-year body depth	0.42	Smíšek (1979A)
2-year body depth	0.69	Smíšek (1979A)
3-year body depth	0.47	Smíšek (1979A)

Table 4.1 Heritabilities (h^2) in Fish (*continued*)

Species Phenotype	$h^2 \pm SE^{a,b}$	Reference
Production Phenotypes (continued)		
4-year body depth	0.63	Smíšek (1979A)
body shape (length:weight)		
high (realized)	0.47	Ankorion (1966)
low (realized)	0.33	Ankorion (1966)
fat content	0.14 to 0.15	Smíšek (1979B)
percent N in dry matter	0.15 to 0.16	Smíšek (1979B)
survival during anoxia	0.51*	Nagy *et al.* (1980)
Tilapia nilotica		
4-week wt	0 to 0.06	Lester *et al.* (1988)
45-day wt	0.04 ± 0.14	Tave and Smitherman (1980)
58-day increased wt (realized)	−0.10 ± 0.02	Teichert-Coddington and Smitherman (1988)
58-day decreased wt (realized)	0.36 ± 0.08	Teichert-Coddington and Smitherman (1988)
8-week wt	0 to 0.21 ± 0.42	Lester *et al.* (1988)
10-week wt	0 to 0.46 ± 0.52	Lester *et al.* (1988)
90-day wt	0.04 ± 0.06	Tave and Smitherman (1980)
12-week wt, ♀	0 to 0.77 ± 0.77	Lester *et al.* (1988)
12-week wt, ♂	0 to 0.39 ± 0.67	Lester *et al.* (1988)
14-week wt, ♀	0 to 1.30 ± 0.82	Lester *et al.* (1988)
14-week wt, ♂	0	Lester *et al.* (1988)
136-day wt, ♀	0.71 ± 0.18	Kronert *et al.* (1987)
136-day wt, ♀	0.37 ± 0.10	Kronert *et al.* (1989)
136-day wt, ♂	0.71 ± 0.17	Kronert *et al.* (1987)
136-day wt, ♂	0.30 ± 0.08	Kronert *et al.* (1989)
210-day wt, ♀	0.58 ± 0.16*	Oldorf *et al.* (1989)
210-day wt, ♂	0.66 ± 0.17*	Oldorf *et al.* (1989)
7-month wt (realized)		
increased wt	0.05 to 0.08	C.-M. Huang and Liao (1990)
decreased wt	0.03	C.-M. Huang and Liao (1990)
wt at maturity, ♀	0.02 ± 0.0*	Uraiwan (1988)
wt at maturity, ♂	0.01 ± 0.03*	Uraiwan (1988)
wt at first spawning	0.46 ± 0.44 to 0.64 ± 0.72	Lester *et al.* (1988)
45-day length	0.10 ± 0.19	Tave and Smitherman (1980)
90-day length	0.06 ± 0.06	Tave and Smitherman (1980)
age at first spawn	0	Lester *et al.* (1988)
age at maturity, ♀	0*	Uraiwan (1988)
age at maturity, ♂	0.10 ± 0.04*	Uraiwan (1988)
fecundity at first spawning	0 to 0.09 ± 0.28	Lester *et al.* (1988)

Table 4.1 Heritabilities (h^2) in Fish (*continued*)

Species Phenotype	$h^2 \pm SE^{a,b}$	Reference
Production Phenotypes (continued)		
136-day gonad wt, ♀	0.15 ± 0.06	Kronert *et al.* (1987)
136-day gonad wt, ♀	0.21 ± 0.06	Kronert *et al.* (1989)
136-day gonad wt, ♂	0.12 ± 0.04	Kronert *et al.* (1987)
136-day gonad wt, ♂	0.08 ± 0.03	Kronert *et al.* (1989)
210-day gonad wt, ♀	0.04 ± 0.03*	Oldorf *et al.* (1989)
210-day gonad wt, ♂	0.06 ± 0.04*	Oldorf *et al.* (1989)
136-day GSI, ♀	0.28 ± 0.09	Kronert *et al.* (1987)
136-day GSI, ♀	0.30 ± 0.08	Kronert *et al.* (1989)
136-day GSI, ♂	0.36 ± 0.10	Kronert *et al.* (1987)
136-day GSI, ♂	0.15 ± 0.04	Kronert *et al.* (1989)
210-day GSI, ♀	0.43 ± 0.13*	Oldorf *et al.* (1989)
210-day GSI, ♂	0.44 ± 0.13*	Oldorf *et al.* (1989)
sex ratio	0.26	Lester *et al.* (1989)
Tilapia mossambica		
5-month wt, ♀ (realized)	0.01 to 0.36	Ch'ang (1971B)
5-month wt, ♂ (realized)	0.10 to 0.76	Ch'ang (1971B)
5-month wt, ♀ + ♂ (realized)	−0.01 to 0.33	Ch'ang (1971B)
Tilapia aurea		
at Tifton, GA		
40-week wt gain, ♀ (realized)	0.38 ± 0.08	Bondari *et al.* (1983)
40-week wt gain, ♂ (realized)	0.20 ± 0.09	Bondari *et al.* (1983)
40-week wt loss, ♀ (realized)	−0.17 ± 0.12	Bondari *et al.* (1983)
40-week wt loss, ♂ (realized)	0.14 ± 0.07	Bondari *et al.* (1983)
40-week increased length, ♀ (realized)	0.87 ± 0.20	Bondari *et al.* (1983)
40-week increased length, ♂ (realized)	0.40 ± 0.19	Bondari *et al.* (1983)
40-week decreased length, ♀ (realized)	−0.09 ± 0.11	Bondari *et al.* (1983)
40-week decreased length, ♂ (realized)	0.07 ± 0.08	Bondari *et al.* (1983)
at Auburn, AL		
49-week wt gain, ♀ (realized)	0.10 ± 0.06	Bondari *et al.* (1983)
49-week wt gain, ♂ (realized)	0.27 ± 0.07	Bondari *et al.* (1983)
49-week wt loss, ♀ (realized)	0.34 ± 0.08	Bondari *et al.* (1983)
49-week wt loss, ♂ (realized)	0.41 ± 0.07	Bondari *et al.* (1983)
Red seabream		
45-day length	0.84	Taniguchi *et al.* (1981)
45-day wt	0.92	Taniguchi *et al.* (1981)
Nibe-croaker		
45-day length	0.53	Taniguchi *et al.* (1981)
45-day wt	0.86	Taniguchi *et al.* (1981)

Table 4.1   Heritabilities (h^2) in Fish (*continued*)

Species Phenotype	$h^2 \pm SE^{a,b}$	Reference
Production Phenotypes (continued)		
Mosquito fish		
42-day wt (realized)	0.01 ± 0.10 to 0.04 ± 0.06	Campton and Gall (1988)
60-day wt, ♂	0.25*	Busack and Gall (1983)
60-day wt, ♀	0.77*	Busack and Gall (1983)
60-day length, ♂	0.25*	Busack and Gall (1983)
60-day length, ♀	0.72*	Busack and Gall (1983)
male age at sexual maturity	0.41*	Busack and Gall (1983)
age at maturity (realized)		
early	0.15 ± 0.08	Campton and Gall (1988)
late	0.67 ± 0.17	Campton and Gall (1988)
brood size	0.16*	Busack and Gall (1983)
Guppy		
wt gain	0	Ryman (1972)
resistance to 7 ppm lead	0.58 ± 0.32 to 0.26 ± 0.17	Burger (1974)
resistance to 10 ppm lead (realized)	0.28 to 0.68	Burger (1974)
Threespine stickleback		
age at first reproduction		
coastal population	0.17 ± 0.23	Snyder (1991)
inland population	0.74 ± 0.64	Snyder (1991)
size at first reproduction		
coastal population	0.55 ± 0.32	Snyder (1991)
inland population	0.07 ± 0.36	Snyder (1991)
initial clutch size		
coastal population	0.30 ± 0.17	Snyder (1991)
inland population	0.13 ± 0.33	Snyder (1991)
egg size (inland population)	0.38 ± 0.29	Snyder (1991)
Zebra danio		
4-week wt	0.26 ± 0.07	von Hertell *et al.* (1990)
12-week wt	0.22 ± 0.65	von Hertell *et al.* (1990)
20-week wt, ♀	0.17 ± 0.11	von Hertell *et al.* (1990)
20-week wt, ♂	0.41 ± 0.15	von Hertell *et al.* (1990)
4-week length	0.34 ± 0.08	von Hertell *et al.* (1990)
12-week length	0.29 ± 0.08	von Hertell *et al.* (1990)
20-week length, ♀	0.07 ± 0.08	von Hertell *et al.* (1990)
20-week length, ♂	0.41 ± 0.14	von Hertell *et al.* (1990)
spawning interval (days)	0.69 ± 0.08	von Hertell *et al.* (1990)
egg no.	0.50 ± 0.07	von Hertell *et al.* (1990)
percent hatch	0.78 ± 0.08	von Hertell *et al.* (1990)
percent survival to 14 days	0.78 ± 0.08	von Hertell *et al.* (1990)

Table 4.1   Heritabilities (h^2) in Fish (*continued*)

Species Phenotype	$h^2 \pm SE^{a,b}$	Reference
Meristic Phenotypes (*continued*)		
Swordtail		
105-day length, ♂		
individually-raised	0.74 ± 0.23*	Campton (1992)
group-raised	0.32 ± 0.17*	Campton (1992)
105-day length, ♀	0.05 ± 0.07*	Campton (1992)
175-day length, ♂		
individually-raised	0.77 ± 0.23*	Campton (1992)
group-raised	−0.05 ± 0.12*	Campton (1992)
175-day length, ♀	−0.04 ± 0.11*	Campton (1992)
length at maturity		
individually raised	0.82 ± 0.23*	Campton (1992)
group raised	−0.17 ± 0.15*	Campton (1992)
Meristic Phenotypes		
Common carp		
dorsal fin ray no.	0.36 ± 0.04 to 0.46 ± 0.12	Nenashev (1970)
lateral line scale count		
right	0 to 0.54 ± 0.05	Nenashev (1970)
left	0.01 ± 0.07 to 0.71 ± 0.06	Nenashev (1970)
no. vertebrae in trunk	0.14 ± 0.08 to 0.67 ± 0.06	Nenashev (1970)
total no. vertebrae	0.10 ± 0.08 to 0.90 ± 0.07	Nenashev (1970)
no. gill filaments		
right	0.20 ± 0.06	Nenashev (1966)
left	0.26 ± 0.06	Nenashev (1966)
Guppy		
dorsal fin ray no.		
19°C	0.41 ± 0.12	Tave (1984A)
25°C	0.77 ± 0.18	Tave (1984A)
caudal fin ray no.	0.4 to 1.0	Beardmore and Shami (1976)
lateral line scale count	−0.57 ± 0.22 to 1.14 ± 0.82	Shami and Beardmore (1978)
Velvet belly shark		
no. whole vertebrae	0.22 ± 0.36	Tave (1984B)
no. half vertebrae	0.22 ± 0.25	Tave (1984B)
total no. vertebrae	0.59 ± 0.21	Tave (1984B)
position of first dorsal fin spine	0.59 ± 0.20	Tave (1984B)
position of second dorsal fin spine	0.24 ± 0.28	Tave (1984B)

Table 4.1   Heritabilities (h^2) in Fish (*continued*)

Species Phenotype	$h^2 \pm SE^{a,b}$	Reference
Meristic Phenotypes (*continued*)		
Threespine stickleback		
no. lateral plates		
21°C	0.50 ± 0.04	Hagen (1973)
16°C	0.83 ± 0.03	Hagen (1973)
asymmetry of lateral plates	0.63 ± 0.16	Hagen (1973)
no. gill rakers	0.58 ± 0.06	Hagen (1973)
Fourspine stickleback		
dorsal spine no.	0.47 ± 0.05 to 0.80 ± 0.13	Hagen and Blouw (1983)
Tilapia nilotica		
dorsal fin ray no.		
hard spines	0.24 ± 0.27	Tave (1986)
soft rays	0.23 ± 0.26	Tave (1986)
total	0.67 ± 0.38	Tave (1986)
pectoral fin ray no.	0.36 ± 0.25	Tave (1986)
anal fin ray no.		
hard	0.0 ± 0.10	Tave (1986)
soft	0.59 ± 0.31	Tave (1986)
total	0.61 ± 0.30	Tave (1986)
no. gill rakers	0.21 ± 0.20	Tave (1986)
Rainbow trout		
no. pyloric caeca	0.75 ± 0.34*	Bergot *et al.* (1976)
no. pyloric caeca	0.46 ± 0.10 to 0.53 ± 0.07	Chevassus *et al.* (1979)
no. pyloric caeca	0.68 ± 0.08	Bergot *et al.* (1981B)
no. vertebrae	0.66	Kirpichnikov (1981)
anal fin ray no.	0.93 ± 0.50	Leary *et al.* (1985B)
dorsal fin ray no.	0.90 ± 0.27	Leary *et al.* (1985B)
no. gill rakers		
lower arch (right + left sides)	0.37 ± 0.21	Leary *et al.* (1985B)
upper arch (right + left sides)	0.67 ± 0.11	Leary *et al.* (1985B)
Brown trout		
no. pyloric caeca	1.10 ± 0.26	Bergot *et al.* (1976)
no. pyloric caeca	0.36 ± 0.15	Blanc *et al.* (1979)
no. vertebrae	0.90	Kirpichnikov (1981)
Viviparous eelpout		
no. vertebrae	0.81	Kirpichnikov (1981)
dorsal fin ray no.	0.79	Kirpichnikov (1981)
pectoral fin ray no.	0.54	Kirpichnikov (1981)
anal fin ray no.	0.60	Kirpichnikov (1981)

strain of *T. nilotica* that were grown in tanks in Germany were 0.6 for males and 0.7 for females, while those for fish that were grown in Kenya were 0.39 for males and 0.34 for females. These studies demonstrate that h^2s must be calculated in the environment in which selection will occur, or the h^2 will give a false impression of the heritable variance that is available for exploitation.

Even factors such as the fish's age when the phenotype is measured may affect h^2. Several studies have shown that h^2s for length and weight change with age (Nenashev 1966; Aulstad *et al.* 1972; Chevassus 1976; Reagan *et al.* 1976; Gall and Gross 1978A; Klupp 1979; Reagan 1979; Smíšek 1979A; Tave and Smitherman 1980; Bailey and Loudenslager 1986; McKay *et al.* 1986; Beacham and Murray 1988A; Gall and Huang 1988A; Beacham 1989). Age can also have an effect on the h^2 of a meristic phenotype (Shami and Beardmore 1978). Campton (1992) found that culture density affected the h^2 for length in male swordtails.

The fact that h^2 is not immutable should be considered when making management decisions. Moav and Wohlfarth (1976) found that the h^2 for increased growth rate in an Israeli strain of common carp was ≈ 0. Many concluded, as a result of this study, that it is impossible to improve productivity in common carp culture via selection, because they have no V_A for increased growth rate. That conclusion should not be extrapolated to include all populations of common carp. It may be true for the population that Moav and Wohlfarth (1976) examined, but Smíšek (1979A) found h^2s of 0.49, 0.15, 0.24, and 0.21 for body weight at 1, 2, 3, and 4 years, respectively, for a Czechoslovakian population of common carp. On a practical level, this means that h^2s calculated in populations other than the one at your station should be used only as guidelines, not as gospel.

It is not mandatory to determine h^2s prior to selection, but it is often advisable because they will enable you to predict whether a breeding program is likely to be worthwhile. The common carp breeding program in Israel is an example of a breeding program that was done before a h^2 was obtained. Moav and Wohlfarth (1976) selected for increased growth rate for five generations without success. Heritability for increased growth in that population was essentially zero. Had they obtained a h^2 for increased growth rate prior to selection, they could have predicted that selection would not have worked, and this would have saved a lot of effort.

An example of a selective breeding program that was discouraged because a small h^2 was detected occurred in the tilapia breeding program at Auburn University. Tave and Smitherman (1980) found that the h^2 for early growth rate in the Auburn University-Ivory Coast strain of *T.*

nilotica was ≤ 0.1, and they predicted that selection in that population would not be an efficient way to improve growth rate. Several aquaculturists questioned the results, because they felt that the h^2s that were calculated were far too small. Consequently, a selection program was initiated to test the validity of the prediction; the response to selection was that which had been predicted, and the realized h^2 was similar to the h^2 calculated by Tave and Smitherman (1980) (Teichert-Coddington and Smitherman 1988).

Realized h^2 is the h^2 that is obtained from a selection program. A realized h^2 gives you the total variance that was available for exploitation in that generation. Although the calculation is a bit more complicated than that shown below, realized h^2 is basically the ratio of response to reach [a rearrangement of Eq. (4.2)]:

$$h^2 = \frac{R}{S}$$

There are three major techniques that are used to calculate h^2s: sib analysis; regression; and realized h^2. The only reason it's important to know what the techniques are and what they measure is so that you can critically evaluate an experiment that estimates a h^2. In sib analysis, V_P is partitioned into its components by analyzing data that are gathered from a series of families. In full-sib analysis, the families are not related. In half-sib analysis, a series of full-sib families are nested within a series of half-sib families, i.e., each male is mated to two or more females (it can be the other way, but the usual procedure is to produce paternal half-sib families). In diallele analysis, groups of males and females are mated so that each male is mated to all females and each female is mated to all males within each group to create a series of overlapping paternal and maternal half-sib families.

In parent–offspring regression, the average value of the progeny from each of a series of families is paired with either the father's or the mother's values, and a regression of the paired data is used to determine h^2. Mid-parent–offspring regression differs from parent–offspring regression only in that the average value of each family is paired with the average of the two parents.

Realized h^2s are determined from a selection program. They are a function of the selection differential and the response to selection.

Half-sib, diallele, parent–offspring, and mid-parent–offspring analyses provide accurate estimates of h^2. Full-sib analysis really does not provide a h^2, because it confounds V_A with V_D. Realized h^2 gives the percentage of V_P that was exploited that generation. Table 4.2 lists what

Table 4.2   Comparison of What the Six Major Techniques That Are Used to Calculate h^2 Measure[a]

Technique	What h^2 measures	Components of V_P that can be quantified
Full-sib	$\dfrac{V_A + V_D}{V_P}$	$(V_A + V_D)$, V_E
Half-sib	$\dfrac{V_A}{V_P}$	V_A, V_D, V_E
Diallele	$\dfrac{V_A}{V_P}$	V_A, V_D, V_E
Parent–offspring	$\dfrac{V_A}{V_P}$	V_A
Mid-parent–offspring	$\dfrac{V_A}{V_P}$	V_A
Realized h^2	$\dfrac{R}{S}$	presumably V_A

[a] Two assumptions were made: V_I and V_M (maternal variance) $= 0$.

the different techniques measure and the components of V_P that can be determined from properly designed experiments.

It is not difficult to calculate h^2s, but a detailed digression about experimental designs that are needed to generate the data that will reveal h^2s is beyond the scope of this book. Excellent sources for information about experimental designs, statistical models, and ways to analyze data that are used to calculate h^2s are Becker's (1985) "Manual of Quantitative Genetics" and Falconer's (1981) "Introduction to Quantitative Genetics."

Selection Programs

Once you decide to employ a selection program in order to exploit a population's V_A and increase productivity, you must decide what type of selection program will give the best results. There are four basic types of selection (Fig. 4.6). All have their uses and are important in either evolution or breeding. In fish culture, two are more important than the others: no selection and directional selection.

No Selection It may appear ridiculous to begin a section about the exploitation of V_A by discussing a program that tries to avoid selection, but the absence of selection is often the most important breeding program that a hatchery manager can employ. An understanding of how to prevent selection gives a better appreciation of broodstock management. When accompanied by goals and plans, selection is desirable, but unprogrammed or unintentional selection is not.

Unintentional selection can alter the gene pools of game fish populations so that they are unable to survive and reproduce in the wild. It can also alter gene pools of food fish populations by eliminating potentially valuable alleles for disease resistance or growth, thus hamstringing future selective breeding programs.

Unintentional selection occurs every time a hatchery manager handles his fish. Subconsciously, hatchery managers and workers who handle fish generally select the larger fish, the fish that are gravid when the manager wants to spawn the fish, and fish that have pronounced secondary sexual characteristics. An example of unintentional selection that eliminated the potential for increased growth rate and, in turn, eliminated selection as a method for improving productivity is what happened in the common carp breeding program in Israel (Moav and Wohlfarth 1976). They attributed the absence of V_A for increased growth rate in the population that they evaluated to the practice of spawning the largest fish to get more eggs per female.

The hatchery itself also selects fish in an unintentional manner. Hatcheries that raise game fish may produce populations with narrow genetic bases because hatcheries select the fish that perform best in raceways or small earthen ponds and the fish that were able to become gravid and reproduce under hatchery conditions. The fish that are not able to survive and reproduce in hatcheries may be the fish that would be best in the wild.

Consequently, a program of no selection may be one of the most important aspects of broodstock management for populations of game fish. Unintentional selection can ruin a population or make the produced stock far too costly. If you do not know why you are culling certain phenotypes, do not do it. Unintentional selection may be a prime reason why it is difficult to establish a self-reproducing population of game fish by stocking hatchery-produced fish. The hatchery environment and the aquaculturists may cull the alleles that enable the population to do well under environmental extremes.

To prevent unintentional selection from lowering the average fitness of the population, you may have to alter your cultural practices. To prevent unintentional selection you should: (1) spawn fish over the

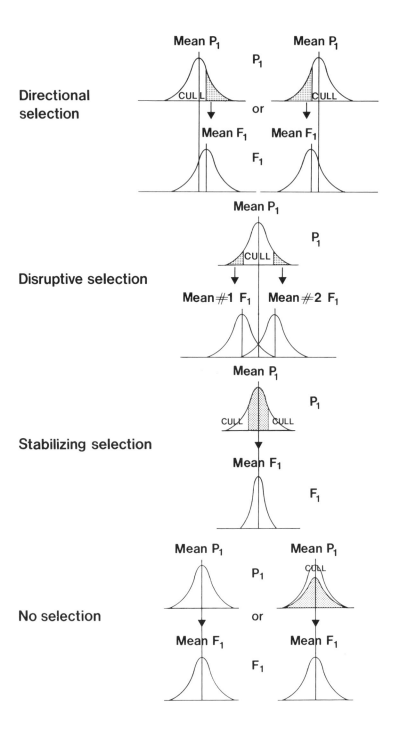

entire spawning season; (2) spawn fish of all sizes; (3) spawn as many fish as possible (you only need to keep a fraction of each spawn); (4) not cull the slow growers or those with poor secondary sexual characteristics, etc. You do not have to spawn every fish in a population; a representative sample is sufficient. Spawn a proportionate sample from each percentile along the normal distribution.

A program of no selection does not mean that you do nothing. Doing nothing is what may ruin the population. To do no selection properly, you must have extensive data about your population so that you can choose your broodstock in a way to include a representative sample from every percentile along the continuum. You must also have records so that you are able to compare generations in order to verify that no change has occurred.

In summary, when you do not know what you want to do or you do not know what phenotype is the best, no selection is the most appropriate form of selection for populations of game fish. Unintentional, unplanned, random changes usually do not improve productivity.

Unintentional selection also alters food fish gene pools and can reduce productivity by inadvertently selecting for poor aquacultural traits. For example, Eknath and Doyle (1985) found that the standard practice of spawning the largest catla and rohu in Indian hatcheries reduced productivity, because by choosing the largest females, hatchery personnel selected for slower growing and late-maturing fish.

While unintentional selection may eliminate some potentially valuable alleles, not all unintentional selection is bad in food fish populations. The elimination of fish that will not reproduce under hatchery conditions, that will not accept artificial feed, that are nervous and frighten easily, that avoid capture, or that do not grow well in captivity is an important aspect of domestication.

Food fish populations lead relatively pampered lives. Hatchery managers try to maximize growth and survival by removing predators, environmental disturbances, etc. Consequently, it is important that farms, management techniques, and aquaculturists select for populations that perform well under specific cultural conditions.

Figure 4.6 Schematic representations of the four basic types of selection. Directional selection shifts the mean either to the right or left. Disruptive selection produces two populations, one with a mean to the right and one with a mean to the left of the mean in the P_1 generation. Stabilizing selection reduces the variance while the mean remains unchanged. No selection changes neither the mean nor the variance.

The selection of food fish populations for hatchery environments (domestication) is why hatchery populations generally outperform wild stocks in yield trials at fish farms. For example, hatchery strains of channel catfish grow faster than wild strains when stocked at commercial rates (7410-9880/ha) and when fed artificial diets (Dunham and Smitherman 1984). Dunham and Smitherman (1983) estimated that the domestication process has increased growth rate by 2–6% per generation in channel catfish.

Doyle (1983) examined the role of domestication on production in aquaculture by analyzing the effects of daily ration, length of generation, age of maturation, size-selective mortality, or stocking density on several species of fish and crustaceans. He found that domestication is beneficial when fitness (the ability to survive and reproduce) is positively correlated with desired aquacultural traits. But if fitness is negatively correlated with desired aquacultural traits or if it is positively correlated with undesired traits, then the domestication process will decrease productivity. For example, if fitness is positively correlated with growth rate, domestication will increase growth rate and yield, as is the case with channel catfish. But if fitness is negatively correlated with age of maturity or is positively correlated with seine escapability, domestication will decrease growth rate because the fish will become sexually mature at a younger age or the population will be more difficult to harvest. For example, selection for early maturity in rainbow trout may decrease yield because growth slows at maturity (Møller *et al.* 1979).

The results of domestication can be predicted if correlations among standard aquacultural practices that lead to domestication and important aquacultural traits are determined (Doyle 1983). Once this is done, domestication can be either accelerated to increase productivity or altered to prevent future depression in productivity and irreparable damage to the gene pool.

Directional Selection   If you wish to improve productivity by altering a population's mean and when you have definite goals and plans, you typically use directional selection to improve your population. You can either increase phenotypic mean or decrease it to suit your purpose. For example, you may want to increase average weight and average length, or you may want to decrease average food conversion and average percentage body fat. Directional selection is used to change the population's phenotype(s) to your preconceived notion of what is desirable in terms of productivity and profits.

The first requirement for a successful program of directional selection (called selection hereafter) is goals. You must establish definite concrete goals so that you can quantify what you are trying to accomplish. Goals are whatever you desire: A game fish farmer may desire more catchable fish, less catchable fish, better survival, better growth, etc.; a food fish farmer may want better growth, better disease resistance, lower food conversion, etc.

The only constraint on your goals is that they must be realistic. You cannot produce a population of 2-year-old bluegills that will average 3 kg during a 3-year selection program. You also have to keep your goals realistic within the constraints of your budget, hatchery size, and time span for the project. It would be difficult to do selection at a hatchery that has only five ponds.

Goals are necessary, but well-conceived plans are mandatory. Plans are the blueprints that you will use to achieve your goals. A directionless program goes nowhere. Breeding programs are expensive because they deal with generations, not growing seasons; whimsical approaches are expensive lessons in futility. Plans are sets of instructions that outline the methods that will be used to grow the fish, tell how the phenotypes will be measured, tell when the phenotypes will be measured, and explain the way the cutoff values will be determined.

For example, if one of the goals of a channel catfish selective breeding program is to increase percent hatchability, the plans needed to achieve that goal can proceed as follows: First, spawn 300 four-year-old females and males that are in good physical shape and that have no obvious deformity. Second, transfer all egg masses to the hatchery and incubate all egg masses in separate containers in the hatchery troughs to maintain the genetic integrity of each family. Third, remove all unfertilized eggs. Fourth, determine the number of eggs in each egg mass. Fifth, count the number of eggs that die or fail to hatch. Cull all families where percent hatch is less than 95%. Sixth, take 100 fry from all families (both selected and culled) and pool them for the control population, so that response to selection can be measured. This plan is a bare-bones outline of those that would actually be needed. The actual set of plans would be fleshed out with culture techniques, stocking rates, feeding rates, number of replicates, etc.

An important aspect of the plan is to decide what phenotypes will be incorporated into the selection program, how they will be measured, and when they will be measured. When deciding what phenotypes will be incorporated into a selection program, a hatchery manager must choose phenotypes that represent real biological traits and not artificial constructs or various aspects of management. For example, selection for survival rate of fry will probably produce little gain, because early viability is strongly influenced by management practices such as feeding, water quality, and predator control.

Selection for phenotypes that are functions of the environment or management will be futile, since differences among individuals will be a function of V_E rather than V_A. For example, selection for increased growth rate during the first month of life will probably be unsuccessful, since early growth rate is mainly a function of V_E—egg size, female age and size, spawning date, feeding practices, and stocking rates—not V_A.

When measuring a phenotype, it is important to measure the trait properly, or selection will act on another phenotype. For example, if a hatchery manager wants to select for increased hatchability, he should not define it as the percentage of eggs that produce fry, but the percentage of fertilized eggs that produce fry. If hatchability is measured as the percentage of eggs that produce fry, selection may not improve hatchability since fertilizing success, which is a function of the males, is confounded with hatchability.

Conversely, hatchery managers who want to improve phenotypes that are difficult to measure can make selection less tedious by selecting for easily measured phenotypes that are highly correlated to the one that they want to improve. For example, selection for increased weight gain at a given age is often measured as length at that age because: (1) the correlations between length and weight at various ages are essentially 1.0 (Table 4.5); (2) it is easy to obtain accurate individual lengths on hundreds of fish, but it is difficult to obtain accurate individual weights on hundreds of fish.

Knowing when to measure a phenotype is often as important as knowing what phenotype will be measured. For example, if a hatchery manager wants to select for increased growth rate, he must define carefully the age at which he wants growth to be improved because growth is not constant; it is different in males and females; and growth rate changes with the onset of sexual maturity. Additionally, h^2s for length and weight are not constant (Table 4.1), which means the response to selection will vary.

Finally, the phenotypes that will be incorporated into a breeding program must represent economically important traits. It makes little sense to increase the number of rays in channel catfish dorsal fins, because consumers buy catfish by the kilogram, not the fin ray. Selection for larger head muscle pads in male channel catfish may be unwise, because if it is successful it will decrease dressing percentage.

An excellent example of goals and the plans to execute these goals in a selective breeding program are those outlined by Gall (1979) for a selective breeding program with rainbow trout. In that paper, he described a selective breeding program to improve egg size, number of eggs per female, hatchability, and weight at 1 year in rainbow trout produced by California state hatcheries.

The first step in any breeding program is to gather data (or use data that exist) to create a quantitative description of the phenotype. You especially want to know the mean, variance, standard deviation (SD), coefficient of variation (CV), range, and h^2. Once you have that information, you can then decide if selection will enable you to achieve your goal. If the desired phenotype is well outside the population's range, it will take many generations to achieve the goal. The information will also help decide how to select fish and where to place the cutoff value. All fish that fall below the value will be culled, and those that fall on or above it will be selected.

Although h^2 is usually considered the factor that acts as a governor and dictates whether selection will be effective, the SD and CV [(SD/mean)100] tell you whether the population has enough phenotypic variation to allow goal achievement via selection. The SD also gives you a good idea as to where to place the cutoff value.

Heritability gives the percentage of reach that can be captured as progress; SD and CV define how much reach is possible. Populations with large SDs and large CVs enable you to have large selection differentials; populations with small SDs and small CVs make selection difficult, because there is little variation to exploit (SD = $\sqrt{\text{variance}}$).

Large SDs and CVs improve the likelihood of success in a breeding program. For example, if two populations of fish have identical mean weights of 400 g at 1 year of age, but one population has an SD of 10 g (CV = 2.5%) while the other has an SD of 100 g (CV = 25%), selection will be far more effective in the population with the 100-g SD, because it has far more variance. Figure 4.7 shows the difference between two populations that have the same mean but different SDs.

Coefficients of variation in most populations of fish are fairly large, and Gjedrem (1975) felt that this is why selection should be an effective way to improve productivity in fish culture. But not all populations have large SDs or CVs; some have so little variance that it will hinder selection (Tave and Smitherman 1980; Tave 1984B, 1986A; Gjerde 1989; Tave et al. 1989B, 1990A). Average CVs for some economically important phenotypes in several species of food fish are shown in Table 4.3.

The SD also gives an indication about the intensity of selection required to achieve your goals. When you establish a cutoff value, you

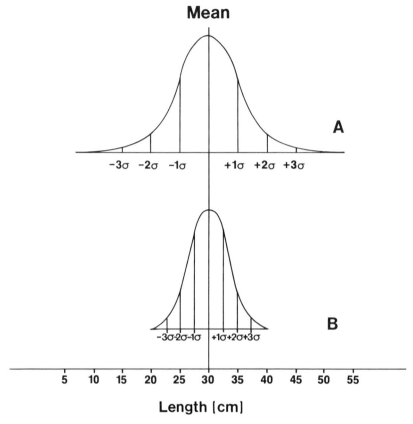

Figure 4.7 Two populations with the same mean length (30 cm) but different standard deviations. Population A has a much larger standard deviation and CV. All other factors being equal, selection will be far easier in population A because there is more variance that can be exploited.

create a cutoff value so many SDs above or below the mean. Even when the cutoff value is labeled as, say, the top 1% or top 10%, what you really have done is establish the cutoff value at 2.33 or 1.28 SDs above the mean, respectively. When you know the SD, you have more control over selection, because you can customize selection intensity in order to maximize the selection differential (reach). The selection differential (S) is the product of the intensity of selection (i) and the standard deviation of the phenotype (σ_P):

$$S = i\sigma_P$$

Table 4.3   Average CVs for Some Important Production Phenotypes in Rainbow Trout, Atlantic Salmon, Common Carp, Channel Catfish, and Tilapia

Phenotype	Rainbow trout	Atlantic salmon	Common carp	Channel catfish	Tilapia
Body weight, juveniles	33	78		46	26
Body weight, adults	22	27	22	27	
Body length, juveniles	14	23			8
Body length, adults	9	8		8	
Mortality/resistance			28		
Meatiness	20	19			
Meat color	23	16			
Percentage fat	10			8	
Dressing percentage	6	4		2	

Source: After Gjedrem (1983).

The intensity of selection is the standardized selection differential in terms of SDs (Fig. 4.8). Intensities of selection for different levels of selection are shown in Table 4.4.

Knowledge about a population's σ_P allows you to adjust i in order to create the value of S that is needed to achieve the desired response per generation:

$$R = i\sigma_P h^2 \tag{4.4}$$

This information also lets you know if the desired response requires too great a selection differential. If the desired response requires an $i \geq 2.67$ or ≤ -2.67 (culling $\geq 99\%$ of the population), the population has too little variance to achieve the desired response per generation without running the risk of lowering the effective breeding number to a value which will cause inbreeding and genetic drift (more on this later).

Figure 4.9 illustrates the effect that SDs and CVs have on a population's response to selection. Selection in a population that has a small SD and CV requires a much higher selection differential to achieve a given response to selection than in a population that has a large SD and CV. The example shown in Fig. 4.9 shows that selection just above the 50th percentile will achieve a response to selection of 0.25 kg when the SD is 0.7 kg (CV = 70%), but that selection must be at the 95th percentile to achieve the same result when the SD is 0.3 kg (CV = 30%).

After you have this baseline information and you decide to proceed, you must choose the selection program that will enable goal achieve-

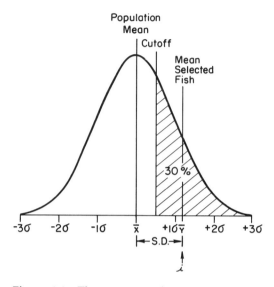

Figure 4.8 The intensity of selection (i) is the standardized selection differential (S.D.) in terms of the standard deviations; i is the number of standard deviations between the population mean ($\bar{x}$) and the mean of the selected fish ($\bar{y}$). In this example, the top 30% of the population was selected (the cutoff value was the 70th percentile, and the mean of the selected fish is the 86.15 percentile. The selection differential is the difference between the population mean (50th percentile) and the mean of the selected fish (86.15 percentile). Because the population mean is, by definition, ±0.0 standard deviations from itself, the mean of the selected fish, when converted to standard deviations from the mean, is i. Therefore, in this example, i = 1.159, because the 86.15 percentile is 1.159 standard deviations above the mean.
Source: After Falconer (1957).

ment. If you are interested in improving only one phenotype, the process is quite simple. Establish a cutoff value, cull all fish that fall below it, and repeat the procedure in succeeding generations. The results of a breeding program to create both a population of guppies with an increased number of dorsal fin rays and one with fewer dorsal fin rays (bidirectional selection) are shown in Figure 4.10.

Table 4.4 Intensities of Selection (i) for Specific
Percentages of a Population Saved During Selection

Up-selection[a]		Down-selection[a]	
Percent saved	i	Percent saved	i
50	0.798	50	−0.798
40	0.966	40	−0.966
30	1.159	30	−1.159
25	1.271	25	−1.271
20	1.400	20	−1.400
15	1.554	15	−1.554
10	1.755	10	−1.755
9	1.804	9	−1.804
8	1.858	8	−1.858
7	1.918	7	−1.918
6	1.985	6	−1.985
5	2.063	5	−2.063
4	2.154	4	−2.154
3	2.268	3	−2.268
2	2.421	2	−2.421
1	2.665	1	−2.665
0.5	2.982	0.5	−2.982
0.1	3.367	0.1	−3.367

[a] In up-selection you save animals above the 50th percentile;
in down-selection you save animals below the 50th percentile.

When conducting selection for traits such as length or weight, you should determine whether sexual dimorphism (one sex is larger) exists. If sexual dimorphism exists, you must establish separate cutoff values for each sex. If you establish a single cutoff value regardless of sex, your select broodfish may be a monosex population. For example, Brooks *et al.* (1982) found that selection for the heaviest 10% in channel catfish would produce a population of broodstock that is predominantly: if not all, male. Thus, in channel catfish breeding programs, it is important to establish separate cutoff values for each sex, to avoid ending up with too few select female broodstock.

Tandem selection To change two or more phenotypes, you have a choice of several different types of selective breeding programs. The simplest is tandem selection. In tandem selection you first select for one phenotype for a number of generations until the goal is reached for that

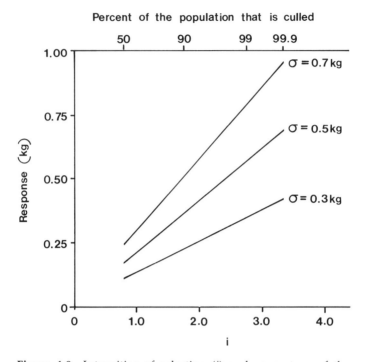

Figure 4.9 Intensities of selection (i) and percentage of the population that must be culled to achieve a desired response to selection in three populations that have the same mean (1.0 kg), the same h^2 (0.4), but different standard deviations (σ = 0.3, 0.5, and 0.7 kg). The regressions show that as the standard deviation decreases, the intensity of selection must increase in order to achieve a desired response to selection.

phenotype, and you then select for the second phenotype until the goal is reached for that phenotype. For a third or fourth phenotype, you simply continue with this pattern.

Unfortunately, tandem selection is very inefficient for two reasons: (1) It takes a very long time to improve two or more traits this way. (2) Selecting for one trait automatically means selecting for others unless the correlations among the traits are zero. If the two phenotypes that you want to change have a negative correlation, you will decrease the mean for one phenotype while improving the other, and all your previous selection may be undone when selecting for the second phenotype. For example, when Millenbach (1950) selected for rainbow trout that matured at year 2 instead of years 3 and 4, he found that the fish in the selected line had a poorer growth rate and also produced fewer eggs.

Ehlinger (1977) found that strains of brook trout that had been selected for increased resistance to furunculosis had become more susceptible to gill disease.

The latter reason illustrates the importance of obtaining correlations among phenotypes. It is helpful to know the correlations among economically important phenotypes, because they give an indication about indirect selection (the effect that selection for one phenotype will have on others). An example of a correlation that produces disastrous results by indirect selection is one that salmon and trout farmers are cursed with: growth rate and age of sexual maturity are negatively correlated (Thorpe *et al.* 1983; Gall 1986; Herbinger and Newkirk 1987; Tofteberg and Hansen 1987; Crandell and Gall 1992). This means that selection for faster growing fish produces fish that mature at a younger age. The production of early-maturing or precocious fish is detrimental because: in species that die shortly after they become sexually mature, precocious fish can cost farmers a sizable portion of their crops; in species that do not die, growth declines as food energy is diverted to gamete production, and flesh quality deteriorates so they cannot be marketed.

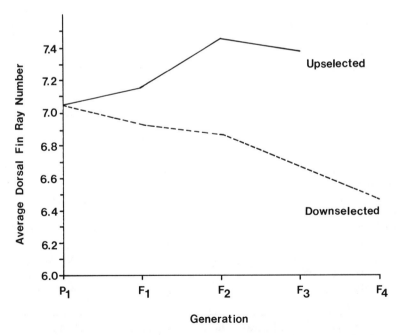

Figure 4.10 Selected data from Svärdson's (1945) bidirectional selection experiment for dorsal fin ray number in the guppy.

Table 4.5   Genotypic and Phenotypic Correlations

Species Phenotype	Correlation[a]		Reference
	Genetic ± SE	Pheno-typic	
Channel catfish			
length–wt at 1 month		0.96	El-Ibiary *et al.* (1979)
length–wt at 30 days	1.01		Reagan (1979)
length–wt at 60 days	0.98		Reagan (1979)
length–wt 12 weeks		0.94	El-Ibiary *et al.* (1979)
length–wt 90 days	1.27		Reagan (1979)
length–wt at 120 days	0.96		Reagan (1979)
length–wt at 150 days	1.60		Reagan (1979)
length–wt at 5 months	1.47 ± 1.49	0.79	Reagan *et al.* (1976)
length–wt at 24 weeks		0.93	El-Ibiary *et al.* (1979)
length–wt at 44 weeks		0.95	El-Ibiary *et al.* (1979)
length–wt at 15 months	1.03 ± 0.53	0.90	Reagan *et al.* (1976)
length–wt at 18 months	1.02		Reagan (1979)
wt 4 weeks–wt 8 weeks		0.78	El-Ibiary *et al.* (1979)
wt 4 weeks–wt 12 weeks		0.92	El-Ibiary *et al.* (1979)
wt 4 weeks–wt 24 weeks		0.72	El-Ibiary *et al.* (1979)
wt 4 weeks–wt 56 weeks		0.53	El-Ibiary *et al.* (1979)
wt 8 weeks–wt 24 weeks		0.61	El-Ibiary *et al.* (1979)
wt 8 weeks–wt 56 weeks		0.46	El-Ibiary *et al.* (1979)
wt 12 weeks–wt 24 weeks		0.81	El-Ibiary *et al.* (1979)
wt 12 weeks–wt 56 weeks		0.62	El-Ibiary *et al.* (1979)
wt 4 weeks–survival 0 to 15 weeks		0.80	El-Ibiary *et al.* (1979)
wt 4 weeks–survival 16 to 40 weeks		−0.22	El-Ibiary *et al.* (1979)
wt 4 weeks–survival 41 to 56 weeks		−0.05	El-Ibiary *et al.* (1979)
length–dress out wt at 18 months	1.05		Reagan (1979)
wt–dress out wt at 18 months	1.03		Reagan (1979)
length–percent fat	1.13		Reagan (1979)
wt–percent fat	1.12		Reagan (1979)
length–dressing percent, ♀		0.28	Bondari (1982)
length–dressing percent, ♂		0.44	Bondari (1982)
wt–dressing percent, ♀		0.28	Bondari (1982)
wt–dressing percent, ♂		0.38	Bondari (1982)

[a] When the standard error (SE) was determined for a genetic correlation it is listed with the correlation.

Table 4.5   **Genotypic and Phenotypic Correlations** (*continued*)

| | Correlation[a] | | |
Species Phenotype	Genetic ± SE	Pheno-typic	Reference
wt–dress out wt, ♀		1.0	El-Ibiary *et al.* (1976)
wt–dress out wt, ♂		1.0	El-Ibiary *et al.* (1976)
wt–dress out percentage, ♀		0.43	El-Ibiary *et al.* (1976)
wt–dress out percentage, ♂		0.31	El-Ibiary *et al.* (1976)
length–dress out wt, ♀		0.89	El-Ibiary *et al.* (1976)
length–dress out wt, ♂		0.90	El-Ibiary *et al.* (1976)
length–dress out percent-age, ♀		0.54	El-Ibiary *et al.* (1976)
length–dress out percent-age, ♂		0.38	El-Ibiary *et al.* (1976)
head wt–dressing percent-age, ♀		−0.37	El-Ibiary *et al.* (1976)
head wt–dressing percent-age, ♂		−0.26	El-Ibiary *et al.* (1976)
wt–lipid percentage at 13 months		0.40	Dunham *et al.* (1985)
girth–body wt at 13 months		0.37	Dunham *et al.* (1985)
body wt–dressing percent-age at 22 months, ♀		−0.04	Dunham *et al.* (1985)
body wt–dressing percent-age at 22 months, ♂		−0.21	Dunham *et al.* (1985)
Tifton strain			
feed consumption–feed conversion at 1 month	−0.79	0.55	Burch (1986)
feed consumption–feed conversion at 2 months	−1.86	−0.19	Burch (1986)
Kansas strain			
feed consumption–feed conversion at 3 months	−0.18	−0.08	Burch (1986)
Rainbow trout			
length 3 months–length 4 months		0.86	Aulstad *et al.* (1972)
length 3 months–length 12 months		0.59	Aulstad *et al.* (1972)
length 3 months–length 16 months		0	Aulstad *et al.* (1972)
length 5 months–length 12 months		0.73	Aulstad *et al.* (1972)
length 5 months–length 16 months		0.20	Aulstad *et al.* (1972)

Table 4.5   Genotypic and Phenotypic Correlations (*continued*)

| | Correlation[a] | | |
| | | | |
Species Phenotype	Genetic ± SE	Pheno-typic	Reference
length 6 months–length 12 months		0.81	Møller *et al.* (1979)
length 6 months–length 18 months		0.67	Møller *et al.* (1979)
length 6 months–length 24 months		0.48	Møller *et al.* (1979)
length 6 months–length 30 months		0.06	Møller *et al.* (1979)
length 12 months–length 16 months		0.57	Aulstad *et al.* (1972)
length 12 months–length 18 months		0.40	Møller *et al.* (1979)
length 12 months–length 24 months		0.25	Møller *et al.* (1979)
length 12 months–length 30 months		−0.18	Møller *et al.* (1979)
length 18 months–length 24 months		0.79	Møller *et al.* (1979)
length 18 months–length 30 months		0.63	Møller *et al.* (1979)
length 24 months–length 30 months		0.74	Møller *et al.* (1979)
length–wt at 9 months		0.92	Klupp (1979)
length–wt at 2 years	0.96 ± 0.03	0.92	Gjerde and Gjedrem (1984)
growth–food conversion	0.99 ± 0.01	0.96	Kinghorn (1983)
growth–O_2 consumption	0.85 ± 0.20	0.21	Kinghorn (1983)
growth–percent fat	−0.47 ± 0.70	−0.23	Kinghorn (1983)
food consumption–percent fat	−0.40 ± 0.69	−0.28	Kinghorn (1983)
14-month wt–spawning as 2-year-old		−0.2	Tofteberg and Hansen (1987)
2-year wt–maturity at 2 years	0.11 ± 0.26	0.11	Gjerde and Gjedrem (1984)
2-year length–maturity at 2 years	0.16 ± 0.43	0.11	Gjerde and Gjedrem (1984)
♀ body wt–no. eggs spawned		0.11 to 0.28	Gall (1972)
no. eggs spawned–egg size		0.17 to 0.36	Gall (1972)

Table 4.5 Genotypic and Phenotypic Correlations (*continued*)

Species Phenotype	Genetic ± SE	Pheno-typic	Reference
no. eggs–no. eggs/kg ♀	0.41 ± 0.14	0.51	Gall (1975)
♀ body wt–egg size		−0.03 to −0.26	Gall (1972)
egg size–egg volume	−0.55 ± 0.12	−0.45	Gall (1975)
egg size–no. eggs	0.09 ± 0.17	0.15	Gall (1975)
egg size–no. eggs/kg ♀	0.47 ± 0.13	0.26	Gall (1975)
egg volume–no. eggs	0.77 ± 0.07	0.78	Gall (1975)
♀ body wt–egg size	−0.46 ± 0.13	−0.19	Gall (1975)
♀ body wt–egg volume	0.56 ± 0.11	0.43	Gall (1975)
♀ body wt–no. eggs	0.39 ± 0.14	0.35	Gall (1975)
♀ body wt–no. eggs/kg ♀	−0.61 ± 0.10	−0.50	Gall (1975)
post spawning wt–egg volume	0.54		N. Huang and Gall (1990)
post spawning wt–egg size	−0.43		N. Huang and Gall (1990)
post spawning wt–egg no.	0.41		N. Huang and Gall (1990)
age at spawning–egg volume	0.37		N. Huang and Gall (1990)
age at spawning–egg size	−0.45		N. Huang and Gall (1990)
egg volume–yearling wt	0.27		N. Huang and Gall (1990)
egg volume–25-month wt	0.47		N. Huang and Gall (1990)
egg size–yearling wt	−0.19		N. Huang and Gall (1990)
egg size–25-month wt	−0.32		N. Huang and Gall (1990)
no. pyloric caeca–body length		0.25	Bergot *et al.* (1976)
no. pyloric caeca–body length		0.29	Bergot *et al.* (1981A)
no. pyloric caeca–body length		0.04	Ulla and Gjedrem (1985)
no. pyloric caeca–fat digestibility		−0.03	Ulla and Gjedrem (1985)
no. pyloric caeca–protein digestibility		−0.10	Ulla and Gjedrem (1985)
2-year wt–gutted wt	1.00 ± 0.01	0.98	Gjerde and Gjedrem (1984)
2-year length–gutted wt	0.84 ± 0.12	0.91	Gjerde and Gjedrem (1984)
2-year wt–meatiness score	0.75 ± 0.16	0.53	Gjerde and Gjedrem (1984)
2-year length–meatiness score	0.75 ± 0.24	0.52	Gjerde and Gjedrem (1984)

Table 4.5   Genotypic and Phenotypic Correlations (*continued*)

Species Phenotype	Genetic ± SE	Pheno-typic	Reference
gutted wt–meatiness score	0.77 ± 0.15	0.53	Gjerde and Gjedrem (1984)
meatiness score–maturity at 2 years	−0.15 ± 0.28	−0.01	Gjerde and Gjedrem (1984)
body wt–percent fat		0.16	Morkramer *et al.* (1985)
3-year body wt–semen production			
first stripping volume		0.25	Gjerde (1984B)
total volume		0.18	Gjerde (1984B)
3-year body length–semen production			
first stripping volume		0.24	Gjerde (1984B)
total volume		0.12	Gjerde (1984B)
gutted body wt–belly thickness	0.36	0.55	Gjerde and Schaeffer (1989)
gutted body wt–K factor	0.16	0.27	Gjerde and Schaeffer (1989)
gutted body wt–dressing percentage	0.07	0.29	Gjerde and Schaeffer (1989)
gutted body wt–abdominal fat	0.19	0.18	Gjerde and Schaeffer (1989)
gutted body wt–gonad wt	0.54	0.54	Gjerde and Schaeffer (1989)
gutted body wt–percent fat	−0.19	0.21	Gjerde and Schaeffer (1989)
gutted body wt–meat color score	0.21	0.07	Gjerde and Schaeffer (1989)
belly thickness–K factor	0.05	0.02	Gjerde and Schaeffer (1989)
belly thickness–dressing percentage	0.79	0.50	Gjerde and Schaeffer (1989)
belly thickness–abdominal fat	−0.54	−0.17	Gjerde and Schaeffer (1989)
belly thickness–gonad wt	−0.20	−0.03	Gjerde and Schaeffer (1989)
belly thickness–percent fat	0.24	0.20	Gjerde and Schaeffer (1989)
belly thickness–meat color score	−0.45	−0.23	Gjerde and Schaeffer (1989)

Table 4.5 **Genotypic and Phenotypic Correlations** (*continued*)

Species Phenotype	Genetic ± SE	Pheno-typic	Reference
	Correlation[a]		
K factor–dressing percentage	−0.09	−0.02	Gjerde and Schaeffer (1989)
K factor–abdominal fat	0.04	0.12	Gjerde and Schaeffer (1989)
K factor–gonad wt	0.25	0.18	Gjerde and Schaeffer (1989)
K factor–percent fat	0.15	0.16	Gjerde and Schaeffer (1989)
K factor–meat color score	−0.08	−0.01	Gjerde and Schaeffer (1989)
dressing percentage–abdominal fat	−0.68	−0.39	Gjerde and Schaeffer (1989)
dressing percentage–gonad wt	−0.36	−0.18	Gjerde and Schaeffer (1989)
dressing percentage–percent fat	0.24	0.13	Gjerde and Schaeffer (1989)
dressing percentage–meat color score	−0.45	−0.22	Gjerde and Schaeffer (1989)
abdominal fat–gonad wt	0.08	−0.07	Gjerde and Schaeffer (1989)
abdominal fat–percent fat	−0.33	−0.06	Gjerde and Schaeffer (1989)
abdominal fat–meat color score	0.44	0.19	Gjerde and Schaeffer (1989)
gonad wt–percent fat	0.0	−0.11	Gjerde and Schaeffer (1989)
percent fat–meat color score	−0.44	−0.23	Gjerde and Schaeffer (1989)
percent water–percent fat	−1.03	−0.84	Gjerde (1986)
body wt–percent fat	−0.51	−0.19	Gjerde (1986)
percent fat–dressing percentage	0.53		Gjerde (1986)
percent fat–gonad wt	−0.49		Gjerde (1986)
percent fat–meat color	−0.98		Gjerde (1986)
percent fat–visceral fat	−0.76		Gjerde (1986)
dressing percentage–belly thickness	0.94		Gjerde (1986)
Brown trout			
no. pyloric caeca–egg size		0.50	Blanc *et al.* (1979)

Table 4.5 Genotypic and Phenotypic Correlations (*continued*)

Species Phenotype	Correlation[a]		Reference
	Genetic ± SE	Pheno-typic	
Coho salmon			
57-day post swim-up wt–84-day post swim-up wt	0.80 ± 0.21	0.49	Iwamoto *et al.* (1982)
57-day post swim-up wt–141-day post swim-up wt	0.62 ± 0.56	0.29	Iwamoto *et al.* (1982)
length–wt at 84 days post swim-up	0.95 ± 0.04	0.96	Iwamoto *et al.* (1982)
length–wt at 141 days post swim-up	1.03 ± 0.06	0.98	Iwamoto *et al.* (1982)
57-day post swim-up wt–141-day post swim-up length	0.74 ± 0.64	0.33	Iwamoto *et al.* (1982)
wt at transfer to saltwater–length at 8 months in salt-water	0.57 ± 0.16 to 0.75 ± 0.08	0.09 to 0.18	Hershberger *et al.* (1990A)
wt at transfer to saltwater–wt at 8 months in saltwater	0.42 ± 0.21 to 0.73 ± 0.09	0.10 to 0.18	Hershberger *et al.* (1990A)
percent survival–length 6 months after transfer to salt water		0.09	Saxton *et al.* (1984)
percent survival–wt 6 months after transfer to salt water		0.09	Saxton *et al.* (1984)
length 8 months in salt water–carotenoid level in .flesh	0.56 ± 0.05	0.85	Iwamoto *et al.* (1990)
wt 8-months in salt water–carotenoid level in flesh	−0.21 ± 0.10	0.58	Iwamoto *et al.* (1990)
percent lipid–carotenoid level in flesh		0.16	Iwamoto *et al.* (1990)
length–dressed wt after 8 months in salt water		0.96	Iwamoto *et al.* (1990)
Atlantic salmon			
length–wt at 12 weeks	0.98	0.93	Bailey and Loudenslager (1986)
length–wt at 6 months	1.0	0.92	Bailey and Loudenslager (1986)
length–wt at 7 months	1.0	0.96	Refstie and Steine (1978)

Table 4.5   Genotypic and Phenotypic Correlations (*continued*)

| Species   Phenotype | Correlation[a] | | Reference |
	Genetic ± SE	Pheno-typic	
length–wt at 1 year	1.0	1.0	Riddell *et al.* (1981)
length–wt at 15 months	1.0	0.95	Bailey and Loudenslager (1986)
length–wt at 2 years	0.99 ± 0.01	0.92	Gjerde and Gjedrem (1984)
length–wt at 3 years	0.99	0.91	Gunnes and Gjedrem (1978)
length 6 months–length 12 months		0.75	Naevdal *et al.* (1975)
length 6 months–length 18 months		0.27	Naevdal *et al.* (1975)
length 6 months–length 24 months		0	Neavdal *et al.* (1975)
length 12 months–length 18 months		0.49	Neavdal *et al.* (1975)
length 12 months–length 24 months		0.32	Neavdal *et al.* (1975)
length 12 months–length 30 months		0.20	Naevdal *et al.* (1975)
length 12 months–length 42 months		0.10	Naevdal *et al.* (1976)
length 18 months–length 24 months		0.32	Naevdal *et al.* (1975)
length 18 months–length 30 months		0.49	Naevdal *et al.* (1975)
length 18 months–length 42 months		0.19	Naevdal *et al.* (1976)
length 24 months–length 30 months		0.69	Naevdal *et al.* (1975)
length 24 months–length 42 months		0.50	Naevdal *et al.* (1976)
length 30 months–length 36 months		0.93	Naevdal *et al.* (1976)
length 30 months–length 42 months		0.72	Naevdal *et al.* (1976)
length 36 months–length 42 months		0.76	Naevdal *et al.* (1976)
length 42 months–maturity		−0.49	Naevdal *et al.* (1976)
2-year wt–maturity at 2 years	0.52 ± 0.13	0.32	Gjerde and Gjedrem (1984)

Table 4.5 Genotypic and Phenotypic Correlations (*continued*)

| Species   Phenotype | Correlation[a] | | Reference |
	Genetic ± SE	Pheno-typic	
2-year length–maturity at 2 years	0.28 ± 0.17	0.13	Gjerde and Gjedrem (1984)
egg size–survival at 6 months		0.29	Naevdal *et al.* (1975)
2-year wt–gutted wt	1.00 ± 0.0	0.99	Gjerde and Gjedrem (1984)
2-year length–gutted wt	0.92 ± 0.03	0.90	Gjerde and Gjedrem (1984)
body length–head length at 1 year	1.0		Riddell *et al.* (1981)
body length–pectoral fin length at 1 year	0.95		Riddell *et al.* (1981)
body length–pelvic fin length at 1 year	0.97		Riddell *et al.* (1981)
wt–head length at 1 year	1.0		Riddell *et al.* (1981)
wt–pectoral fin length at 1 year	0.95		Riddell *et al.* (1981)
wt–pelvic fin length at 1 year	0.97		Riddell *et al.* (1981)
2-year wt–meatiness score	0.88 ± 0.08	0.49	Gjerde and Gjedrem (1984)
2-year length–meatiness score	0.89 ± 0.08	0.49	Gjerde and Gjedrem (1984)
gutted wt–meatiness score	0.98 ± 0.08	0.50	Gjerde and Gjedrem (1984)
maturity at 2 years–meatiness score	0.22 ± 0.18	0.01	Gjerde and Gjedrem (1984)
4-year body wt–semen production			
first stripping volume		0.36	Gjerde (1984B)
total volume		0.63	Gjerde (1984B)
4-year body length–semen production			
first stripping volume		0.34	Gjerde (1984B)
total volume		0.56	Gjerde (1984B)
2-year survival–2-year maturity		−0.45 to 0.13	Standal and Gjerde (1987)
2-year survival–2 year wt		0.16 to 0.22	Standal and Gjerde (1987)

Table 4.5 Genotypic and Phenotypic Correlations (*continued*)

| | | Correlation[a] | | |
		Genetic ± SE	Pheno-typic	Reference
Species	Phenotype			
	2-year survival–2 year length		0.03 to 0.25	Standal and Gjerde (1987)
	2-year survival–2-year K factor		−0.04 to 0.38	Standal and Gjerde (1987)
Pink salmon				
	60-day wt–150-day wt	1.06 ± 0.08		Beacham and Murray (1988B)
	60-day wt–315-day wt	1.03 ± 0.09		Beacham and Murray (1988B)
	60-day wt–500-day wt	0.82 ± 0.17		Beacham and Murray (1988B)
	75-day wt–245-day wt	0.95 ± 0.09		Beacham (1989)
	75-day wt–420-day wt	0.02 ± 0.28		Beacham (1989)
	150-day wt–315-day wt	0.82 ± 0.11		Beacham and Murray (1988B)
	150-day wt–500-day wt	0.63 ± 0.23		Beacham and Murray (1988B)
	215-day wt–315-day wt	0.94 ± 0.05		Beacham and Murray (1988B)
	215-day wt–500-day wt	0.73 ± 0.17		Beacham and Murray (1988B)
	315-day wt–500-day wt	0.95 ± 0.08		Beacham and Murray (1988B)
	410-day wt–500-day wt	0.98 ± 0.05		Beacham and Murray (1988B)
	315-day wt–GSI, ♀	0.97 ± 2.20		Beacham and Murray (1988B)
	500-day wt–GSI, ♀	0.90 ± 1.96		Beacham and Murray (1988B)
	315-day wt–GSI, ♂	0.91 ± 0.15		Beacham and Murray (1988B)
	500-day wt–GSI, ♂	0.93 ± 0.10		Beacham and Murray (1988B)
	embryo survival–fry length			
	4°C	0.98 ± 4.13		Beacham (1988)
	16°C	0.80 ± 0.24		Beacham (1988)
	alevin length–fry length			
	4°C	1.15 ± 0.58		Beacham (1988)
	16°C	0.33 ± 0.38		Beacham (1988)

Table 4.5 **Genotypic and Phenotypic Correlations** (*continued*)

| Species Phenotype | Correlation[a] | | Reference |
	Genetic ± SE	Pheno-typic	
Chum salmon			
embryo survival–fry length			
3°C	0.34 ± 0.52		Beacham (1988)
8°C	−0.01 ± 0.90		Beacham (1988)
alevin length–fry length			
3°C	0.71 ± 0.40		Beacham (1988)
8°C	0.71 ± 0.37		Beacham (1988)
Amago salmon			
percent eyed eggs–hatch-ability	0.78		Sato and Morikawa (1982)
Tilapia nilotica			
length–wt 45 days	1.09 ± 1.43	0.94	Tave and Smitherman (1980)
length–wt 90 days	1.12 ± 1.61	1.01	Tave and Smitherman (1980)
total rays of dorsal fin–total rays of anal fin		0.23	Tave (1986)
total rays of dorsal fin–pectoral fin rays		0.05	Tave (1986)
136-day wt–gonad wt, ♀	0.0 ± 0.27	−0.35	Kronert *et al.* (1987)
136-day wt–gonad wt, ♀	0.05	0.33	Kronert *et al.* (1989)
136-day wt–gonad wt, ♂	−0.37 ± 0.22	0.51	Kronert *et al.* (1987)
136-day wt–gonad wt, ♂	−0.40	0.56	Kronert *et al.* (1989)
136-day wt–GSI, ♀	−0.28 ± 0.23	0.16	Kronert *et al.* (1987)
136-day wt–GSI, ♀	−0.06	0.16	Kronert *et al.* (1989)
136-day wt–GSI, ♂	−0.78 ± 0.10	0.26	Kronert *et al.* (1987)
136-day wt–GSI, ♂	−0.64	0.33	Kronert *et al.* (1989)
gonad wt–GSI, ♀	0.94	0.99	Kronert *et al.* (1989)
gonad wt–GSI, ♂	0.93	0.95	Kronert *et al.* (1989)
Tilapia aurea			
length–wt at 48 weeks, ♀		0.92	Bondari (1982)
length–wt at 48 weeks, ♂		0.82	Bondari (1982)
48-week wt–dressing per-centage, ♀		0.23	Bondari (1982)
48-week wt–dressing per-centage, ♂		0.33	Bondari (1982)
48-week length–dressing percentage, ♀		0.03	Bondari (1982)
48-week length–dressing percentage, ♂		0.30	Bondari (1982)

Table 4.5 Genotypic and Phenotypic Correlations (*continued*)

Species Phenotype	Genetic ± SE	Pheno-typic	Reference
Mosquitofish			
length–wt at 2 months, ♀	0.96	0.96	Busack and Gall (1983)
length–wt at 2 months, ♂	0.90	0.95	Busack and Gall (1983)
Velvet belly shark			
position anterior dorsal fin spine–no. whole vertebrae		0.21	Tave (1984B)
position posterior dorsal fin spine–no. whole vertebrae		0.74	Tave (1984B)
no. whole vertebrae–no. half vertebrae		−0.45	Tave (1984B)

Examples of phenotypic and genetic correlations are shown in Table 4.5. Genetic correlations give the best prediction about indirect selection, but it is difficult to obtain accurate genetic correlations when working with the population sizes used in most quantitative genetic analyses. Genetic correlations determined from data that are gathered from small populations are often greater than 1.0, and that is a biological impossibility. Examples of such correlations are found in Reagan *et al.* (1976), Tave and Smitherman (1980), Gjerde (1986), Beacham (1988), and Beacham and Murray (1988B). Because of the difficulties in obtaining accurate genetic correlations, phenotypic correlations can be used as good approximations.

Independent culling Independent culling is a selection program that can be employed when you wish to select simultaneously for two or more phenotypes. Here you establish minimal performance levels for each phenotype, and a fish must pass all minima in order to be selected. An example of independent culling is shown diagrammatically in Fig. 4.11.

Independent culling is more efficient than tandem selection, but it has two disadvantages: First, to be selected, a fish has to be outstanding in all phenotypes. This means that a fish that is great in one phenotype but only average in another will be culled. This may eliminate some valuable alleles. This problem can be circumvented by modifying independent culling as shown diagrammatically in Fig. 4.12. A second disadvantage is that when you select for several phenotypes and establish

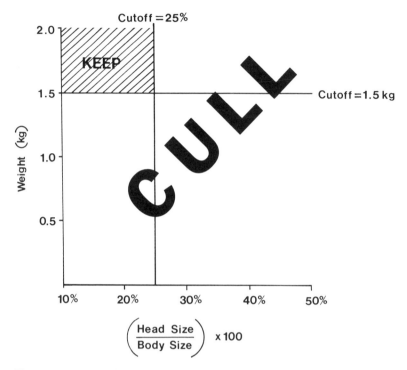

Figure 4.11 A schematic diagram of independent culling for body weight and percent head size (an approximation of dressing percentage). Minimal performance levels are 1.5 kg and 25%, respectively. Fish that weigh at least 1.5 kg and have a percentage head size no larger than 25% are kept. Fish that do not meet both minima are culled.

minimal levels for each independent of the others, you may be able to keep only a few fish:

	Percentage of the population that is saved	
No. phenotypes	Each phenotype culled at 50th percentile	Each phenotype culled at 90th percentile
1	50.0	10.0
2	25.0	1.0
3	12.5	0.1
4	6.25	0.01
5	3.12	0.001
6	1.56	0.0001
7	0.78	0.00001

As you can see, the overall percentage of the population that is saved drops precipitously. If you select at the 50th percentile and you select four phenotypes simultaneously, you will save only 6.25% of the population. Culling at the 50th percentile is not very intense. But even with this low cutoff value, the simultaneous selection of many traits at this level of selection will restrict the population quite severely.

Very intense selection can cause two problems: One, you may not be able to maintain a constant population. You need to retain a certain percentage of the population in order to maintain a constant population size. This is a big problem with livestock, but it is of less importance with fish because of their great fecundity. Second, severe reductions in the effective breeding number can alter the gene pool through genetic drift and inbreeding (more about this later). To prevent independent culling from restricting the population, you have to relax your standards. For

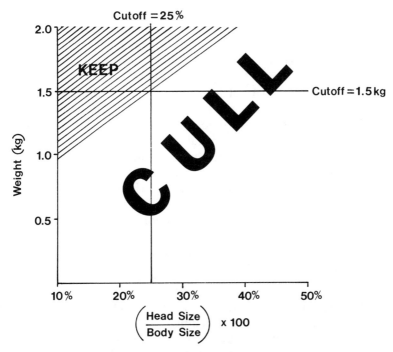

Figure 4.12 A schematic diagram of modified independent culling for weight and percent head size (an approximation of dressing percentage). Minimal performance levels are 1.5 kg and 25%, respectively. Unlike the selection program diagrammed in Fig. 4.11, fish that are outstanding in one phenotype are kept instead of being culled.

example, if you want to save 10% of the population, and you are going to select for two equally important phenotypes, your cutoff value is at the $\sqrt{0.1}$ level for each phenotype [$(\sqrt{0.1})(\sqrt{0.1}) = 0.1$] or the 31.6% level. You can, of course, adjust the cutoff values for the different phenotypes if you want to select one more intensely than the other. For example, cutoff values of 40% and 25% for two phenotypes will still retain 10% of the population [$(0.4)(0.25) = 0.1$].

Selection index   A selection index is the most efficient selection program when selecting simultaneously for two or more phenotypes, because all phenotypes are entered into a formula that will produce an overall value for each fish. Fish are then selected according to their scores, e.g., top 10% of the scores. A fish's numerical score is determined by the following formula:

$$I = b_1 X_1 + b_2 X_2 + b_3 X_3 + \cdots + b_n X_n \qquad (4.5)$$

where I is an individual's score, the bs are multiple regression coefficients, and the Xs are an individual's phenotypes. The bs are obtained from h^2s, correlations, and the economic importance of the phenotypes. One of the ultimate goals of any breeding program is to develop a selection index, because it is the most efficient way to improve productivity. Reagen *et al.* (1976) developed a selection index for channel catfish, Gall (1979) developed one for rainbow trout, and Hershberger *et al.* (1990A) developed one for coho salmon. Reagen *et al.*'s (1976) selection index for channel catfish was constructed to select for increased length and weight at both 5 and 15 months, and was based on the h^2s for length and weight at 5 and 15 months and on the correlations among those traits that they calculated in their population of channel catfish:

$$I = 1.578(\text{wt at 5 months}) - 4.6135(\text{length at 5 months})$$

$$+ \ 0.2122(\text{length at 15 months}) + 26.5624(\text{wt at 15 months})$$

This selection index should not be considered to be the definitive channel catfish selection index by all channel catfish farmers, because it may be valid only for the population in the Reagan *et al.* (1976) study.

Although this type of selection program is the most efficient, the baseline data needed to calculate the bs do not exist for the vast majority of cultured fish.

Despite the fact that the data needed to generate selection indexes do not exist for most populations of fish, you can approximate a selection index by assigning your own values to the bs. These values (importance factors) are based on your ideas about the importance of each phenotype and its relative value.

For example, if a catfish farmer wants to select for increased weight at 18 months of age, for decreased dressing percentage (based on head size to body size ratio), and for increased gain per day during August in his population of channel catfish, the first thing he needs to do is to assign relative values to the three phenotypes. The following are the values that he assigns for this project:

Phenotype	Population mean	Relative importance
weight	454 g	50%
dressing percentage	60%	30%
gain/day	3 g/day	20%

Importance factors are determined by using the following equation:

$$\text{Importance factor } (I) = \frac{\text{relative importance}}{\text{population mean}} \tag{4.6}$$

The importance factors for weight, dressing percentage, and gain per day during August in this selection program are

$$I_{\text{weight}} (I_W) = \frac{50\%}{454 \text{ g}} = 0.1101322$$

$$I_{\text{dressing }\%} (I_D) = \frac{30\%}{60\%} = 0.5$$

$$I_{\text{gain/day}} (I_G) = \frac{20\%}{3 \text{ g/day}} = 6.6666667$$

The approximated selection index for this breeding program is

$$I = (I_W)(\text{weight}) + (I_D)(\text{dressing }\%) + (I_G)(\text{gain/day})$$

A fish that ranks at the 50th percentile in all three phenotypes (I_{mean}) will have an $I = 100.0$:

$$I_{\text{mean}} = (0.1101322)(454 \text{ g}) + (0.5)(60\%) + (6.6666667)(3 \text{ g/day}) = 100.0$$

Now that the farmer has created his approximated selection index, the next step is to compute the I values for every channel catfish in the population. Those channel catfish that have Is greater than 100 are above average, while those with Is less than 100 are below average. If the farmer wishes to retain the top 10% of the population, he simply keeps those channel catfish above the 90th percentile in terms of I values. He can even use the Is to choose which of two brooders is the better fish. For example, which channel catfish is better, AU-23 or AL-22?

	AU-23	AL-22
weight	544 g	589 g
dressing %	63%	62%
gain/day	4.0 g/day	3.2 g/day

$I_{AU-23} = (0.1101322)(544 \text{ g}) + (0.5)(63\%) + (6.6666667)(4.0 \text{ g/day}) = 118.079$

$I_{AL-22} = (0.1101322)(589 \text{ g}) + (0.5)(62\%) + (6.6666667)(3.2 \text{ g/day}) = 117.201$

AU-23 beats AL-22.

Family selection All of the programs that have been discussed in the previous sections are based on individual merit and are called "mass selection" or "individual selection." Each individual, regardless of family, is compared to all others, and the best are used as broodstock in the next generation. Individual selection is effective if h^2 is large.

When h^2 is small, individual selection is not an efficient way to change the population mean. If the phenotypes in question are so valuable that improving them is of the utmost importance, family selection can be used to change population means more efficiently. Family selection will work only if V_A exists for a particular phenotype. If no V_A exists, neither individual nor family selection will be able to change the phenotype.

In individual selection, each fish stands or falls on its own merit and is selected or culled without regard to familial relationships. All families are pooled; family averages are not determined; and individuals are judged against the population mean.

An alternate form of selection is based on familial ties. Individuals are not pooled but are kept isolated by families (they can be grown together if they can be branded or tagged). The population average is not as important as family averages, and selection is based on relationships to family means.

There are two basic types of family selection: the first is called either "family selection" or "between-family selection" (I will call it "between-family selection"; the term "family selection" will be used generically to describe all forms of family selection). In this type of breeding program, family means are compared, and whole families are either culled or selected, based on family means. The whole family does not have to be selected; a random sample will suffice. The sample must be a random sample, or it will not be between-family selection.

Between-family selection is useful if selection must be done on characteristics, such as carcass traits, which cannot be measured in live fish. If cryopreservation of gametes has been perfected for a species, dead fish can be used as broodfish. On the other hand, if this technique has not been developed, the only way to select for these phenotypes is to use a breeding program called "sib selection."

Sib selection is a form of between-family selection, but the fish that are selected or are culled are not measured. The decision to select or to cull the families is based on the phenotypic values of the fish that were slaughtered. If family size is large (which is the case with fish) and if the fish that are measured are chosen randomly from each family, the means and variances of the fish that are measured should be the same as those that are not; consequently, phenotypic means of the fish that aren't measured are known.

Another form of between-family selection is called "progeny testing." This type of breeding program assesses the genetic value of broodstock by examining their offspring; this program is similar to that discussed in Chapter 3 for qualitative phenotypes. This breeding program can produce better results than individual selection, because an individual fish's offspring present a fairly accurate estimate of its breeding value. The performance of families will differentiate parents whose phenotypes are produced by heritable variance (V_A) from those whose phenotypes are due to non-heritable variance (V_D, V_I, V_{G-E}, or V_E).

Progeny testing has one major drawback—it takes an inordinate amount of time; in some cases it nearly doubles the generation interval. In some species, such as salmon, progeny testing is impossible unless gametes can be cryopreserved, because the fish die after they spawn.

The second type of family selection is called "within-family selection." In this type of breeding program, each family is considered a temporary sub-population, and selection occurs simultaneously and independently within each family. Individuals within each family are selected or culled, based solely on their relationship to their family mean. Figure 4.13 illustrates the differences among individual, between-family, and within-family selection.

Individual selection is generally more effective than family selection if h^2 is large. There is no magic h^2 value that identifies whether individual selection will be more efficient; however, phenotypes with h^2s $\lesssim 0.15$ are difficult to improve by individual selection, and family selection is more effective.

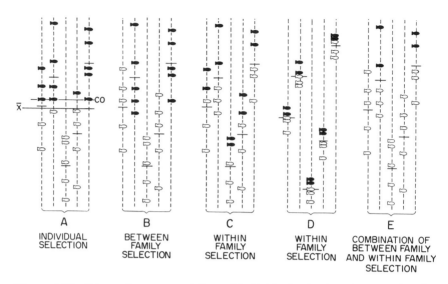

A	B	C	D	E
INDIVIDUAL SELECTION	BETWEEN FAMILY SELECTION	WITHIN FAMILY SELECTION	WITHIN FAMILY SELECTION	COMBINATION OF BETWEEN FAMILY AND WITHIN FAMILY SELECTION

Figure 4.13 Schematic representation of individual selection, between-family selection, within-family selection, and a combination of between-family and within-family selection. Populations A, B, C, and E are identical. There are five families (represented by the vertical dashed lines). Arranged along each dashed line are individual fish that comprise that family, and the position of each fish represents its phenotypic value. The small horizontal bar on each dashed line is the family mean. The black fish represent fish that are selected, while the white fish represent fish that are culled. In individual selection (A), family means are ignored and individual fish are compared to the population mean ($\overline{X}$). A cutoff value (CO) is created, based on the distribution of individuals around the population mean, and all fish that are equal to or larger than the cutoff value are selected, while those that fall below the cutoff value are culled. In between-family selection (B), the population mean is ignored and selection is based on family means. Population B shows that the two families with the largest means are selected, while the other three families are culled. No fish from the culled families are saved, even if they are outstanding fish. In within-family selection (C and D) the population mean is once again ignored. This time, family means are not compared. Fish within each family are compared only to their family mean. In populations C and D, the top two fish in each family are saved. All other fish in each family are culled. Populations C and D illustrate the effects of two different types of environmental variance. In population C, the effect is felt more on an individual basis, and increases individual deviations from family means. In population D, the effect is felt at the family level and increases family mean deviations from the population mean. Within-family selection is very effective in the situation illustrated in population D. Population E illustrates the results of a combination of between-family and within-family selection. Initially, the best two families are selected; then the top two fish within each of the two selected families are selected.
Source: after Falconer (1957)

Between-family selection is more effective than individual selection when h^2 is small, because most of the phenotypic variance observed among individuals is due to non-heritable sources of variance. As a result, an individual's phenotypic value does not accurately represent its breeding value.

When family means are compared, much of the non-heritable variance, particularly environmentally-induced phenotypic variance, is reduced, since deviations of individual values from family means negate each other. Thus, the family phenotypic means approximate their mean breeding values. This is particularly true if environmental variance has a greater effect on the individual level than on the family level.

Within-family selection is more efficient than individual or between-family selection when the major sources of non-heritable variance, particularly environmental variance, are common to all individuals of a family but are different among families; i.e., environmentally-induced variations are at the family level. Examples of this type of non-heritable variance are time of birth and age of females (these sources of phenotypic variance will be discussed later). Figure 4.13 illustrates the results of environmental variance that cause individuals to deviate from family means (Fig. 4.13 A, B, C, E) vs that produced when environmental variance is common to all fish within families but is different among families (Fig. 4.13 D). If a large component of V_P is due to environmental variance at the family level, within-family selection will be more efficient than between-family selection because the best families may be best simply because of V_E.

When individual selection is used, all fish that are selected are phenotypically the best (equal to or greater than the cutoff value), and if h^2 is large, phenotypic values approximate breeding values. For example, if individual selection for weight is conducted and the cutoff value is 1.0 kg, all select broodfish will be ≥ 1.0 kg. When family selection is used, the select broodfish will consist of individuals with a broad range of phenotypic values. When using between-family selection, whole families are selected, so runts tag along with their families. When within-family selection is used, the best fish in one family may be half as large as fish that were culled in other families. The way to prevent this from becoming too pronounced is to combine between-family and within-family selection (Fig. 4.13 E). This is a two-stage selection program: the first step is to use between-family selection to select the best families; the second step is to use within-family selection to select only the best fish from each of the selected families.

The progress that can be made when using family selection can be computed in a manner similar to that described earlier for individual selection. If between-family selection is used, predicted response to selection is

$$R_{b\text{-}f} = i\sigma_{pf}h_f^2 \qquad (4.7)$$

which is a modified version of Eq. (4.4), where: $R_{b\text{-}f}$ is the response to between-family selection; i is the intensity of selection; σ_{pf} is the standard deviation of the family means of the phenotype; h_f^2 is the heritability of the family means, which is the proportion of phenotypic variance of family means that is due to V_A.

If within-family selection is used, predicted response to selection is

$$R_{w\text{-}f} = i\sigma_{pw}h_w^2 \qquad (4.8)$$

which is also a modified form of Eq. (4.4), where: $R_{w\text{-}f}$ is the response to within-family selection; i is the intensity of selection; σ_{pw} is the within-family standard deviation; h_w^2 is the heritability of within-family phenotypic deviations, which is the proportion of phenotypic variance of within-family phenotypic deviations that is due to V_A. A detailed discussion on how these h^2s are calculated can be found in Falconer (1981).

The relative efficiencies of individual selection, between-family selection, within-family selection, sib selection, progeny testing, and a combination of between-family and within-family selection can be calculated once the h^2, SD of the phenotype, correlation of phenotypic values of members of the families, family size, and type of family (full-sib or half-sib) are known. Falconer (1981) provides an excellent discussion of this, as well as the formulae that are needed to determine which program is most efficient.

The efficiencies of individual vs family selection has been estimated in *T. nilotica* and in rainbow trout. In *T. nilotica*, within-family selection was predicted to be more efficient than individual selection as a way of improving growth rate, because of large familial-level environmental effects (Uraiwan and Doyle 1986). In rainbow trout, individual selection was predicted to be more efficient than either between-family or within-family selection as a way of improving body weight; a combination of between-family and within-family selection was predicted to be best (Gall and Huang 1988A). On the other hand, between-family selection was predicted to be more effective than either within-family or individual selection as a way of improving female post-spawning weight, egg volume, egg size, and number of eggs; the most efficient breeding programs to improve these traits were progeny testing and a combination of between-family and within-family selection (Gall and Huang 1988A).

Control population A key aspect of any breeding program is to use a control population to enable you to evaluate the genetic gain. If no control population exists, you will be unable to separate improvements that occurred because of your exploitation of V_A from those which occurred because of an improved environment (better handling skills, better rations, etc.). Figure 4.14 shows that the F_1-generation control population's mean is larger than the parental generation's mean, which indicates that environmental gains occurred.

Examples of breeding programs that had no controls are those that were done by Donaldson and Olson (1957) and Savost'yanova (1969).

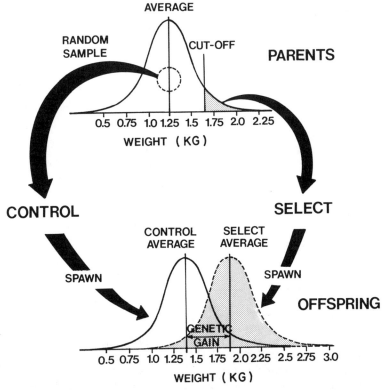

Figure 4.14 Schematic diagram of individual selection for weight. Mean weight in the parental generation was 1.25 kg. The cutoff weight for the selected broodfish was 1.65 kg. A random sample of fish around the mean was used to choose the control broodfish. The mean weights of the F_1 select and F_1 control lines were 1.88 and 1.39 kg, respectively. Weight was improved by 36% over that of the control line.
Source: after Tave (1987)

There is no doubt that they improved their strains of rainbow trout genetically; unfortunately, it is impossible to separate improvements that were due to the exploitation of V_A from those that were due to V_E.

Kincaid *et al.* (1977) and Kincaid (1983A) maintained a control population in their rainbow trout selection program. Because of that, they were able to determine that 35.7% of the increased growth rate was genetic gain, while 64.3% was due to improved culture conditions. If no selection had been done, average 147-day weight of the even-year line would have increased from 2.02 to 6.44 g over the six generations simply because of improvements in V_E (Kincaid 1983A); see Fig. 4.15. Obviously, a program which fails to incorporate a control will produce results which cannot be interpreted properly. It is difficult to make correct management decisions when you have incomplete information.

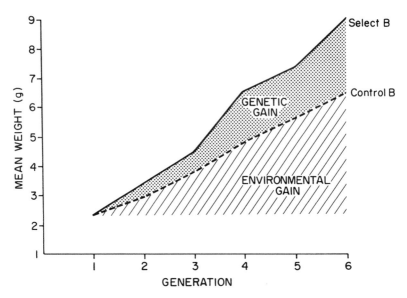

Figure 4.15   Results from Kincaid *et al.*'s (1977) and Kincaid's (1983A) selection program for increased 147-day weight in rainbow trout. Data are for the even-year line (see Fig. 4.16). The data show that average weight of the select line went from 2.02 g to 8.89 g, while that for the control line went from 2.02 to 6.48 g. Had no selection been conducted, average weight would have been improved by 4.46 g because of better handling skills, etc. The control population enables us to determine that the genetic improvement was 2.41 g, which means that the genetic improvement was 119.3%.

Source: after Kincaid (1983C)

There are several ways to create a control population. The first is to take a random sample of the population before selection is conducted. A second way is to select fish from around the mean (Fig. 4.14). A third is to respawn the parental generation. The third approach is limited in that it will generally work for only one or two generations, and it is impossible with fish that die after spawning. If there are age-related (V_E) effects on offspring growth (V_P), this approach will not work. The best approach is to obtain a random sample before selection.

Once the control population is created, it is perpetuated every generation and raised under the same culture conditions as the select population. The only difference is that random samples are taken in the control population every generation to create the next generation's control broodfish.

Selective Breeding Programs with Fish   A number of selective breeding programs have been conducted to change length, weight, time of spawning, viability, disease resistance, meristics, or sex ratio in channel catfish (Bondari 1982, 1983A, 1986; Dunham and Smitherman 1983), rainbow trout (R.C. Lewis 1944; Millenbach 1950; Donaldson and Olson 1957; Savost'yanova 1969; von Limbach 1970; Bridges 1973, 1974; Kincaid et al. 1977; Bergot et al. 1981B; Kincaid 1983A; Ihssen 1986; Langholz and Hörstgen-Schwark 1987; Siitonen and Gall 1989; Fevolden et al. 1991; Sadler et al. 1992), steelhead trout (Tipping 1991), brook trout (Embody and Hayford 1925; Hayford and Embody 1930; Ehlinger 1964, 1977), brown trout (Ehlinger 1964, 1977) Atlantic salmon (Gjedrem 1979; Friars et al. 1990; Fevolden et al. 1991), chinook salmon (Donaldson and Menasveta 1961), coho salmon (Hershberger 1988; Hershberger et al. 1990A), Siberian whitefish (Andrijasheva 1981; Andrijasheva et al. 1983), common carp (Kirpitschnikow and Faktorowitsch 1969; Kirpichnikov 1972; Kirpichnikov and Factorovich 1972; Kirpichnikov et al. 1972A, 1972B, 1979; Moav and Wohlfarth 1976; Babouchkine 1987), T. mossambica (Ch'ang 1971B), T. nilotica (Hulata et al. 1986; Abella 1987; Teichert-Coddington and Smitherman 1988; Uraiwan 1988; Abella et al. 1990; Wedekind et al. 1990; C.-M. Huang and Liao 1990), T. aurea (Bondari et al. 1983), hybrid tilapia (Jarimopas 1986; Behrends et al. 1988), mosquitofish (Busak 1983; Campton and Gall 1988), guppy (Schmidt 1919; Svärdson 1945; Burger 1974), and fourspine stickleback (Hagen and Blouw 1983).

Some of these programs incorporated crossbreeding in addition to selection. For example, Kirpichnikov's (1972) and Kirpichnikov et al.'s (1972A) selection program for increased weight gain in the Ropsha strain of common carp incorporated individual and family selection for weight gain, crossbreeding, and progeny testing to cull all s alleles.

A detailed examination of every selective breeding program that has been conducted with fish is beyond the scope of this book. Several programs will be described to illustrate different ways breeding programs can be conducted to improve productivity.

The first is a simple program by which individual selection was used to improve a single phenotype. The best selection program that achieved its goal with a warmwater fish is the one done at Auburn University to improve growth rate in channel catfish. The only goal was to improve size at 18 months of age (after the second summer), which is the age when catfish were traditionally harvested. At that age they weigh between 340 and 568 g. Consequently, the goal was to improve weight at 18 months of age and thus decrease time needed to produce a marketable-sized fish.

A preliminary study had determined h^2 for length and weight (Reagan et al. 1976), and these h^2s indicated that considerable amounts of heritable variance for growth existed. Consequently, individual selection was chosen to improve growth. Three strains were entered into this breeding program: Marion, Rio Grande, and Kansas.

Six pairs of fish in each of three lines were spawned to produce the parental generation. Fish produced from the six spawns within each strain were pooled and raised to fingerling size during the first summer. At that point, 900 fish from each strain were randomly selected and stocked in three 0.04-ha ponds at 7,410/ha. Fish were grown for a second summer and were harvested at the end of the second summer. At harvest, fish from each pond were separated by sex and were graded by inch (2.45 cm) groups. The sexed inch-groups from the three ponds were then combined. The top 10% of the males and the top 10% of the females were selected and were weighed. Each select line consisted of 45 males and 45 females. After the select fish were removed, the remaining males and females were each combined, and 45 of each sex were randomly chosen to form a control line for each strain (Chappell 1979).

The select and control broodfish were then grown until mature. The select fish were "held back" by underfeeding, and the control fish were allowed to grow so that they would be the same size when they were spawned (Dunham 1981). When sexually mature, 3 to 8 males and females from each select line and each control line were spawned. This created the F_1 generations. The fish were raised as above described, and when the fish were harvested at the end of the second summer: average weights of the Marion strain were 486 g for the select line and 413 g for the control line, a genetic gain of 18%; average weights of the Rio

Grande strain were 431 g for the select line and 368 g for the control line, a genetic gain of 17%; average weights of the Kansas strain were 513 g for the select line and 459 g for the control line, a genetic gain of 12% (Dunham and Smitherman 1983).

At harvest, the fish were measured and selected as described above to produce the F_2 select broodstock and F_2 control broodstock. These fish were raised until mature, and the process was repeated for a second generation of selection. The only differences were: one, the Rio Grande select line went extinct; two, the Marion control line was not spawned; the Kansas control was used as the control for both the Kansas and Marion select lines.

Six to ten males and females were spawned in the three groups, and the F_2-generation fish were raised as described above. When the fish were harvested at the end of the second summer: average weights of the F_2-generation Kansas strain were 363 g for the select line and 309 g for the control line, a total genetic gain of 17% (5% additional gain in the second generation); average weight of the Marion select line was 337 g, a total genetic gain of 21% (3% additional gain in the second generation) when compared to the Kansas control (Brummett 1986).

The F_2-generation average weights were less than the F_1-generation weights. The reason was that the F_2 fingerlings were smaller when they were stocked for grow-out. This occurred because F_2 fry were stocked and cultured at a greater density and because they were held at that density for a greater period (Dunham and Smitherman 1983; Brummett 1986). This clearly shows why a control line is valuable. The control showed that the select lines were still diverging from the control line. Had no control been maintained, the only conclusion would have been that something had gone wrong.

A second example of a breeding program that improved weight gain was done with rainbow trout by the U.S. Fish and Wildlife Service (Kincaid et al. 1977; Kincaid 1983A). The objective was to improve weight gain. Selection was a combination of between-family and within-family selection. Preliminary studies suggested that family selection would be efficient, because large family differences were observed at 147 days. Secondly, there was a large correlation between 147-day weight and family weight at 1 year. By selecting the best families when they were 147 days old, the fish would be small, which meant that the breeding program would require less grow-out space, water, food, and labor. The goals were to improve 147-day weight by direct selection and to improve 1-year weight by both direct and indirect selection.

Between 43 and 78 families were spawned each generation. Family size was equalized at 600 eggs, 300 fry, 150 fish at 105 days, and 110 fish at 147 days in the last four generations. Fish within each family were chosen by random samples at all stages. The fish were raised using standard rainbow trout culture techniques. On day 147, the fish were weighed, and the 8 families with the greatest mean weights were selected, while all other families were culled. One constraint was placed on the selection process: each parental line had to contribute one of the selected families, but it could not contribute more than three families.

A control population was created by taking a random sample from the base population from which the lines were selected. Each generation, 250 parents were randomly selected and mated to perpetuate the control lines.

Family size in the 8 selected families was reduced to 110 fish, and they were grown until they were one year old. At that age, within-family selection was performed in each of the eight selected families. The fish in each family were weighed, and the 20 heaviest fish in each family were retained for broodstock.

After three generations, both the select and the control lines were split into odd-year and even-year lines. The process was repeated for six generations. After six generations, average weight at 147 days increased from 2.02 to 5.84 g in the odd-year select line and from 2.02 to 4.57 g in the odd-year control line. After six generations, average weight at 147 days increased from 2.02 to 8.89 g in the even-year select line and from 2.02 to 6.48 g in the even-year control line. Genetic gain was 0.401 g/generation for a total of 2.41 g in the even-year line, an increase of 119.3%, and was 0.217 g/generation for a total of 1.27 g in the odd-year line, an increase of 62.9% (Figure 4.16).

Selection also increased 1-year weight in the select lines. At the end of the project, average 1-year weights were: 188.0 g in the odd-year select line and 100.8 in the odd-year control line; 220.5 g in the even-year select line and 166.4 g in the even-year control line. The genetic improvement was about 40%.

This breeding program is an excellent example of how genetic correlations can be used to improve a trait. In this breeding program, weight at 1 year was improved by both direct selection (within-family selection at 1 year) and by indirect selection (between-family selection at day 147). By selecting at day 147, 1-year weight was improved by indirect selection and less space, feed, and labor were needed than if selection were only conducted at 1 year. This breeding program was both biologically efficient and economically prudent.

Another example with rainbow trout was the breeding program done at the University of California, Davis to change time of spawning (Siitonen and Gall 1989). Modern rainbow trout farms are almost like

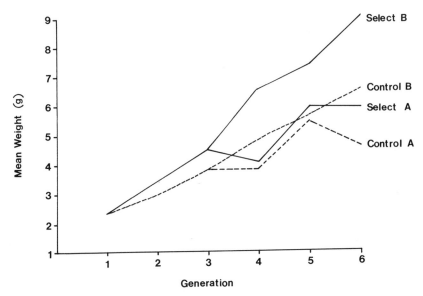

Figure 4.16 Results from Kincaid *et al.*'s (1977) and Kincaid's (1983A) selection experiment for increased 147-day weight in rainbow trout. After the third generation, both the select and control populations were split into odd-year spawners (A) and even-year spawners (B).
Source: After Kincaid (1983C)

factories in that they are capable of producing market-sized fish on a continual basis. Because of this, these farms require fingerlings on a continual basis, which means that they also require fry and eggs continually.

Because rainbow trout are seasonal spawners, the production system may be constrained by the brief, once-a-year spawning season. This can be circumvented by maintaining several strains that will spawn at different times, but this is expensive, can be a logistic headache in that the strains must be isolated because of disease concerns, and because different strains have different growth rates, production schedules would change constantly.

A second option is to import eggs from around the world as rainbow trout mature and spawn in different regions. This too is expensive and it also creates health problems. As with the first option, the culture of different strains would alter production schedules because each strain has its own growth rate.

A third option is to stockpile fingerlings and maintain them until they are needed. This approach creates some logistic headaches, and it is expensive because maintaining a fish for a year costs about as much as raising one for market.

A fourth approach is to alter time of spawning by manipulating the photoperiod in indoor facilities. This approach is effective, but the facilities and effort needed for this endeavor are costly.

A final approach is to use selection to develop lines that spawn at different times. Obviously, farmers would have to develop and maintain facilities to accommodate the different lines, but each farm would only have to maintain one strain and no new fish would be imported which would minimize disease problems.

The goal of this breeding program was to make the fall-spawning Hot Creek strain spawn earlier by selecting for early spawning date. Previous research had suggested that this approach would work, because h^2 for age at spawning in rainbow trout had been estimated to be 0.38 (Gall *et al.* 1988). Additionally, rainbow trout broodstocks maintained by California Department of Fish and Game hatcheries spawn over an 8-month period. The original stocks for these hatcheries were probably spring spawners, which suggested that domestication and unintentional selection had altered time of spawning in these hatchery populations.

Selection for spawning was done by individual selection in both the odd-year and even-year lines of the Hot Creek strain of rainbow trout. Selection was based on female spawning time, because females spawn once a year; thus, a single discrete date could be quantified for each female. Males, on the other hand, are capable of spawning over a longer period, so it was difficult to quantify a single discrete value for each male, something which is required in a breeding program. This obviously meant that progress would not be as rapid as was theoretically possible, because selection would be conducted on only half the population.

Historical data on the spawning history of the strain were used to create a quantitative description of the phenotype: mean spawning date, variance, range, etc. This information was then used to establish the cutoff value for the P_1 generation, delineating females which would be saved from those that would be culled. Females were checked weekly to determine maturity. No fish that matured during the first two weeks of the spawning season were saved, because the eggs were generally of poor quality. Females that spawned before the projected mean spawning date were saved and became the select broodstock. The projected mean spawning date for subsequent generations was based on the previous generation's data. Those that spawned before the projected mean spawning date were saved, while those that spawned on or after the projected mean spawning date were culled. This process was repeated for six generations (13 years).

The control population was the Mt. Whitney strain of rainbow trout, a spring spawning strain. This wasn't a true control population, because it was a different strain, because it was maintained at a different hatchery, and because it spawned at a different time. However, it could be used to monitor progress, because the times that the two strains spawned could be compared.

After six generations, average date of spawning had been changed by 69 days in the even-year line and by 67 days in the odd-year line (Fig. 4.17). The response to selection was −6.63 and −6.87 days/generation for the odd-year and even-year lines, respectively (the sign is negative because the fish were spawning earlier in the year). These results clearly indicate that selection to change time of spawning in rainbow trout was successful.

The most impressive selection program that has been conducted with farm-raised fish was conducted by Domsea Farms, Inc. and the University of Washington with coho salmon (Iwamoto *et al.* 1982; Saxton *et al.* 1984; Hershberger 1988; Hershberger *et al.* 1990A, 1990B). Domsea raised pan-sized coho salmon (300–350 g) in marine net-pens. The breeding program began when Domsea decided to develop their own

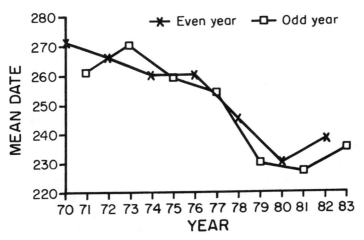

Figure 4.17 Mean spawning date during six generations of selection for early spawning date in rainbow trout. Spawning date is expressed in Julian dates (January 1 = 1 . . . December 31 = 365).
Source: after Siitonen and Gall (1988)

broodstock. When Domsea made this decision, they realized that they could use selection to direct genetic improvements instead of hoping that domestication would improve the fish.

Their first step was to list the phenotypes that they wished to improve. Their initial list included: 1) growth during the saltwater phase of production; 2) growth during the freshwater phase of production; 3) feed conversion during the saltwater phase; 4) percentage smoltification; 5) survival during the saltwater phase; 6) dressing percentage.

These goals were then examined biologically and economically. Based on these assessments, feed conversion during the saltwater phase was eliminated because of the inability to obtain a satisfactory measurement of the trait. Dressing percentage was eliminated because it was not considered as important economically as the remaining traits. Additionally, the small amount of genetic variability for dressing percentage would have made it difficult to improve the trait significantly.

Goals were then made for improving the remaining phenotypes: 1) increase body weight at the end of the freshwater phase of production—at seven months post-fertilization; 2) increase body weight at the end of the eighth month of the saltwater phase of production; 3) increase percentage smoltification; 4) increase survival during the first 3.5 months of the saltwater phase.

This list was shortened once again, because in the culture system used by Domsea, smoltification was an instantaneous event, in that fingerlings were transferred directly from freshwater to saltwater (as opposed to the gradual change which occurs naturally). Consequently, percentage smoltification was simply the percentage that survived this transfer.

Because of the way smoltification was defined, the geneticists cleverly realized that they did not have to include this goal in their breeding program. If the trait were heritable, the percentage would automatically increase as part of domestication, because those that died could not pass on their genes. Additionally, preliminary studies found that growth rate during the freshwater phase was positively correlated with survival during transfer to saltwater. This meant that selection for increased freshwater growth would increase percent smoltification by indirect selection.

Consequently, the breeders had to incorporate only three goals into the breeding program: growth in the freshwater phase, survival in the saltwater phase, and growth in the saltwater phase. They then decided to select the fish at two ages. The first selection occurred eight months after transfer to saltwater, which is the normal time for harvesting the coho salmon. This selection was based on weight at the end of the freshwater period, survival at 3.5 months in the saltwater phase, and weight at eight months in the saltwater phase. To do this, the breeders created a selection index where survival at 3.5 months in the saltwater phase was 3 times more important than freshwater weight and weight at eight months in the saltwater phase was five times more important than freshwater weight; thus, the relative values for these traits were 1 : 3 : 5, respectively. The traits were given these weighted values to reflect their relative economic values.

The second selection occurred at 14 months in the saltwater phase, about four months prior to sexual maturation. The information used during the second selection included all data collected to this point and also incorporated some adult size traits.

Selection at these two stages was conducted by using between-family selection. Family selection was chosen because of low heritabilities and because family size (about 2,000) was large enough to make family selection an effective approach.

The selection program is illustrated in Fig. 4.18. Each generation

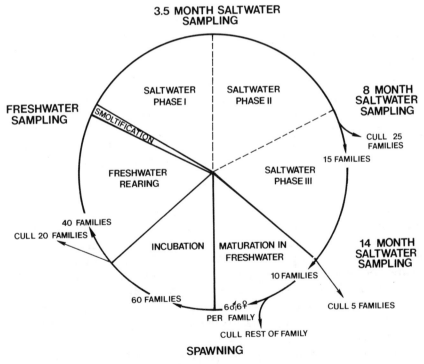

Figure 4.18 Schematic diagram of the selective breeding program used to improve growth rate in coho salmon.
Source: after Hershberger (1988)

started with 60 full-sib families. At the end of incubation, the number was reduced to 40. The reason that 20 families were culled was because of facility constraints and because some families had poor percentage fertilization or percentage hatch, and the geneticists didn't want growth influenced by differential stocking rates as a result of poor survival.

When fish were large enough, fish in each family were marked by fin clip and by freeze brand. Family size was reduced to 600 during the freshwater phase. Fish were grown in freshwater for seven months and were then transferred to marine net-pens (the families were combined), where they grew until just prior to spawning.

After eight months in saltwater, the weight at the end of the freshwater period, survival, and 8-month saltwater weight were entered into the selection index to make the first selection. The top 15 families were saved, while the other 25 were culled.

At the end of the 14th month in the saltwater phase, the second selection was made: the 10 best families were saved, while the other five were culled. The remaining 10 families were then grown to maturity in salt water, and one month prior to spawning they were transferred to fresh water so they could undergo final maturation. The heaviest six males and six females from each family were selected by within-family selection to be the broodfish to produce the next generation.

This process was repeated for four generations. There were two lines: an odd-year line and an even-year line. After four generations of selection, average freshwater weight increased from 14.6 to 19.2 g in the odd-year line and increased from 17.6 to 23.8 g in the even-year line (Figure 4.19 A). Body weight after eight months in salt water increased from 239 to 430 g in the odd-year line and from 296 to a density corrected value of 406 g (the actual value of 668 g is shown in Figure 4.19 B) in the even-year line.

It is difficult to quantify the genetic gain exactly, because control lines were not maintained from the beginning and because husbandry techniques were constantly improving. However, impressive gains were made. Growth was improved an average of 6.7%/generation during the freshwater period and an average of 10.1%/generation during the saltwater period. This decreased time needed to produce a marketable fish from 11 months to just 6 to 8 months in the select lines. In contrast, time needed to produce marketable fish decreased just one month in the control line.

The goal of improving percentage smoltification was also achieved. Survival during smoltification increased from 40% to 56–72%, an increase of 35–40%.

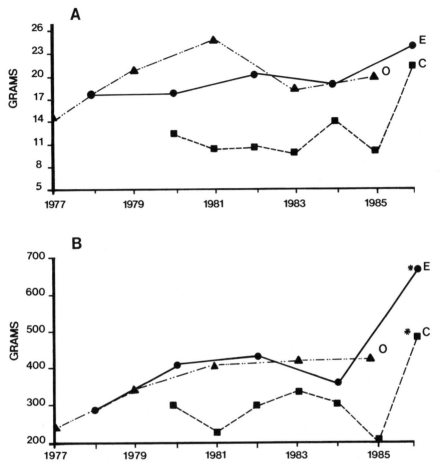

Figure 4.19 Results of the selective breeding program to improve weight gain in coho salmon. Graph A shows average weight in the freshwater period, and graph B shows average weight after 14 months in salt water: O = odd-year line; E = even-year line; C = control; * = the fish were grown at 1/4 normal stocking density.
Source: after Hershberger (1988)

Coho salmon that were culled during selection were not destroyed or sold for food. These fish were placed with the production broodstock in order to improve growth of the fish that were grown for market. Even though these fish were culls, they came from a breeding program, and their growth had been improved over that of the unselected production broodstock. This procedure transferred some the improved growth rate to the production fish, without compromising the efficiency of the selective breeding program.

Perhaps the biggest benefit of this program was the development of a prototype selective breeding program that not only improved production traits, but also demonstrated that such a program can complement, not complicate, husbandry and production efforts at a commercial aquaculture facility.

DOMINANCE GENETIC VARIANCE AND HYBRIDIZATION

When little or no V_A exists and it is either difficult or impossible to improve a phenotype by selection, the breeding technique that can be used to improve productivity is hybridization (crossbreeding). Hybridization improves productivity by exploiting V_D. You can, of course, use hybridization to exploit V_D even when a large amount of V_A exists.

If you recall, V_D is the genetic variance produced by the interaction of alleles at each locus. Because this form of genetic variance depends on interactions, it is disrupted during meiosis and cannot be transmitted from a parent to its offspring. Dominance genetic variance is created anew and in different combinations each generation, and its effects are basically those based on luck. Fish that are superior because of certain interactions are superior because of fortuitous combinations of alleles. Unfortunately, these fish cannot transmit their superiority to their progeny. Consequently, the goal in a crossbreeding program is to discover which combination of parents produces the combination of alleles that produces the desirable interaction in the progeny, and thus improves productivity.

Additive genetic variance and V_D are diametrically opposed; consequently, selection and crossbreeding are also diametrically opposed. In selection, you choose animals based on individual or family merit in the hope that the next generation will closely approximate those you have chosen. In crossbreeding, the next generation does not have to approximate the brooders, and unless the cross was made previously, you cannot predict the outcome of a mating to produce F_1 hybrids. The production of superior F_1 hybrids is a hit-or-miss proposition. Sometimes you are lucky; sometimes you are not. You can mate two superior fish and produce nothing but culls, or you can mate two culls and produce a grand champion. Basically, what you are trying to do is to discover a cross that nicks (produces superior hybrid progeny). When you have found the right combination, you use that cross to improve productivity.

Uses of Hybridization

Hybridization can be used in one of several ways to improve productivity. It can be used as a "quick and dirty" method before selection will be employed. Exploitation of V_D is independent of V_A, so hybridization can be used to improve productivity whether h^2 is large or small. When h^2 is small, hybridization is often the only practical way to improve productivity, because selection will be too inefficient. Hybridization can be incorporated into a selection program as a final step to produce animals for grow-out. In this case, you select fish in two lines that have been shown to produce good hybrids; after the selection programs, you hybridize the select lines. A third use of hybridization is for the production of new breeds or strains. A fourth use is for the production of uniform products. Processing plants and consumers often desire uniform products, and hybridization is the most efficient method of producing uniform progeny. A fifth use is for the production of monosex populations. And a sixth use is for the production of hybrids to be stocked in natural bodies of water that are unable to maintain self-reproducing populations.

One thing that hybridization generally does not do is produce good broodstock. By good broodstock, I am referring, not to eggs/kg female, but to the ability of the hybrids to produce above average progeny. F_1 hybrids do not produce above average progeny (unless they exhibit maternal heterosis; more about this later), because their superiority is due to V_D, and that is disrupted during gametogenesis. Because hybrid superiority is due to interactions, when hybrids reproduce, their progeny exhibit a wide range of interaction effects. Although hybridization can be used to create new breeds and thus provide new pools of V_A that can be exploited by selection, hybridization is mainly used to produce superior animals and plants for grow-out. Selection is used to produce superior broodstock.

Because improvement by hybridization is a hit-or-miss proposition, few definite patterns exist in the literature. While some experiments show that hybridization has improved productivity, others have recorded no improvement. Some hybrids have even lessened productivity.

Rather than list the dozens of studies that were or were not successful, I shall present examples of the different ways hybridization has been used to improve productivity:

Table 4.6 Selected Data from Chappell's (1979) Comparison of Four Interspecific Hybrid Catfish, Two Intraspecific Hybrid Catfish, Four Strains of Channel Catfish, White Catfish, and Blue Catfish[a]

Group	Seinability (%)	Feed conversion	Gain (kg/ha) (unadjusted)	Percent marketable	Dressing percentage	Survival (%)
Interspecific hybrids						
Channel ♀ × Blue ♂	64.6[ab]	1.21[a]	4,018[a]	99.8	62.0[b]	95.6[ab]
Blue ♀ × Channel ♂	52.3[b]	1.41[bc]	3,485[bcd]	98.3	59.0[c]	93.4[b]
White ♀ × Blue ♂	73.1[a]	2.42[d]	2,193[e]	47.8	59.0[c]	100.0[a]
Channel ♀ × White ♂	56.4[ab]	1.49[c]	3,313[cd]	67.6	56.5[d]	95.0[ab]
Intraspecific hybrids						
Marion ♀ × Kansas ♂	25.6[cd]	1.22[a]	3,998[a]	99.4	60.0[c]	98.6[a]
Auburn ♀ × Rio Grande ♂	20.8[d]	1.27[ab]	3,728[abc]	96.0	61.5[d]	99.2[a]
Strains of channel catfish						
Kansas	24.3[cd]	1.26[ab]	4,025[a]	99.5	59.3[c]	98.6[a]
Marion	35.4[c]	1.26[ab]	3,880[ab]	99.1	59.3[c]	95.9[ab]
Rio Grande	29.2[cd]	1.42[bc]	3,390[cd]	93.0	64.0[a]	97.3[a]
Auburn	18.5[d]	1.36[abcd]	3,610[abcd]	95.6	63.3[a]	99.9[a]
Blue catfish	68.4[a]	1.51[a]	3,210[d]	99.5	64.3[a]	98.1[a]
White catfish	29.8[cd]	1.99[d]	2,475[e]	70.7	55.0[e]	84.1[c]

[a] Means followed by the same letter are not significantly different ($P = 0.05$).

Hybridization has been used to improve productivity in catfish culture as a stop-gap method until selection could be employed to create better strains of channel catfish. Plumb *et al.* (1975) showed that hybridization improved resistance to channel catfish virus disease. Giudice (1966), Yant *et al.* (1976), and Chappell (1979) found that some hybrids improved yield by 10 to 18%. Chappell (1979) also found that hybridization improved seinability and food conversion. Some of Chappell's (1979) data are shown in Table 4.6. Horn (1981) and Dunham *et al.* (1983) found that hybridization improved egg production.

Hybridization has also been used to produce new strains of rainbow trout, brown trout, brook trout, cutthroat trout (Kincaid 1981), common carp (Kirpichnikov 1981), and channel catfish (Dunham and Smitherman 1985). For example, Dunham and Smitherman (1985) created the AU-MK strain of channel catfish from Marion × Kansas hybrids. A population of Marion × Kansas F_1 hybrids were spawned to produce Marion × Kansas F_2 hybrids, and the F_2 hybrids were selected for improved body weight. The F_3 generation (AU-MK-3) had a faster growth rate, had a greater spawning rate as 3-year-olds, and produced more fingerlings/kg female than other select strains. Figure 4.20 shows the pedigree of the AU-MK strain of channel catfish.

Some hybrid crosses produce monosex populations. Monosex populations have been produced by interspecific hybridization (hybridization of two species) in sunfishes (Childers 1967) and tilapia (Hickling 1960; Pruginin *et al.* 1975; Majumdar and McAndrew 1983). The most famous and most important example is that which occurs in tilapia. Hickling (1960) discovered that progeny produced by the hybridization of two species were all males. (At the time, Hickling did not realize that

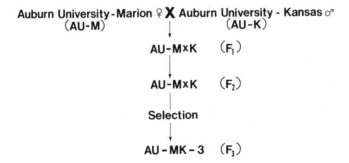

Figure 4.20 Pedigree of the AU-MK strain of channel catfish that was created by hybridizing the Marion and Kansas strains.
Source: After Dunham and Smitherman (1985)

he had hybridized two species. He thought that he had hybridized two strains of the same species.) The explanation for this phenomena is very interesting: Some species of tilapia have the XY sex-determining system, while others have the WZ sex-determining system (F. Y. Chen 1969). The proper combination of sex chromosomes in the parents will result in all-male progeny. That combination is produced by the hybridization of XX females with ZZ males:

XY System (XX females)	WZ System (ZZ males)
T. nilotica	T. hornorum
T. mossambica	T. aurea

For example, the hybridization of a T. nilotica ♀ × a T. hornorum ♂ will produce a monosex population:

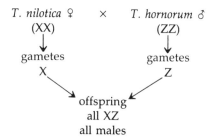

Unfortunately, this technique is not 100% successful in producing all-male populations; many matings produce some females.

There are two reasons why this technique is not 100% successful. The first is the fact that many tilapia culturists cannot or will not maintain pure species. Some even brag that their stocks are "12/16-" or "15/16-pure" and claim that because their stock contains only a "small percentage" of genes from another species it can be considered "pure" for breeding purposes. That is nonsense. It is like being almost pregnant—you either are or you are not; there is no in-between. A mongrelized population of tilapia will not breed as expected, because it can contain as many as four sex chromosomes. The inability of a mongrelized tilapia population to breed true means that some individuals will transmit the "incorrect" sex chromosomes to their offspring, which means that some hybrid offspring that are expected to be males will, in fact, be females.

A second reason that this technique is not 100% successful is because of autosomal sex-influencing or sex-modifying genes (Avtalion and Hammerman 1978; Hammerman and Avtalion 1979; Majumdar and McAndrew 1983; W. L. Shelton *et al.* 1983). These genes turn some males into females.

The autosomal sex-modifying genes can be eliminated by a breeding program called recurrent selection (more about this later). Lahav and Lahav (1990) performed one such program and produced tilapia which bred true and which produced 100% male hybrid offspring.

Hybridization has been used to improve fishing success in put-and-take situations. Donaldson *et al.* (1957) found that hybrid cutthroat trout were far more catchable than the parental strains. Tave *et al.* (1981) and Dunham *et al.* (1986) found that F_1 hybrid catfish were more catchable than parental strains. Channel catfish ♀ × blue catfish ♂ hybrids were extremely vulnerable to capture (Table 4.7).

Hybrids can also be used to improve a wild fishery. Moav *et al.* (1978, 1979) and Wohlfarth (1986) outlined programs in which hatchery strains would be stocked to hybridize with local populations in order to produce faster growing hybrids that could be harvested by commercial fishermen. Splake (brook trout × lake trout) and white bass × striped bass hybrids are examples of hybrids that are being stocked to replace or supplement natural stocks.

Planning Crossbreeding Programs

Because the results of hybridization cannot be predicted, how do you initiate a crossbreeding program? Although the discovery of a cross that

Table 4.7 Relative Abundance in the Population vs Proportion Caught by Angling for Blue Catfish, Channel Catfish, and Their Reciprocal Hybrids

Group	Relative abundance (%)		Proportion in catch (%)	
	Number	Weight	Number	Weight
Channel catfish	9.07	9.23	2.67	1.53
Blue catfish	32.82	28.65	22.67	17.32
Channel ♀ × Blue ♂	29.54	37.44	57.33	63.85
Blue ♀ × Channel ♂	28.57	24.68	17.33	17.30
Total	100.00	100.00	100.00	100.00
Parent species	41.89	37.88	25.34	18.85
Hybrids	58.11	62.12	74.66	81.15
Total	100.00	100.00	100.00	100.00

Source: Tave *et al.* (1981).

will produce superior progeny is basically fortuitous, you can improve your chances by judicious planning. The principle is the same as that in cards. What you are dealt is luck, but those with skill and an understanding of basic probability seem to win more pots than those who draw two cards to fill inside straights.

Certain hybrid matings can be eliminated before you initiate your program. You must pay attention to the phylogenetic tree. If the species are so distantly related that they belong to different families or orders, the odds are you will have little success. For example, it makes little sense to hybridize rainbow trout and common carp. Basically, it is best to stay within a family, and your luck will improve if you stay within a genus. Chevassus (1979) reviewed interspecific hybridization among three genera of salmonids; his findings are shown schematically in Fig. 4.21. (Some taxonomists are rearranging the salmonid family tree and are renaming some species, so the taxonomy used in Fig. 4.21 may not meet with universal approval.) The data in Fig. 4.21 show that hybridization between genera in salmonids is not as successful as hybridization within a genus. By starting with logical matings, you can achieve a higher degree of success.

An important bit of information that can often tell you whether an interspecific (between two species) cross will produce large numbers of viable F_1 offspring is the species' karyotypes (chromosome numbers and relative sizes and the morphology of the chromosomes). Successful crosses between species with dissimilar chromosome numbers are rare. An example of the problem created by trying to hybridize fish with dissimilar chromosome numbers is Chappell's (1979) hybridization study with channel catfish, blue catfish, and white catfish. He found that channel × blue hybrids were viable and grew well, but that hybrids with white catfish were very difficult to produce, and of those that were produced, many were abnormal. Subsequent karyotypic analysis of the three species showed that both blue catfish and channel catfish have 58 chromosomes, but white catfish have only 48 chromosomes (LeGrande *et al.* 1984). Additionally, the morphology of some of the chromosomes is different; white catfish have several pairs of very large chromosomes (Fig. 4.22). Prior knowledge of the karyotypes would have enabled Chappell (1979) to predict that hybridization between white catfish and either channel catfish or blue catfish would not be successful.

On the other hand, dissimilarities in the karyotypes between species may enable fish farmers to create sterile hybrids and thus produce a new commodity. Golden shiner is an important cultured bait fish in the U.S. Some bait fish farmers would like to grow a similarly shaped species, but one that is hardier. Rudd, a European cyprinid, was proposed, but

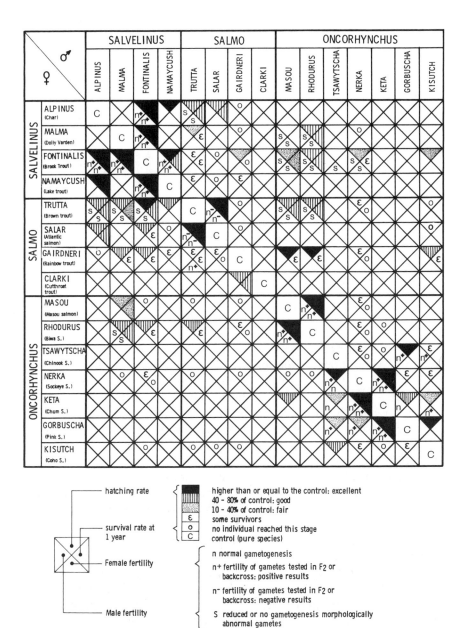

Figure 4.21 Viability and fertility of interspecific salmonid hybrids among the genera *Salvelinus, Salmo,* and *Oncorhynchus.*
Source: Chevassus (1979)

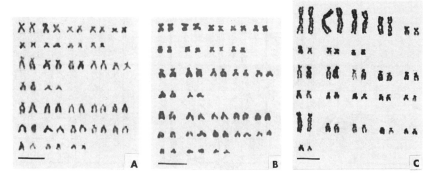

Figure 4.22   Karyotypes of channel catfish (A), blue catfish (B), and white catfish (C). Channel catfish and blue catfish have 58 chromosomes, while white catfish have only 48 chromosomes. In addition, the first four pairs of chromosomes in white catfish are large metacentric chromosomes which have no counterpart in either channel catfish or blue catfish. The bar in each picture represents 5 μm.
Source: after LeGrande et al. (1984)

because it is not native to the U.S., many states banned its importation and sale. Although both species have 50 chromosomes, Gold *et al.* (1991) and Y. Li *et al.* (1991) found a number of differences in chromosome morphology and also found that the nucleolar organizing region (NOR), which is where the genes for ribosomal DNA are encoded, is on different chromosomes. These results suggest at least partial sterility in golden shiner-rudd F_1 hybrids. Preliminary observations of hybrid gonads revealed that they are abnormal, and attempts to spawn the hybrids have been unsuccessful. These results suggest that if sufficient numbers of hybrids can be produced and if they grow well, bait fish farmers may be able to exploit the chromosomal differences in these species to produce a sterile interspecific hybrid bait fish.

Another important clue that can provide information about the potential success of interspecific hybridization is the biology and reproductive behavior of the two species that you wish to hybridize. If one species spawns in March and the other in June, it will be difficult to produce hybrids unless you can cryopreserve the gametes. Such a procedure was used to hybridize red sea bream and crimson sea bream, species that spawn six months apart (Kurokura *et al.* 1986), and to hybridize even- and odd-year pink salmon (Gharrett and Smoker 1991). If one spawns in running water and the other in stagnant water, the gametes may never meet unless they can be stripped. There are many behavioral blocks to hybridization, and they can be very frustrating because they are often difficult to quantify. The two species may need different photoperiods, temperatures, or light intensities to induce spawning. Even if these requirements are the same, the different species may not recognize each other's courtship behavior. Some behavioral problems can be circumvented through the use of hormone injections. For example, Tave and Smitherman (1982) used human chorionic gonadotrophin to improve spawning success when hybridizing channel catfish and blue catfish. The use of hormones can, however, be expensive, so if it is possible to avoid their use, fingerling production will be cheaper. Hybridization below the species level should cause fewer problems in terms of behavioral roadblocks, because the reproductive behavior of the strains should be quite similar.

The results of interspecific hybridization can be improved by evaluating different strains. For example, Ramboux and Dunham (1991) found that the spawning rate between channel catfish ♀ and blue catfish ♂ varied among strains. Since the lack of spawning success in this cross has hindered its commercial development, these results are encouraging and demonstrate the value of this type of research.

Hybridization below the species level will give you the greatest likelihood of success. Many people erroneously think that interspecific hybridization (hybridization of two species) produces better hybrids than intraspecific hybridization (hybridization within a species). This is not so. The quality of the hybrid is simply a matter of luck. Chappell (1979) found that certain intraspecific hybrid catfish were as good as their more famous cousin, the channel ♀ × blue ♂ hybrid (Table 4.6). One advantage of intraspecific hybridization is the fact that you should have little trouble producing progeny, something which is often a problem with interspecific hybridization. Again, the production of superior hybrids is fortuitous, but you can be confident that you will produce hybrids.

One way to improve the likelihood of producing intraspecific F_1 hybrids that exhibit hybrid vigor (positive heterosis) on fish farms is to hybridize hatchery strains rather than wild stocks. For example, hybridization studies with channel catfish showed that 80% of hatchery × hatchery F_1 hybrids exhibited positive heterosis, but only 30% of hatchery × wild F_1 hybrids exhibited positive heterosis (Smitherman and Dunham 1985).

The final area that can determine the difference between success and failure in the production of hybrids is human error. I have noticed that when hatchery personnel try to produce hybrids, they usually choose the best brooders for the normal matings in order to meet fingerling production quotas and then use the culls in an effort to produce hybrids. This practice all but dooms the program. The broodstock that were used in the hybridization program may not have spawned under any condition, but their failure to spawn will be blamed on hybridization. This practice is understandable because hatchery managers must first achieve their fingerling production quotas. But if hybridization is to be attempted, the broodstock must be top quality fish.

If your first attempt at hybridization fails, the next step is to try, try, and try again. The literature will give you information about what crosses have been attempted, but other than that, you simply have to spawn all possible combinations.

Not only do you have to make all possible combinations, but you must also make the reciprocal matings. Reciprocal crosses are the two possible matings between two groups: female A × male B and female B × male A. This is necessary because reciprocal hybrids are seldom the same. For example, the channel ♀ × blue ♂ hybrid catfish is superior to its reciprocal: It is more uniform (Brooks 1977), grows faster, is more seinable, exhibits greater feeding vigor, has a better dressing percentage, has a better food conversion (Chappell 1979), is more catchable by hook and line (Tave *et al.* 1981; Dunham *et al.* 1986), and is easier to produce (Tave and Smitherman 1982). If you do not make the reciprocal crosses, you may miss the better hybrid.

A crossbreeding program does not have to stop with the production of F_1 hybrids. Heterosis is generally assumed to be controlled by V_D. If this is so, heterosis of F_2 hybrids should be half as great as that of the F_1 hybrids. If heterosis is controlled solely by V_D, F_1 hybrids will be the best; however, recent research has indicated that heterosis is also influenced by additive genetic effects, maternal genetic effects, maternal heterosis, epistatic genetic effects, and egg cytoplasmic effects, which means that F_2 or other types of hybrids can be better than F_1 hybrids. Consequently, the production of F_2, backcross hybrids, or other types of hybrids could produce outstanding fish for grow-out.

Jayaprakas *et al.* (1988) compared growth of Auburn University-Egypt and Auburn University-Ivory Coast strains of *T. nilotica* and their F_1, F_2, and backcross hybrids and found that heterosis of the F_2 and backcross hybrids was greater than that of the F_1 hybrids. Tave *et al.* (1990D) determined that the reason why the F_2 and backcross hybrids were better was because of maternal heterosis.

Maternal heterosis is produced when crossbred mothers are spawned. Maternal heterosis does not refer to increased egg production or other traits expressed by the mother. Those traits are part of heterosis for the F_1 hybrids. Maternal heterosis is expressed in the progeny of F_1 hybrid mothers—the F_2 and the backcross hybrids.

Even though V_D effects in the F_2 hybrids were only half as great as those in the F_1 hybrids, maternal heterosis was expressed in the F_2 and backcross hybrids, and this made them grow faster than the F_1 hybrids (Tave *et al.* 1990D). These results suggest that crossbred mothers make better broodstock in *T. nilotica*. The universal application of these results is unknown, so the results should not be used to claim that all crossbred mothers produce faster-growing offspring.

Once the various genetic parameters that contribute to heterosis are known, they can be used to predict the outcome of other hybrid matings. The information gathered in the tilapia crossbreeding program indicated that Egypt ♀ × Ivory Coast ♂ F_2 hybrids and Egypt ♀ × Ivory Coast ♂ F_1 hybrid ♀ × Egypt ♂ backcross hybrids would be better than any other possible Egypt-Ivory Coast *T. nilotica* hybrid (Tave *et al.* 1990D).

Types of Crossbreeding Programs

There are several types of crossbreeding programs. The most common is the two-breed cross. Here two breeds, species, etc., are hybridized to produce F_1 hybrids for grow-out:

$$A \times B$$
$$\downarrow$$
$$AB\ F_1\ hybrids$$

The production of the F_1 hybrids is a terminal cross. You seldom want to incorporate hybridization into a line that is undergoing selection, because it will undo some of your work on the exploitation of V_A. Basically, what you are trying to do is exploit V_D. This is the crossbreeding program that has been used in most fish hybridization studies. If you hybridize two lines that have undergone selection, you will exploit both V_A and V_D.

Topcrossing is a variation on the two-breed cross in which an inbred line is mated to a noninbred line or strain. R. H. Davis (1976) found that some topcrossed rainbow trout, brown trout, and brook trout grew faster than parental lines.

Backcrossing is another type of hybridization. Here the F_1 hybrid is mated back to one of its parents or parental lines. This is done to produce a hybrid with a greater percentage of one group:

$$A \times B$$
$$\downarrow$$
$$AB\ F_1\ hybrids$$

$$AB\ F_1\ hybrids \times A$$
$$\downarrow$$
$$AB\text{-}A\ backcross\ hybrid$$

The genome of the AB-A backcross hybrid is 75% A and 25% B.

This technique can be used to transfer desirable alleles from one strain or species to another. Behrends and Smitherman (1984) back-crossed cold-tolerant *T. aurea* to hybrid red tilapia to produce a population of cold-tolerant red tilapia.

This technique can also be used to improve growth by exploiting maternal heterosis. To do this, F_1 hybrid females are backcrossed to parental strain males.

A three-breed cross is used to produce various combinations of three different groups:

	%A	%B	%C
A × B ↓			
AB F_1 hybrids	50	50	0
AB F_1 hybrids × C ↓			
ABC F_2 hybrids	25	25	50
ABC F_2 hybrids × A ↓			
ABC-A F_3 hybrids	62.5	12.5	25

When one group is brought back (A was brought back and mated to the F_2 hybrid in the above example), the program is also called a rotational cross.

Recurrent Selection Although you cannot select for hybrid vigor, you can select for combining ability or for certain specific combinations which are most desirable. If you find that certain individuals hybridize more easily than the rest of the population, you can initiate a breeding program called recurrent selection in order to improve reproductive success during hybridization. Basically, all you do is select those that are willing to spawn, mate them within their own group (breed, strain, or species), and use their progeny to produce hybrids in the following generation. This should improve reproductive success during hybridization. This is repeated until reproductive success reaches the desired level. Recurrent selection is diagrammed schematically in Fig. 4.23. If recurrent selection is done with both sexes, the program is called reciprocal recurrent selection.

Recurrent selection can also be used to improve the results of hybridization. For example, a major goal in tilapia culture is the production of monosex male populations in order to prevent reproduction during

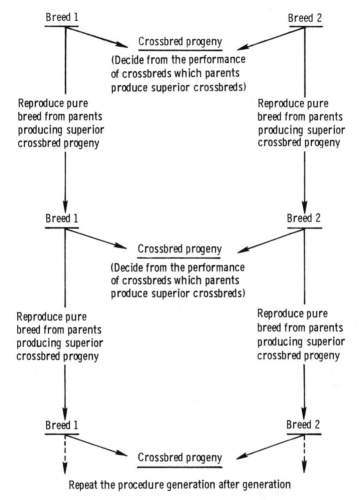

Figure 4.23 Schematic diagram of recurrent selection. If this breeding program is done for only Breed 1 or Breed 2, it is called recurrent selection. If it is done with both breeds, it is called reciprocal recurrent selection.

Source: John F. Lasley, Genetics of Livestock Improvement, 3rd edition, © 1978, p. 200. Reprinted by permission of Prentice-Hall, Inc., Englewood Cliffs, NJ

grow-out. Hybridization in tilapia often produces 5–15% females, which negates the effort to produce a monosex population. Individual tilapia that produce 100% hybrid males should be placed in a reciprocal recurrent selection program to eliminate tilapia that produce hybrid females (Hulata *et al.* 1983). Lahav and Lahav (1990) used this procedure to identify and produce broodfish that produce 100% male hybrid tilapia. Ramboux and Dunham (1991) found that they could improve the spawning success rate between channel catfish and blue catfish by using different strains. The production of these hybrids could be improved further by entering the two strains which hybridize most readily into a reciprocal recurrent selection program. McKay *et al.* (1992A) proposed that reciprocal recurrent selection be used to improve viability of tiger trout (brown trout ♀ × brook trout ♂).

Heterosis

The superiority or inferiority of hybrids is measured as heterosis. Heterosis (H) can be determined by using the following formula:

$$H = \left(\frac{\text{mean reciprocal F}_1 \text{ hybrids} - \text{mean parents}}{\text{mean parents}}\right) 100 \qquad (4.9)$$

For example, say a hatchery manager spawns and raises channel catfish, blue catfish, and their reciprocal hybrids in order to evaluate the relative growth of these catfishes at his hatchery. He harvests the four groups when they are 18 months old and records the following average weights:

Group	mean wt (g)
channel catfish	460
blue catfish	440
channel catfish ♀ × blue catfish ♂	600
blue catfish ♀ × channel catfish ♂	462

What is the heterosis in this experiment?

Step 1.   Calculate the average weight of the parental groups:

$$\text{mean wt parental groups} = \frac{460 \text{ g} + 440 \text{ g}}{2}$$

$$\text{mean wt parental groups} = 450 \text{ g}$$

Step 2.   Calculate the average weight of the hybrids:

$$\text{mean wt hybrids} = \frac{462 \text{ g} + 600 \text{ g}}{2}$$

$$\text{mean wt hybrids} = 531 \text{ g}$$

Step 3.   Calculate heterosis using Eq. (4.9):

$$H = \left(\frac{531 \text{ g} - 450 \text{ g}}{450 \text{ g}}\right) 100$$

$$H = 18\%$$

Note that both parental groups and the two reciprocal hybrids are needed in order to calculate heterosis. If all four groups are not measured, you cannot calculate heterosis. You can say that one or both hybrids were better or worse than one or both parents, but you cannot calculate heterosis. If $H > 0\%$, hybrid vigor was produced.

In summary, hybridization is the breeding technique that is used to exploit V_D. The production of better fish by hybridization is a hit-or-miss proposition. You cannot predict which cross will nick. You can only try, and if you are successful you will exploit V_D and improve productivity. If you are not successful, you simply have to try again.

INBREEDING

Inbreeding is the third major breeding program that can have a tremendous impact on productivity. Inbreeding is one of those concepts that everyone knows but few really understand. The term usually conjures up images of deformed and depraved individuals, and it is used as a cruel joke to explain virtually all developmental or behavioral defects, although inbreeding usually has nothing to do with these problems. Most people are aware of inbreeding because of legal and moral laws against certain consanguineous (blood-related) marriages. As recently as 200 years ago, there were laws forbidding consanguineous matings even in livestock, because it was considered immoral and against the laws of God and nature. But breeders quickly discovered that inbreeding is one of the most important breeding techniques; without its use, agricultural productivity would decline precipitously.

Inbreeding is simply the mating of related individuals; nothing more, nothing less. Inbreeding does not imply, nor does the definition mention, anything about viability, growth, or productivity. Inbreeding is neither good nor bad; it can, however, like any breeding program, be used either wisely or stupidly.

Genetically, all inbreeding does is create homozygosity. Related individuals share alleles through one or more common ancestor(s). When related individuals mate, the alleles that they share as a result of their common ancestor(s) can be paired. This produces progeny that are homozygous at one or more loci, and the progeny are inbred.

The mating of unrelated individuals can also produce offspring that are homozygous at one or more loci, so how can you distinguish homozygosity that is produced by inbreeding from that which is produced without inbreeding? Genetically, how do these two forms of homozygosity differ? The answers are: (1) you cannot distinguish between the two; (2) genetically, there is no difference. They are the same. The only difference is that inbred animals are homozygous because they have alleles that are alike by descent, whereas noninbred individuals are homozygous because they have alleles that are alike in kind. Again, there is no chemical or physical way to distinguish between the two, other than the way in which the alleles are inherited.

Since there is no genetic difference, why make a distinction? The reason is that related individuals are genetically more alike than unrelated individuals, i.e., they share more alleles in common, so the mating of related individuals will, on the average, produce offspring that are more homozygous than those produced by unrelated parents. Thus, inbreeding increases homozygosity over that which would occur if the mates were unrelated. The measure of inbreeding (F) quantifies the percentage increase in homozygosity over the population average.

That is all inbreeding does. Inbreeding does not change gene frequencies; selection, genetic drift, migration, and mutation change gene frequencies, but inbreeding does not. Because inbreeding increases homozygosity, it does change genotypic frequencies by increasing the homozygotes at the expense of the heterozygotes. This is demonstrated in Table 4.8.

Because genotypic variance increases, phenotypic variance also increases. Genotypic and phenotypic variances increase because the population is being split into separate homozygous lines, and that changes a normal distribution into a bimodal distribution (Fig. 4.24). If the homozygous lines are separated into distinct populations, genotypic and phenotypic variances will be reduced. If that happens gene frequencies and genetic variance will change too.

So, what does inbreeding do that causes it to have bad connotations? Almost every organism carries deleterious recessive alleles that

Table 4.8 **Effects of Inbreeding on Genotypic Frequencies and Allelic Frequencies at a Given Locus. The Following Matings Occur Each Generation:** $AA \times AA; Aa \times Aa; aa \times aa$

Generation	Genotypic frequency[a]			Allelic frequency	
	$f(AA)$	$f(Aa)$	$f(aa)$	$f(A)$	$f(a)$
P_1	0.25	0.5	0.25	0.5	0.5
F_1	0.375	0.25	0.375	0.5	0.5
F_2	0.4375	0.125	0.4375	0.5	0.5
F_3	0.46875	0.0625	0.46875	0.5	0.5
F_4	0.48437	0.03125	0.48437	0.5	0.5
F_5	0.49218	0.015625	0.49218	0.5	0.5
F_6	0.49609	0.007812	0.49609	0.5	0.5
F_7	0.49804	0.003906	0.49804	0.5	0.5
F_8	0.49902	0.001953	0.49902	0.5	0.5
F_9	0.49951	0.000976	0.49951	0.5	0.5
F_∞	0.5	0.0	0.5	0.5	0.5

[a] Due to rounding error, genotypic frequencies do not always sum to 1.0 for a given generation.

are hidden in the heterozygous state. If these alleles are expressed, they will produce abnormal or lethal phenotypes. Related individuals are likely to have the same detrimental recessive alleles. Because inbreeding creates homozygosity by pairing alleles that are alike by descent, rare deleterious recessive alleles have a greater likelihood of being paired and expressed when relatives mate than when unrelated individuals mate. The probability of pairing these deleterious recessive alleles increases as the relationship between the parents increases. The pairing of these alleles and their subsequent expression in inbred animals is what gives inbreeding its bad reputation and produces the social and legal taboos against inbreeding.

There is no certainty that inbred offspring will be abnormal. However, the probability of producing abnormal or subviable progeny increases when the parents are related, and the closer the relationship, the greater the probability.

The pairing of detrimental recessive alleles produces a general trend toward lowered viability, survival, growth, egg production and, simultaneously, increases the percentage of abnormalities. Generally, the greater the inbreeding, the more pronounced the depression in productivity.

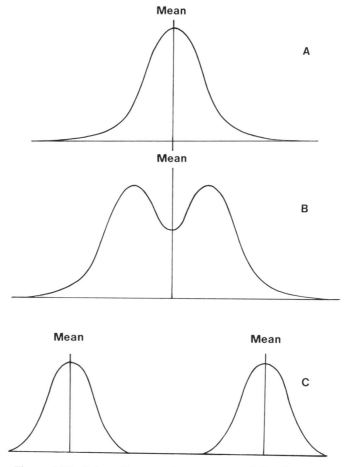

Figure 4.24 Inbreeding can change a normal distribution (A) into a bimodal curve (B) as inbreeding increases the percentage of homozygous individuals. This increases phenotypic variance. If inbreeding continues and splits the population into two separate lines (C), the variance in each line is suddenly reduced.

There have been relatively few inbreeding studies of fish. Some work has been done in rainbow trout (Calaprice 1969; Aulstad and Kittelsen 1971; Bridges 1973; Kincaid 1976A, 1976B, 1983B; R. H. Davis 1976; Gjerde et al. 1983; Gjerde 1988), Atlantic salmon (Ryman 1970), brook trout (Cooper 1961; R. H. Davis 1976), brown trout (R. H. Davis 1976), common carp (Moav and Wohlfarth 1968), *T. mossambica* (Ch'ang

1971A), channel catfish (Bondari 1984B; Bondari and Dunham 1987), zebra danio (Piron 1978; Mrakovčić and Haley 1979), eastern mosquito-fish (Leberg 1990), and convict cichlids (Winemiller and Taylor 1982). With few exceptions, the studies showed similar trends: Inbreeding de-pressed production phenotypes such as growth, viability, and survival, and increased the number of abnormalities. Kincaid has done the most detailed studies, and some of his results are shown in Table 4.9. Ryman (1970) found that inbred Atlantic salmon had a significantly lower return rate when stocked in natural waters. His study suggests that posthatch-ery mortality, as a by-product of inbreeding, may be one reason why it is difficult to reclaim a resource by stocking hatchery-produced fish.

These studies have generally been based on levels of inbreeding in the range of 25–60%, although Kincaid (1976A) looked at $F = 12.5\%$. Future studies should examine lower levels of inbreeding, because low levels can actually enhance productivity. Some of Kincaid's (1976A) results (Table 4.9) suggest that this may be true in rainbow trout. Kin-

Table 4.9 Inbreeding Depression in Rainbow Trout[a]

Phenotype	F (%)					
	12.5	18.5	25.0	37.5	50.0	59.4
Hatch[b]	−4.5	1.7	10.1	−5.5	9.0	
Survival[b]	−2.5	−14.4	−7.7	−4.0	−11.0	
77-day wt[b]	0.0	−15.4	−16.7	−23.0	−29.0	
91-day wt[b]	4.4	−16.3	−17.5	−33.0	−30.0	
105-day wt[b]	6.8	−14.2	−15.6	−32.0	−30.0	
126-day wt[b]	3.8	−7.9	−15.7	−23.0	−20.0	
150-day wt[b]	8.5	−1.8	−11.6	−21.0	−21.7	
1-year wt, ♂[c]			−23.3	−16.4	−12.8	−34.8
1-year wt, ♀[c]			−20.9	−27.2	−21.9	−41.8
2-year wt, ♂[c]			−26.2	−32.2	−33.7	−41.8
2-year wt, ♀[c]			−18.0	−38.1	−28.8	−51.2
2-year length, ♂[c]			−9.2	−13.0	−15.2	−19.0
2-year length, ♀[c]			−9.1	−16.8	−13.9	−21.8
Wt egg mass[c]			−18.1	−33.9	−40.3	−57.0
Percent cripples[d]			−2.9	−10.0		

[a] Inbreeding depression is expressed as percent depression when compared with a control population ($F = 0.0$). A positive value means that the inbred group was better than the control.
[b] Kincaid (1976A).
[c] Kincaid (1983B).
[d] Kincaid (1976B).

caid (1977) estimated that the critical level of inbreeding in rainbow trout is about 18%; below 18%, inbreeding produced few problems, but above 18%, productivity was depressed significantly.

McKay *et al.* (1992B) used computer simulation to determine the amount of inbreeding that would occur as a result of breeding programs that are used by many salmonid farmers. They found that the programs that are typically used by farmers would produce 3–5% *F*/generation. Such levels of inbreeding will probably cause inbreeding depression within 1 to 3 generations and could counteract the benefits of selective breeding programs.

Uses of Inbreeding

Inbreeding studies in fish have generally shown that inbreeding reduces productivity, so how could it ever be used to improve a population? One of the major uses of inbreeding is in a breeding program called line-breeding. Linebreeding occurs when an outstanding individual (usually a male) is brought back into the line to mate with a descendant. This is done because the animal is so outstanding that you want to increase his contribution to each descendant and to increase his contribution to the gene pool. Two types of linebreeding are shown in Fig. 4.25.

A second major use of inbreeding is for the creation of inbred lines that will be hybridized to produce F_1 hybrids for grow-out. Here, two or more selected lines are inbred to fix certain alleles. When the inbred lines are mated, the hybrids will be identical at the desired loci and will be uniform, which is often one of the goals in a crossbreeding program. Inbreeding in two or more lines followed by hybridization is the classic way of producing uniform progeny for grow-out. Gjerde (1988) developed inbred lines of rainbow trout for crossbreeding programs. Although his inbred lines produced crossbreds which exhibited hybrid vigor, he concluded that this was not profitable, because the time and effort needed to produce the inbred lines was worth more than the hybrid vigor gained by creating the crossbreds.

Inbreeding is also used by many researchers to produce animals that will be used in various experiments. In some cases, the scientists do not realize that they are creating inbreeding, do not consider the effects that inbreeding will have on their stock, and do not realize that inbreeding may confound the variable that they are examining in their experiment. The natural consequence of inbreeding sometimes surprises researchers, who then draw incorrect conclusions. One example is a study by Piron (1978). He produced high levels of inbreeding in zebra danio that were to be used in toxicity tests. The inbreeding produced some fish

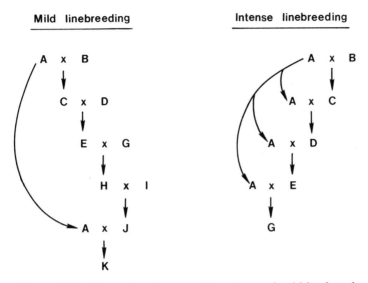

Figure 4.25 Schematic diagram of two types of mild linebreeding. The goal in both is to increase individual A's contribution to his descendants' genome. In the example of mild linebreeding, individual A contributed 53.12% of individual K's alleles. In the example of intense linebreeding, individual A contributed 93.75% of individual G's alleles (a parent normally contributes 50% of an individual's alleles).

with skeletal deformities, and this led Prion (1978) to conclude that this species was not suitable for use in toxicity tests. The production of abnormalities as a result of inbreeding was not a valid reason to reach this conclusion.

Until all detrimental recessive alleles are culled, almost every population will produce some abnormalities as a result of inbreeding. As a matter of fact, many species that are used for biomedical research are highly inbred in order to produce homozygous populations so that all animals will react in a similar manner to experimental variables. The production of highly inbred populations for such research minimizes individual variation, which can be a significant portion of total variance in an experiment.

Even when used improperly and when inbreeding depression does occur, inbreeding can still produce good offspring. The depressions seen for various phenotypes are populational means. Outstanding individuals can be and are produced even though the population average may decline. Outstanding inbred animals can be valuable as brood stock because they breed true for many phenotypes and will not perpetuate undesirable phenotypes.

The genetics of inbreeding is similar to that of crossbreeding. Both depend on the interactions of alleles. Inbreeding depresses production by pairing detrimental recessive alleles. Thus, inbreeding is basically a function of V_D.

Calculating Inbreeding

Individual inbreeding values can be calculated by using a technique called path analysis. In path analysis, you convert a pedigree to a path diagram and determine the inbreeding of an individual by adding the different possible paths to one or more common ancestor(s). For example:

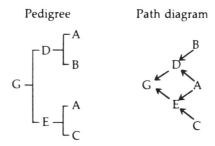

Each arrow in the path diagram represents a gamete and 50% of an individual's genome. Note that F was not used to designate an individual; F is never used because it is the symbol for inbreeding.

Individual inbreeding values are determined by using the following formula:

$$F_X = \Sigma[(0.5)^N(1 + F_A)] \tag{4.10}$$

where F_X is the inbreeding of an individual, Σ is the symbol for "sum of" or "add," N is the number of individuals in a given path, and F_A is the inbreeding of the common ancestor. If $F_A = 0$, Eq. (4.10) becomes

$$F_X = \Sigma[(0.5)^N] \tag{4.11}$$

Individual G in the preceding pedigree is inbred because one of his ancestors appears on both the maternal and paternal side of the pedigree (the definition of a common ancestor). Individual A is the common ancestor of G. The inbreeding of G is determined by tracing a path from G to A. When you trace the path, what you will do is determine how A's genes ended up in G. To do this, you start with one of G's parents, trace a path to A, and then trace the path from A to G's other parent:

To calculate F_G, trace a path from D to E, through G's common ancestor:

Common ancestor of G: A

Path from G to A: D–A–E

There are three individuals in this path, so $N = 3$. Individual A is not inbred, so Eq. (4.11) can be used to calculate F_G:

$$F_G = (0.5)^3$$

$$F_G = 0.125$$

If more than one common ancestor exists you simply add the products of each path. For example:

Pedigree   Path diagram

Individuals G and I are inbred because both have common ancestors.

To calculate F_G, trace the paths from C to D, through G's common ancestors:

Common ancestors of G: A and B

Path from G to A: C–A–D

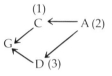

Path from G to B: C–B–D

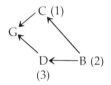

There are two paths, and there are three individuals in each path; $N = 3$ in both paths. Individuals A and B are not inbred, so Eq. (4.11) can be used to calculate F_G. To calculate F_G, add the products of the two separate paths:

$$F_G = (0.5)^3 + (0.5)^3$$

$$F_G = 0.25$$

To calculate F_I, trace the paths from G to H, through I's common ancestors:

Common ancestors of I: C, A, and B

Path from I to C: G–C–H

Path from I to A: G–D–A–C–H

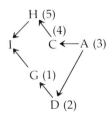

Path from I to B: G–D–B–C–H

H (5)

I   C (4)

G (1)

D←—B (3)
(2)

There are three paths. There are three individuals in the path from I to C and five individuals in the paths from I to A and I to B. Consequently, N is 3, 5, and 5, respectively, for the three paths. As before, none of the common ancestors are inbred, so Eq. (4.11) can be used to calculate F_I. To calculate F_I, the products of the three separate paths are added:

$$F_I = (0.5)^3 + (0.5)^5 + (0.5)^5$$

$$F_I = 0.1875$$

There may even be more than one path between an individual and a common ancestor. If there is a second path, you simply calculate it and add it to the total.

There is one important rule about determining a path between an individual and a common ancestor: You cannot retrace a path, i.e., you cannot go through an individual twice in a given path. Thus, you cannot make the path G–D–A–C–G for the path from I to A in the preceding example, because you would go through individual G twice.

The second path diagram illustrates an important concept. Inbreeding can be reduced to zero if two unrelated individuals mate. Individual K is not inbred, because K has no common ancestor. Two of K's ancestors are inbred, but since K's parents are not related, $F_K = 0$. Consequently, if you can identify individual fish and produce pedigrees, you can prevent inbreeding simply by mating unrelated individuals. The classic way to eliminate any inbreeding in animals that will be used for grow-out is to produce hybrids. If strains or lines are kept pure, F_1 hybrids will always have $F = 0$.

What do the F values mean? F is a measure of the increase in homozygosity as a result of inbreeding. Thus, a fish with $F = 25\%$ is 25% more homozygous than the average fish in the population. F says nothing about the exact level of homozygosity or the population average. F is a value relative to the population mean.

Calculating individual F values has one huge liability in fish culture: To do so, you must know the individual's pedigree. This information is sadly lacking, because most hatcheries are not equipped or are unable to give fish individual identification marks. Because of this, it is not possible to calculate individual levels of inbreeding for most fish.

Does that mean that inbreeding cannot be measured and should be ignored? The answer is a definite no. You may not be able to determine individual pedigrees and calculate inbreeding values for individual fish, but because inbreeding occurs in any population, it is crucial to calculate the average inbreeding for the individuals in a population if you are to manage it properly.

Effect of Population Size on Inbreeding and Genetic Drift

Unintentional inbreeding and genetic drift occur in hatchery populations because they are small and closed. This combination can quickly destroy a population's genetic variance and increase inbreeding, which will lower productivity and increase production costs. Unintentional inbreeding and genetic drift caused by the small populations of broodfish that are usually maintained at hatcheries may be among the leading factors that lower productivity and increase the cost of achieving production quotas.

This problem is probably more critical for game fish populations. Fish that are raised for food or bait lead relatively pampered lives, in that hatchery managers do all they can to keep each fish alive. Fish that are stocked in natural bodies of water go from a relatively mild hatchery environment to a very harsh one where survival is tenuous. Consequently, the loss of genetic variance and inbreeding may adversely affect populations that are stocked in the wild far more than populations that never leave a hatchery.

Genetically, ideal populations are infinitely large. Unfortunately (or fortunately depending on your point of view), hatchery managers cannot work with infinitely large populations. Hatchery managers must work with small finite populations. When a population is finite, the best way to describe it is not by total population but by the effective breeding number. Effective breeding number depends on several factors; the most critical are total number of breeding individuals, sex ratio, mating system, and variance of family size.

When no selection is occurring, there are two mating systems that hatchery managers can use: random mating or pedigreed mating. Random mating is almost exclusively used in aquaculture. Effective breeding number in a population where random mating is used is calculated by using the following formula:

$$N_e = \frac{4(\female)(\male)}{(\female) + (\male)} \tag{4.12}$$

where N_e is the effective breeding number, $\female$ is the number of females that produce viable offspring, and $\male$ is the number of males that produce viable offspring. An examination of the preceding formula shows that N_e can be increased in one of two ways: increase the number of breeding individuals or bring the population closer to a 50:50 sex ratio (Fig. 4.26).

Effective breeding number is one of the most important concepts in the management of a population, in that it gives an indication about the genetic stability of the population because N_e is inversely related to both inbreeding and genetic drift. As N_e decreases, inbreeding and the variance of changes in gene frequencies as a result of genetic drift increase. The inbreeding produced by a single generation of mating in a closed population is

$$F = \frac{1}{2N_e} \tag{4.13}$$

The inverse relationship between F and N_e clearly indicates that as N_e decreases, F increases (Fig. 4.27). The F calculated in Eq. (4.13) is the average inbreeding value for every fish in the population.

Why does inbreeding occur in a closed population, and why is it inversely related to N_e? The reason is simple chance encounters. In a closed population, there is a probability that when fish are mated at random, related individuals will be mated; the smaller the N_e the greater the likelihood that this will occur.

For example, if you live in a closed city of 3,000,000 and if you have 20 relatives living in that city, the odds are 10/1,500,000 that you will marry a relative if your mate is chosen at random from citizens of that city (assuming that both populations have a 50:50 sex ratio); consequently, the probability of marrying a relative in that situation is 0.0000067. On the other hand, if you live in a closed town of 500 and if you have 20 relatives living in that town, the odds of marrying a relative is 10/250 if your mate is chosen at random from citizens of that town; consequently, the odds of marrying a relative is 0.04. In this example, the odds of marrying a relative and of producing inbred offspring is 5,970 times greater in the town, simply because it is smaller.

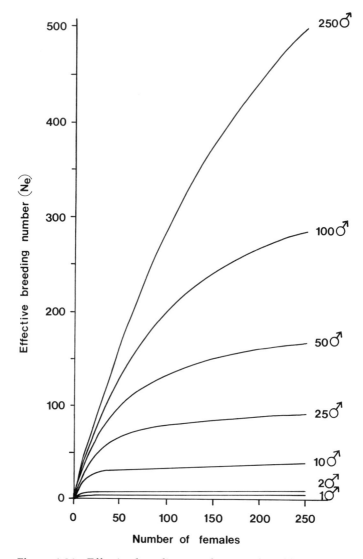

Figure 4.26 Effective breeding numbers produced by the mating of various combinations of males and females. The effective breeding numbers were calculated using two assumptions: (1) random mating was in effect; (2) all parents contributed equally to the next generation.

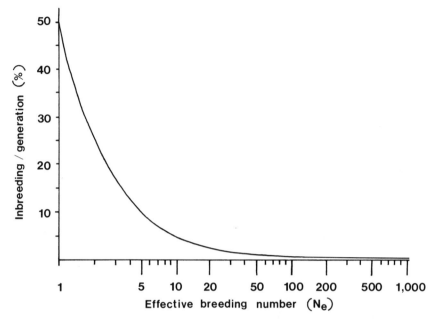

Figure 4.27 Relationship between effective breeding number and inbreeding. *Source: Tave (1990B)*

Genetic drift is random changes in gene frequency created by sampling error. The sampling error can be natural, such as that which occurs when portions of populations are isolated by earthquakes, floods, etc., or it can be manmade by inaccurate collections.

When a population is sampled—e.g., broodstock selection or shipment to another station—there is a chance that the sample is an inaccurate reflection of the makeup of the population. The smaller the sample, the greater the likelihood that inaccuracies in sampling will occur. These inaccuracies extend to all characteristics of the population, including the alleles. Changes in gene frequency due to sampling error are called genetic drift. Genetic drift is expressed as the variance of the change in gene frequency, and it is inversely related to N_e:

$$\sigma^2_{\Delta q} = \frac{pq}{2N_e} \tag{4.14}$$

where $\sigma^2_{\Delta q}$ is the variance of the change in gene frequency, and p and q are the frequencies of alleles p and q at a given locus.

As with inbreeding, the inverse relationship between genetic drift and N_e means that as N_e decreases, genetic drift increases. The ultimate effect of genetic drift is the loss of some alleles and the fixation of others. Rare alleles (alleles whose frequencies are low; generally <0.01) are easily lost, but more common alleles can also be lost via genetic drift.

The overall effect of a small N_e is homozygosity produced by inbreeding and through the loss of alleles as a result of genetic drift. Thus, reductions in N_e can irreversibly damage a gene pool by eliminating alleles and creating homozygosity. Once homozygosity is increased via inbreeding, alleles may be lost more rapidly because small samples may collect a disproportionate number of fish that are homozygous at a given locus. As a result, there is a lowering of overall fitness, viability, and productivity, since the population becomes unable to adjust to changes in the environment, because some of its genetic potential has been lost.

Genetically uniform hatchery populations may be one reason why it is difficult to create self-reproducing populations in lakes and oceans by stocking hatchery-produced fish. Genetically uniform lines may be adequate in controlled environments like hatcheries, but they are detrimental in the wild where the environment is anything but uniform. This is the reason that a major goal in the ocean ranching of salmon is to maintain as much genetic diversity as possible.

Once decreases in N_e have produced inbreeding, the inbreeding in turn lowers future N_e in a positive feedback cycle. Once inbreeding has occurred, N_e becomes

$$N_{eF} = \frac{N_e}{1 + F} \tag{4.15}$$

where N_{eF} is N_e in an inbred population. Thus, restrictions in N_e and F tend to feed each other, and the situation can steadily deteriorate.

Once a population has had its N_e reduced, subsequent increases do not correct the damage that has been done. Restrictions in N_e often occur during the transfer of stocks from one hatchery to another. It is expensive and difficult to move large populations, so many hatcheries are started with a handful of fish, especially if the fish are very prolific. When this occurs, the population is said to have gone through a bottleneck. The dramatic lowering of N_e in a bottleneck has long-term consequences on the genetics of the stock and on productivity. The N_e over a series of generations is the harmonic mean of the N_e in each generation. Over t generations, overall N_e can be determined from the following formula:

$$\frac{1}{N_{e\ overall}} = \frac{1}{t} \left(\frac{1}{N_{e1}} + \frac{1}{N_{e2}} + \cdots + \frac{1}{N_{et}} \right) \tag{4.16}$$

where $N_{e\ overall}$ is the overall effective breeding number, and N_{e1}, N_{e2}, and N_{et} are the N_es in generations 1, 2, and t, respectively. The preceding formula shows that the generation with the smallest N_e has the greatest impact on overall N_e.

A bottleneck has long-lasting effects on the genetics of a population. Nei *et al.* (1975) showed that bottlenecks reduce genetic variance and average heterozygosity, and depending on the N_e of the bottleneck and the increase in population size thereafter, a population may not recover its genetic diversity for hundreds of generations, other than through the introduction of new stock.

This has important practical implications for the management of hatchery stocks. If a hatchery population is started with only a few individuals, with progeny from only a few spawns, or if the population is decimated by disease, subsequent increases in N_e only keep inbreeding and genetic drift from getting worse; they do not rectify the problems that have occurred. Thus, it is crucial to know the N_e for each generation, both at your hatchery and before the stock arrives.

The N_e of a population can also go through a bottleneck during selection. If you establish an extreme cutoff value, you may be able to save too few fish, and N_e will be reduced precipitously in the selected population. Another way that selection can create a bottleneck is if the selected fish come from only a few families. Fish from a few families may be superior to all others due either to V_A, V_D, V_I, V_{G-E}, or V_E, so when you select the fish, you may retroactively reduce N_e in the previous generation if only a few parents produced the selected population. While both types of bottlenecks may be equally damaging, the latter is an invisible one, because you will probably not realize that you drastically reduced N_e.

A final factor that can reduce N_e is family size. If all brooders produce the same number of offspring, N_e is calculated as described earlier. However, gametic production and offspring viability are highly variable. This has a negative impact on N_e because unequal contributions to the next generation will lower N_e from what it would have been had all brooders made an equal contribution. When there is unequal production of offspring, N_e in a population with random mating is

$$N_{eUR} = \frac{8(N_e)}{V_{k♀} + V_{k♂} + 4} \tag{4.17}$$

where N_{eUR} is N_e with unequal production of offspring and $V_{k♀}$ and $V_{k♂}$ are the variances of female and male offspring production (family size), respectively. Family size often assumes what is called a Poisson distribution. In a Poisson distribution, the mean and the variance are the same, so when this occurs, mean family size can be substituted for variance (Latter 1959).

The inadvertent loss of genetic variance via reductions in N_e can be devastating. Genetic variance is the raw material with which both nature and geneticists work, and its loss is damaging because it is usually irreversible. The loss of genetic variance can reduce productivity, increase developmental anomalies, and make future improvements via selection difficult.

The loss of genetic variance as a result of small N_es is becoming a major concern in conservation ecology. The management and survival of an endangered population or species may hinge on our ability to manage its N_e. Meffe (1987) called N_e the conservation geneticists' "nemesis," because if it cannot be properly managed, efforts to save the population may be for naught. In fact, improper management of an endangered species' N_e may hasten its demise.

A number of studies have suggested that reductions in N_e had adverse effects on the gene pools of hatchery stocks (Allendorf and Utter 1979; Allendorf and Phelps 1980; Ryman and Ståhl 1980; Tave and Smitherman 1980; Cross and King 1983; Ståhl 1983; Taniguchi et al. 1983; Sugama et al. 1988; Verspoor 1988; Macaranas and Fujio 1990; Cross and Challanain 1991).

For example, Teichert-Coddington and Smitherman (1988) were unable to improve growth rate in the Auburn University–Ivory Coast strain of *T. nilotica* by selection. This may have been due to reductions in N_e during several transfers of this strain which might have eliminated most of the V_A for increased growth rate (Tave and Smitherman 1980). An electrophoretic examination of the strain found no detectable heterozygosity (no detectable genetic variance), which suggests that this was the case (Brummett 1982).

Edds and Echelle (1989) found that rare alleles had been lost in hatchery populations of Leon Springs pupfish, Comanche Springs pupfish, and Pecos gambusia, all of which are endangered species, after 6–8 years in captivity. They attributed these losses to small (30–80 fish) founder populations.

Hershberger (1992) reviewed genetic variability in rainbow trout populations and found that most of the genetic variation was within, not between, populations. He felt that bottlenecks, and subsequent genetic drift, were responsible for much of this variability.

Leary et al. (1985A) found high frequencies of two morphological deformities and an unusually large number of individuals who were asymmetrical at bilateral meristic phenotypes in a hatchery population of cutthroat trout. They attributed these developmental anomalies to a reduction in heterozygosity, as a result of genetic drift. Several studies have shown that heterozygosity is positively correlated with egg size, growth rate, developmental rate, and population size, and it is negatively correlated with developmental problems (Leary *et al.* 1984, 1985A, 1985C; Ferguson et al. 1985; Danzmann *et al.* 1988, 1989; Blanco *et al.* 1990; Kartavtsev *et al.* 1990; Leberg 1990). It isn't known if these relationships exist in most species, but some studies have not found any relationship between heterozygosity and growth or development (Beacham and Withler 1985A, 1985B; McAndrew *et al.* 1986).

Hatching rate has declined in the Donaldson strain of rainbow trout (Hershberger 1985). This may be due to reductions in N_e; N_e was reduced to about 20/generation over the past 40 years. This has produced inbreeding of between 40 and 60% (Hershberger 1983), and electrophoretic data from Allendorf and Utter (1979) suggest that genetic drift has robbed the population of many valuable alleles.

Perhaps the best (or worst, depending on your point of view) example is the extinction of the Auburn University-Rio Grande Select strain of channel catfish (AU-RG-S) (Tave 1988). One generation of selection improved weight in this strain by 17% (Dunham and Smitherman 1983). Although this population was founded with 45 males and 45 females, only three males and three females were spawned to produce the F_1 generation ($N_e = 6$; it may have been lower because of previous unintentional inbreeding). Growth rate was improved, but genetic variance was reduced dramatically. The F_1 generation was examined electrophoretically, and there was no detectable heterozygosity (Hallerman *et al.* 1986). F_1 broodstock were selected, marked, and overwintered in ponds with broodstock from other strains of channel catfish. The winter was particularly harsh, and every AU-RG-S fish died. The other strains in the pond, including the AU-RG control line, which had greater amounts of heterozygosity, survived. The likely reason that AU-RG-S went the way of the dodo was that its genetic variance, which would have enabled the strain to be pre-adapted to the harsh conditions, was lost as a result of the bottleneck during the production of the F_1 = generation select fish.

As bad as this is for hatchery stocks that are used in fish culture, it can be ruinous when it occurs in populations that are used to stock natural bodies of water. The U.S. Fish and Wildlife Service (1982) warned that the loss of genetic variance in hatchery stocks of lake trout that are used to stock Lakes Michigan and Ontario may hinder the restoration of that species. Rasmuson (1981), Ryman (1981), and Johansson (1981) all expressed concern that broodstock management practices were reducing genetic variance in Scandanavian hatchery stocks of salmon that are used to supplement natural production. Ryman (1981) warned that many management programs may actually destroy the populations' gene pools instead of saving them.

Ryman (1991) warned that stocking practices could adversely affect natural populations by lowering their N_e. After they have been stocked and have become sexually mature, if fish from only a few hatchery spawns contribute a disproportionately large percentage of sexually mature fish to the spawning population, they will lower the natural population's N_e. If this occurs, the stocking program could destroy the resource it was designed to save.

Edds and Echelle (1989) found that reductions in N_e resulted in the loss of rare alleles in three species of endangered fish that were being maintained at a hatchery. Great care must be taken to prevent such losses, because the loss of genetic variance in endangered populations may mean that they can survive only in controlled hatchery environments.

Preventing Reductions in Effective Breeding Number   How do you prevent restrictions in N_e from ruining the genetic potential of a hatchery population? First and foremost, keep N_e as large as possible every generation. When acquiring broodstock, you should determine, in advance, the pedigree of the fish that you are purchasing. You need to know how many parents were used to produce the fish that you are buying. This is an important concern in aquaculture, because of the fecundity of fish. I once saw someone fill a request for a channel catfish foundation population by shipping 2000 fry that came from a single spawn. The N_e that produced the 2000 channel catfish was only 2 (less if the parents were related and inbred). The inbreeding that this foundation population would produce in the first generation was 25%, but the loss of many alleles via genetic drift was even more damaging.

How large should N_e be to prevent inbreeding and genetic drift? Kincaid recommended that N_e be at least 200 (1976A) and 500 (1979); Ryman and Ståhl (1980) recommended that N_e be at least 60; Food and Agriculture Organization of the U.N. (FAO) recommended that N_e be at least 50 for short term work and 500 for long term work (FAO/UNEP 1981); U.S. Fish and Wildlife Service (1984) recommended that N_e be 1000; Allendorf and Ryman (1988) recommended that N_e be 200 (100 males and 100 females); and Tave (1990B; [recommendation made in 1984 but not published until 1990]) recommended N_es between 263 and 344 for food fish and bait fish populations and N_es between 424 and 685 for populations to be stocked in natural bodies of water. Tave (1988) reduced his recommended N_es for food fish populations to between 45 and 250. Allendorf and Ryman (1988) recommended N_es of between 200 and 400 for populations to be stocked in rivers and lakes.

Which recommendation is correct? Unfortunately, there is no single number that every hatchery manager can use to prevent genetic drift- or inbreeding-related problems from occurring in his stocks. Tave (1990A, 1990B) outlined procedures to determine minimum N_es that can be used to prevent inbreeding- and genetic drift-related problems.

To calculate the N_e that is needed to prevent inbreeding from reaching levels that depress productivity, you need two pieces of information. The first is the level of inbreeding at which inbreeding depression occurs. Unfortunately, this information does not exist for fish; a single universal value probably will never exist, because it will be different for different phenotypes and different populations. Kincaid's (1976A, 1976B, 1983B) studies with rainbow trout are the most complete, but the smallest level of inbreeding that he produced was 12.5%. Kincaid (1977) estimated that $F = 18\%$ is the level at which inbreeding depression becomes significant in rainbow trout. However, some of his data (Kincaid 1976A; Table 4.9) showed inbreeding depression at $F = 12.5\%$. Data from other animals showed that inbreeding depression can occur at any level of inbreeding (Falconer 1981). Because no critical values of F exist for fish, hypothetical values must be used. Tave (1990A) suggested using 5% as a conservative value and 10% as a liberal estimate.

The second bit of information that is needed is the number of generations that you wish to incorporate into the breeding program before F reaches the critical value. Once that is determined, you simply calculate the minimum constant N_e that will produce the critical level of inbreeding in the prescribed number of generations. For example, if you choose $F = 5\%$ as your critical level of inbreeding and you do not want to reach that level until generation 15, N_e is calculated as follows:

Step 1. Calculate the F/generation needed to produce $F = 0.05$ at generation 15:

$$F/\text{generation} = \frac{0.05}{15 \text{ generations}}$$

$$F/\text{generation} = 0.0033333/\text{generation}$$

Step 2. Calculate N_e needed to produce $F = 0.0033333/\text{generation}$ using Eq. (4.13):

$$F = \frac{1}{2N_e}$$

$$0.0033333 = \frac{1}{2N_e}$$

$$(0.0033333)(2) = \frac{1}{N_e}$$

$$N_e = \frac{1}{(0.0033333)(2)}$$

$$N_e = 150$$

Minimum constant N_es that will produce $F = 5\%$ and $F = 10\%$ after a prescribed number of generations are listed in Table 4.10.

How large should N_e be to prevent genetic drift? It is difficult to prevent genetic drift, because any change in gene frequency as a result of sampling error is genetic drift. If the frequency of an allele changes from 0.500 to 0.499 as a result of sampling error, genetic drift has occurred. But that is not as important as the loss of alleles, so the crucial question becomes: How large does N_e have to be to prevent the loss of rare alleles? The reason that you are interested in the loss of rare alleles

Table 4.10   Effective Breeding Numbers (N_e) Needed to Produce $F = 5\%$ and $F = 10\%$ after Certain Numbers of Generations

	N_e	
No. generations	$F = 5\%$	$F = 10\%$
1	10	5
2	20	10
3	30	15
4	40	20
5	50	25
6	60	30
7	70	35
8	80	40
9	90	45
10	100	50
20	200	100
30	300	150
40	400	200
50	500	250
60	600	300
70	700	350
80	800	400
90	900	450
100	1000	500

Source: Tave (1990A).

is that it is more likely that rare alleles (alleles with low frequencies) will be lost than those with high frequencies. The answer to this question depends on two decisions that you must make: (1) how valuable are the rare alleles, i.e., how rare an allele will you try to save (e.g., $f = 0.1$ or 0.01 or 0.000001); (2) what probability level is desired (e.g., $P = 0.05$ means that you have a guarantee of 95% of saving the allele; $P = 0.01$ means that you have a guarantee of 99% of saving the allele). (Note: I am going to use the term "guarantee" instead of "probability" when I am discussing the likelihood of saving an allele. I realize that this will cause statisticians to develop high blood pressure, but I feel that my use of "guarantee" will make this section easier to understand.)

If you want to save one of the rarest alleles ($f \approx 0.000001$) and you want a 100% guarantee, you will need an N_e so large that it will never fit in a hatchery. Consequently, a compromise must be made between what is ideal and what will cause problems. The N_e that you need can be determined by calculating the number of fish that are needed to ensure that allele q, at a given frequency, is present at a given guarantee. The probability (P) of not keeping an allele in a random sample is

$$P = (1.0 - q)^{2Ne} \qquad (4.18)$$

where P is the probability of losing the allele in a single sample (one generation or one transfer of broodstock from one station to another) and q is the frequency of the allele. The exponent is $2N_e$ because fish are diploids and have 2 alleles per locus. Table 4.11 shows the probabilities of losing an allele for various N_es at various gene frequencies. For example, the information in Table 4.11 shows that you need an N_e of 150 to produce a $P = 0.04904$ for an allele whose $f = 0.01$ (a probability of 4.9% of losing the allele, and a guarantee of 95.1% of keeping it).

The probabilities that are listed in Table 4.11 are the probabilities of losing an allele via genetic drift for only a single generation or single transfer. The probability of losing an allele during the course of several generations or transfers to other hatcheries is the product of the probability for each generation or transfer. For example, if you maintain a constant N_e of 150 for 10 generations, the probability of losing an allele whose $f = 0.01$ after 10 generations is calculated as follows:

Step 1. Calculate the probability of losing the allele in a single generation using Eq. (4.18):

$$P = (1.0 - q)^{2Ne}$$
$$P = (1.0 - 0.01)^{2(150)}$$
$$P = 0.04904$$

Table 4.11 Probabilities of Losing an Allele via Genetic Drift for Eight Allelic Frequencies at Various Effective Breeding Numbers (N_e)[a]. The Guarantee of Keeping the Allele is: 1.0 − The Probability of Losing It.

N_e	Allelic frequency							
	0.5	0.4	0.3	0.2	0.1	0.05	0.01	0.001
2	0.06250	0.12960	0.24010	0.40960	0.65610	0.81451	0.96060	0.99601
3	0.01562	0.04666	0.11765	0.26214	0.53144	0.73509	0.94148	0.99402
4	0.00391	0.01680	0.05765	0.16778	0.43047	0.66342	0.92274	0.99203
5	0.00098	0.00605	0.02825	0.10737	0.34868	0.59874	0.90438	0.99004
6	0.00024	0.00218	0.01384	0.06872	0.28243	0.54036	0.88638	0.98807
7	0.00006	0.00078	0.00678	0.04398	0.22877	0.48768	0.86875	0.98609
8	0.00002	0.00028	0.00332	0.02815	0.18530	0.44013	0.85146	0.98412
9	4×10^{-6}	0.00010	0.00163	0.01801	0.15009	0.39721	0.83451	0.98215
10	1×10^{-6}	0.00004	0.00080	0.01153	0.12158	0.35849	0.81791	0.98019
14		6×10^{-7}	0.00005	0.00193	0.05233	0.23783	0.75472	0.97237
15			0.00002	0.00124	0.04239	0.21464	0.73970	0.97043
20			6×10^{-7}	0.00013	0.01478	0.12851	0.66897	0.96077
25				0.00001	0.00515	0.07694	0.60501	0.95121
30				2×10^{-6}	0.00180	0.04607	0.54716	0.94174
31				1×10^{-6}	0.00146	0.04158	0.53627	0.93985
35					0.00063	0.02758	0.49484	0.93236
40					0.00022	0.01652	0.44752	0.92308
45					0.00008	0.00989	0.40473	0.91389
50					0.00003	0.00592	0.36603	0.90479
55					9×10^{-6}	0.00354	0.33103	0.89578
60					3×10^{-6}	0.00212	0.29938	0.88687
66					9×10^{-7}	0.00115	0.26537	0.87628
70						0.00076	0.24487	0.86930
75						0.00046	0.22145	0.86064
80						0.00027	0.20028	0.85208
85						0.00016	0.18113	0.84359
90						0.00010	0.16381	0.83520
95						0.00006	0.14814	0.82688
100						0.00004	0.13398	0.81865
125						3×10^{-6}	0.08106	0.77870
135						1×10^{-6}	0.06630	0.76328
150							0.04904	0.74071
175							0.02967	0.70456
200							0.01795	0.67019
225							0.01086	0.63748
230							0.00982	0.63114
250							0.00657	0.60638
275							0.00398	0.57679
300							0.00240	0.54865
325							0.00146	0.52188
350							0.00088	0.49641
375							0.00053	0.47219
400							0.00032	0.44915

Table 4.11 (*continued*)

			Allelic frequency					
N_e	0.5	0.4	0.3	0.2	0.1	0.05	0.01	0.001
425							0.00019	0.42723
450							0.00012	0.40639
475							0.00007	0.38656
500							0.00004	0.36770
685							1×10^{-6}	0.25393
1498								0.04991
2302								0.00999
6880								1×10^{-6}

 These probabilities are the probabilities of losing an allele for a single generation.

Step 2. Calculate the guarantee of keeping the allele in a single generation. To calculate the guarantee of keeping the allele, subtract the probability of losing the allele (P) from 1.0:

$$\text{guarantee of keeping the allele} = 1.0 - 0.04904$$

$$\text{guarantee of keeping the allele} = 0.95096$$

Step 3. Calculate the guarantee of keeping the allele after 10 generations. This is the product of the guarantee of keeping the allele for each generation:

$$\text{guarantee of keeping the allele} = (0.95096)^{10}$$

$$\text{guarantee of keeping the allele} = 0.60481$$

Step 4. Calculate the probability of losing the allele (P) after 10 generations. To calculate the probability of losing the allele, subtract the guarantee of keeping the allele from 1.0:

$$P = 1.0 - 0.60481$$

$$P = 0.39519$$

Thus, if you maintain an N_e of 150/generation for 10 generations, you have a guarantee of 60.5% of keeping an allele whose $f = 0.01$ after 10 generations, despite having a guarantee of 95.1% of keeping the allele in each generation.

If you have a fluctuating population, the overall guarantee of keeping an allele is, again, the product of the guarantee for each generation.

For example, what is the guarantee of keeping an allele whose $f = 0.01$ after 10 generations with the following N_es/generation?

$$N_e s: 230, 100, 200, 50, 30, 200, 10, 20, 25, 230$$

Step 1.   Calculate the probability of losing an allele (P) and the guarantee of keeping an allele whose $f = 0.01$ for each generation:

N_e	Probability of losing the allele	Guarantee of keeping the allele
	Use Eq. (4.18).	
	$P = (1.0 - q)^{2N_e}$	$1.0 - P$
230	$P = (1.0 - 0.01)^{2(230)} = 0.00982$	0.99018
100	$P = (1.0 - 0.01)^{2(100)} = 0.13398$	0.86602
200	$P = (1.0 - 0.01)^{2(200)} = 0.01795$	0.98205
50	$P = (1.0 - 0.01)^{2(50)} = 0.36603$	0.63397
30	$P = (1.0 - 0.01)^{2(30)} = 0.54716$	0.45284
200	$P = (1.0 - 0.01)^{2(200)} = 0.01795$	0.98205
10	$P = (1.0 - 0.01)^{2(10)} = 0.81791$	0.18209
20	$P = (1.0 - 0.01)^{2(20)} = 0.66897$	0.33103
25	$P = (1.0 - 0.01)^{2(25)} = 0.60501$	0.39499
230	$P = (1.0 - 0.01)^{2(230)} = 0.00982$	0.99018

Step 2.   Calculate the guarantee of keeping the allele after 10 generations. It is the product of the guarantee of keeping the allele for each generation:

guarantee of keeping the allele $= (0.99018)(0.86602)(0.98205)(0.63397)$

$\times (0.45284)(0.98205)(0.18209)$

$\times (0.33103)(0.39499)(0.99018)$

guarantee of keeping the allele $= 0.00560$

Step 3.   Calculate the probability of losing the allele after 10 generations. To calculate the probability of losing the allele (P), subtract the guarantee of keeping the allele from 1.0:

$$P = 1.0 - 0.00560$$

$$P = 0.99440$$

This example clearly shows the detrimental impact that reductions in N_e can have on genetic drift. Even if N_e is reduced for only one generation, that bottleneck can have a long-lasting impact on the genetics of a population.

For example, if N_e for the foundation (first) generation is 10, and it is then increased and maintained at 230 for the next 9 generations, the probability of losing an allele whose $f = 0.01$ after the 10th generation is

Step 1. Calculate the probability of losing an allele (P) and the guarantee of keeping an allele whose $f = 0.01$ for each generation:

N_e	Probability of losing the allele	Guarantee of keeping the allele
	Use Eq. (4.18).	
	$P = (1.0 - q)^{2N_e}$	$1.0 - P$
10	$P = (1.0 - 0.01)^{2(10)} = 0.81791$	0.18209
230	$P = (1.0 - 0.01)^{2(230)} = 0.00982$	0.99018

Step 2. Calculate the guarantee of keeping the allele after 10 generations. It is the product of the guarantee of keeping the allele for each generation:

$$\text{guarantee of keeping the allele} = (0.18209)[(0.99018)^9]$$

$$\text{guarantee of keeping the allele} = 0.16661$$

Step 3. Calculate the probability of losing the allele (P) after 10 generations. To calculate the probability of losing the allele, subtract the guarantee of keeping the allele from 1.0:

$$P = 1.0 - 0.16661$$

$$P = 0.83339$$

The overall probability of losing an allele gives you the probability of losing the allele via genetic drift after a certain number of generations. If the allele is actually lost, N_e can be any number thereafter, and allelic frequency will still be zero. Once lost, an allele can only be reintroduced by mutation or by the importation of new broodstock. This explains why bottlenecks, particularly small founder populations, can permanently cripple a population's genetic variance.

What are some guideline N_es that hatchery managers should use to prevent genetic drift from eliminating rare alleles? To calculate this, you must determine three things. The first is the frequency of the rare alleles that you want to save. Tave (1990B) recommended that the goal should be to keep alleles whose $f = 0.01$, because population biologists and population geneticists generally consider that alleles whose $f \geq 0.01$ contribute to polymorphism, and the objective is to keep polymorphic loci in the polymorphic state. Meffe (1986) recommended that the goal should be to keep alleles whose $f = 0.05$, because alleles that are rarer than that contribute little to overall genetic variance.

Meffe's (1986) recommendation is acceptable for food fish and bait fish husbandry, because rare alleles probably do not govern productivity on fish farms. In addition, most rare alleles are probably lost during domestication selection unless they are important in the domestication process, in which case the frequency will increase dramatically. But Tave's (1990B) recommendation is probably more appropriate for game fish and ocean ranching programs, as well as for those that maintain standard reference lines, because the major goal of these programs is conservation genetics; to conserve the genetic variance you must save as many alleles as is practicable. In such programs it is better to err on the side of conservatism and be safe.

The second piece of information that is needed is the probability (P) that you wish to use (the probability of losing the allele). In biology, two probabilities are generally used, and these are acceptable; $P = 0.05$ and 0.01; $P = 0.05$ and 0.01 means that there is a 5 and 1% probability of losing the allele and a 95 and 99% guarantee of keeping the allele, respectively.

The third piece of information that is needed is the number of generations that will be incorporated into the breeding program before P reaches the desired level. Once you determine this, you simply back-calculate the constant N_e/generation needed to produce the desired P for the given alleles in the predetermined number of generations.

For example, say you want $P = 0.01$ (a probability of 1% of losing the allele and the guarantee of 99% of keeping the allele) after 10 generations for an allele whose $f = 0.01$. What N_e/generation should be maintained in order to achieve this goal? The answer is calculated as follows:

$$q = 0.01; \qquad P = 0.01 \text{ after 10 generations}$$

Step 1.   Calculate the guarantee/generation of keeping the allele that will produce a guarantee of 0.99 of keeping the allele ($P = 0.01$) after 10 generations:

$$0.99 = (\text{guarantee/generation})^{10}$$

$$\text{guarantee/generation} = (0.99)^{1/10}$$

$$\text{guarantee/generation} = 0.9989955$$

Step 2. Calculate P/generation (P) by subtracting the guarantee of keeping the allele from 1.0:

$$P = 1.0 - 0.9989955$$

$$P = 0.0010045$$

Step 3. Calculate the N_e needed to produce $P = 0.0010045$ by using Eq. (4.18):

$$P = (1.0 - q)^{2N_e}$$

$$0.0010045 = (1.0 - 0.01)^{2N_e}$$

The formula must be converted to logarithms to solve for N_e:

$$\log 0.0010045 = \log (0.99)^{2N_e}$$

$$\log 0.0010045 = (2N_e)(\log 0.99)$$

$$\frac{\log 0.0010045}{\log 0.99} = 2N_e$$

$$\frac{-2.9980377}{-0.0043648} = 2N_e$$

$$2N_e = 686.867$$

$$N_e = \frac{686.867}{2}$$

$$N_e = 343.43$$

$$N_e \text{ is rounded up to } 344$$

In the preceding example, you need a constant N_e of 344/generation to produce $P = 0.01$ (a probability of 1% of losing the allele and a guarantee of 99% of keeping the allele) after 10 generations for an allele whose $f = 0.01$. Constant N_es needed to produce $P = 0.05$ and 0.01 for alleles whose $f = 0.05$ or 0.01 after various numbers of generations are listed in Table 4.12.

As was the case with inbreeding, there is no universal N_e that can be recommended for every fish breeding program. The N_es listed in Table 4.12 should be used only as guidelines. They are not absolute values. Most food fish farmers should maintain constant N_es between 68 and 90. A constant $N_e = 68$ is sufficient for short-term work (≤ 10 generations), because that N_e will produce a 99% guarantee of keeping an allele whose

Table 4.12   Effective Breeding Number per Generation (N_e) Needed to
Produce $P = 0.01$ and 0.05 (Guarantees of Keeping the Alleles of 99% and
95%) after Various Numbers of Generations for Alleles Whose $f = 0.05$ or
0.01^a

	$f = 0.05$		$f = 0.01$	
No. generations	$P = 0.05$	$P = 0.01$	$P = 0.05$	$P = 0.01$
1	30	45	150	230
5	45	61	229	309
10	52	68	263	344
15	56	72	283	364
20	59	75	297	378
25	61	77	308	390
30	63	78	318	399
35	64	80	325	406
40	65	81	332	413
45	67	82	338	419
50	68	83	343	424
75	72	87	363	444
100	74	90	377	458

Source: After Tave (1990B).
[a] Effective breeding numbers were rounded to the next higher whole number.

$f = 0.05$ after 10 generations. A constant $N_e = 90$ should be sufficient for long-term work (≥ 10 generations), because that N_e will produce a guarantee of 99% of keeping an allele whose $f = 0.05$ after 100 generations.

Aquaculturists who are maintaining standard reference lines of fish or who wish to conserve a greater number of alleles should try to maintain N_es between 263 and 343/generation. A constant N_e of 263 should be sufficient for short-term work (≤ 10 generations), because this N_e will produce a guarantee of 95% of keeping an allele whose $f = 0.05$ after 10 generations. An N_e of 343 is more appropriate for long-term work (10–50 generations), because this N_e will produce a guarantee of 95% of keeping an allele whose $f = 0.01$ after 50 generations.

Effective breeding numbers should be at least 424/generation for fish culture programs where the population will be used for fisheries management programs or for ocean ranching. The recommended N_e is larger for these programs for two reasons: First, one of the major goals, if not the major goal, should be to maintain as much genetic variance as possible. These fish will be stocked in the wild, and no one knows which alleles enhance survival or which alleles preadapt fish to an altered environment. The loss of genetic variance via genetic drift may be a prime reason why it is difficult to restore a natural resource with hatchery-produced populations. Second, when working with these programs, a hatchery manager must incorporate long-term planning into broodstock management, and 50 generations is a good minimum because it will encompass anywhere from 50 to 200+ years. A constant N_e of 424/generation will produce guarantees of 99% and 95% of keeping alleles whose $f = 0.01$ after 50 and 257 generations, respectively.

Although it would be ideal to keep alleles that are rarer than 0.01 for these types of programs, the N_e required for this goal is almost unmanageable. For example, the N_e required to produce a guarantee of 95% of keeping an allele whose $f = 0.001$ for a single generation is 1498. Obviously, it is not practical to try and prevent the loss of alleles much rarer than 0.01.

Effective breeding numbers needed to prevent inbreeding from reaching undesired levels and to produce desired guarantees of keeping rare alleles will seldom be the same. To achieve both goals, choose the larger number. Overall recommended minimum constant N_es are listed in Table 4.13. The N_es in Table 4.13 should be considered bottleneck numbers; if N_e goes below the given number for just one generation, it will be impossible to achieve the desired goals.

Table 4.13 Minimum Constant N_es for Food Fish Farming and for Fisheries Management Programs[a]

No. generations	Food fish farming N_e	Fisheries management N_e
1	45	
5	61	
10	100	344
15	150	364
20	200	378
25	250	390
50		500
60		600
70		700
80		800
90		900
100		1000

[a] Food fish N_es were generated using the following assumptions: you want to keep $F \leq 5\%$ and want to produce a 99% guarantee of keeping an allele whose $f = 0.05$. Effective breeding numbers for fisheries management programs were generated using the following assumptions: you want to keep $F \leq 0.05$ and want to produce a 99% guarantee of keeping an allele whose $f = 0.01$.

The N_es that have been recommended to prevent inbreeding from reaching levels that depress productivity and to prevent genetic drift from losing alleles are simply guidelines; they should not be considered as gospel. The N_es that are presented are based on the assumptions that are described in this section. The key aspect of this section is the procedures that are used to generate N_es. If you know how to use the procedures, you can generate your own N_es based on your goals and your assumptions about what is desirable.

What should be done if the N_e of a population drops (or hatchery records indicate that it was reduced previously) and begins to lower productivity? The only way to correct such a problem is to acquire new broodstock. If this is done, make sure that you acquire broodstock which has not gone through a bottleneck. However, as simple as this prescription is, it is often difficult medicine to swallow. Many hatchery managers loathe the idea of bringing in new broodstock because they fear the importation of new diseases more than they fear genetic problems. Diseases are visible and understandable, but genetic drift is invisible. Moreover, hatcheries that propagate threatened or endangered species often cannot import new broodstock because they are not available.

When the importation of new broodstock is not feasible, several options can be used to prevent inbreeding and genetic drift from getting worse. The first option is to increase N_e as much as possible; the larger the N_e, the better. The usual constraints here are the physical limitations of space at the hatcheries and of the budget. Most hatcheries tend to spawn as many fish as practicable, so it is often difficult to increase N_e this way.

A second approach is to spawn a more equal sex ratio, providing it is not already 50:50. The effect of a skewed sex ratio on inbreeding can be demonstrated by this formula:

$$F = \frac{1}{8(\female)} + \frac{1}{8(\male)} \qquad (4.19)$$

where $\female$ is the number of females that produce offspring and $\male$ is the number of males that produce offspring.

When breeding populations are small, skewed sex ratios can lower N_e (Fig. 4.26) and increase inbreeding dramatically. The following example demonstrates this fact:

Population 1: 25 females and 25 males

$$F = \frac{1}{8(25)} + \frac{1}{8(25)}$$

$$F = 1\%/\text{generation}$$

Population 2:   250 females and 10 males

$$F = \frac{1}{8(250)} + \frac{1}{8(10)}$$

$$F = 1.3\%/\text{generation}$$

Population 2 has over 5 times as many breeding individuals, but the N_e in population 1 is 50 while the N_e in population 2 is only 38.5. As a result, the inbreeding produced by population 2 is 30% greater because of the skewed sex ratio. Table 4.14 gives F produced in a single generation by various combinations of males and females.

There is often a great temptation to use skewed sex ratios because they optimize fingerling production in terms of the fewest broodfish needed to achieve fingerling production quotas. Bondari (1983B) demonstrated that catfish farmers who use the open pond spawning technique can skew the sex ratio of channel catfish broodfish up to 4 females : 1 male and not affect fry production. This practice may be beneficial for the economics of fingerling production, but any genetic problems that exist will only get worse.

For example, if a catfish farmer needs 50 egg masses, the inbreeding produced by the two sex ratios is

1 : 1 sex ratio

50♀ : 50♂

$$F = \frac{1}{8(50)} + \frac{1}{8(50)}$$

$$F = 0.5\%/\text{generation}$$

4 : 1 sex ratio

50♀ : 12.5 or 13♂

$$F = \frac{1}{8(50)} + \frac{1}{8(13)}$$

$$F = 1.21\%/\text{generation}$$

A third approach to maximizing N_e is to switch from random mating to pedigreed mating. Pedigreed mating differs from random mating in that each female leaves one daughter and each male leaves one son to be used as broodstock in the following generation. A male that spawns with ten females leaves as many sons as a male that spawns with one female—one (in reality, each parent can leave 10 or 100 offspring—the major criterion is that each parent leaves the same number of offspring and an equal number reproduce). The sons and daughters are chosen in

Table 4.14 Percent Inbreeding per Generation Produced in a Population with Random Mating[a]

No. females	No. males																					
	1	2	3	4	5	6	7	8	9	10	20	25	50	75	100	125	150	175	200	225	250	∞
1	25.00	18.75	16.67	15.62	15.00	14.58	14.29	14.06	13.89	13.75	13.12	13.00	12.75	12.67	12.62	12.60	12.58	12.57	12.56	12.56	12.55	12.50
2	18.75	12.50	10.42	9.38	8.75	8.33	8.04	7.81	7.64	7.50	6.88	6.75	6.50	6.42	6.38	6.35	6.33	6.32	6.31	6.31	6.30	6.25
3	16.67	10.42	8.33	7.29	6.67	6.25	5.95	5.73	5.56	5.42	4.79	4.67	4.42	4.33	4.29	4.27	4.25	4.24	4.23	4.22	4.22	4.17
4	15.62	9.38	7.29	6.25	5.62	5.21	4.91	4.69	4.51	4.38	3.75	3.62	3.38	3.29	3.25	3.22	3.21	3.20	3.19	3.18	3.18	3.12
5	15.00	8.75	6.67	5.62	5.00	4.58	4.29	4.06	3.89	3.75	3.12	3.00	2.75	2.67	2.60	2.58	2.57	2.56	2.56	2.55	2.55	2.50
6	14.58	8.33	6.25	5.21	4.58	4.17	3.87	3.65	3.47	3.33	2.71	2.58	2.33	2.25	2.18	2.17	2.15	2.15	2.14	2.13	2.13	2.08
7	14.29	8.04	5.95	4.91	4.29	3.87	3.57	3.35	3.17	3.04	2.41	2.29	2.04	1.95	1.91	1.89	1.87	1.86	1.85	1.84	1.84	1.79
8	14.06	7.81	5.73	4.69	4.06	3.65	3.35	3.12	2.95	2.81	2.19	2.06	1.81	1.73	1.69	1.66	1.65	1.63	1.62	1.62	1.61	1.56
9	13.89	7.64	5.56	4.51	3.89	3.47	3.17	2.95	2.78	2.64	2.01	1.89	1.64	1.56	1.51	1.49	1.47	1.46	1.45	1.44	1.44	1.39
10	13.75	7.50	5.42	4.38	3.75	3.33	3.04	2.81	2.64	2.50	1.88	1.75	1.50	1.42	1.38	1.35	1.33	1.32	1.31	1.31	1.30	1.25
20	13.12	6.88	4.79	3.75	3.12	2.71	2.41	2.19	2.01	1.88	1.25	1.12	0.88	0.79	0.75	0.72	0.71	0.70	0.69	0.68	0.68	0.62
25	13.00	6.75	4.67	3.62	3.00	2.58	2.29	2.06	1.89	1.75	1.12	1.00	0.75	0.67	0.62	0.60	0.58	0.57	0.56	0.56	0.55	0.50
50	12.75	6.50	4.42	3.38	2.75	2.33	2.04	1.81	1.64	1.50	0.88	0.75	0.50	0.42	0.38	0.35	0.33	0.32	0.31	0.31	0.30	0.25
75	12.67	6.42	4.33	3.29	2.67	2.25	1.95	1.73	1.56	1.42	0.79	0.67	0.42	0.33	0.29	0.27	0.25	0.24	0.23	0.22	0.22	0.17
100	12.62	6.38	4.29	3.25	2.62	2.21	1.91	1.69	1.51	1.38	0.75	0.62	0.38	0.29	0.25	0.22	0.21	0.20	0.19	0.18	0.18	0.12
125	12.60	6.35	4.27	3.22	2.60	2.18	1.89	1.66	1.49	1.35	0.72	0.60	0.35	0.27	0.22	0.20	0.18	0.17	0.16	0.16	0.15	0.10
150	12.58	6.33	4.25	3.21	2.58	2.17	1.87	1.65	1.47	1.33	0.71	0.58	0.33	0.25	0.21	0.18	0.17	0.15	0.15	0.14	0.13	0.08
175	12.57	6.32	4.24	3.20	2.57	2.15	1.86	1.63	1.46	1.32	0.70	0.57	0.32	0.24	0.20	0.17	0.15	0.14	0.13	0.13	0.12	0.07
200	12.56	6.31	4.23	3.19	2.56	2.15	1.85	1.62	1.45	1.31	0.69	0.56	0.31	0.23	0.19	0.16	0.15	0.13	0.12	0.12	0.11	0.06
225	12.56	6.31	4.22	3.18	2.56	2.14	1.84	1.62	1.44	1.31	0.68	0.56	0.31	0.22	0.18	0.16	0.14	0.13	0.12	0.11	0.11	0.06
250	12.55	6.30	4.22	3.18	2.55	2.13	1.84	1.61	1.44	1.30	0.68	0.55	0.30	0.22	0.18	0.15	0.13	0.12	0.11	0.11	0.10	0.05
∞	12.50	6.25	4.17	3.12	2.50	2.08	1.79	1.56	1.39	1.25	0.62	0.50	0.25	0.17	0.12	0.10	0.08	0.07	0.06	0.06	0.05	0.00

Source: Tave (1990A).

[a] These levels of inbreeding assume that each parent contributes an equal number of viable offspring. Inbreeding values were rounded to the nearest hundredth.

a random manner from within each family. This breeding system can double N_e without increasing population size. N_e with pedigreed mating is

$$N_e = \frac{16(♀)(♂)}{3(♀) + (♂) \text{ or } (♀) + 3(♂)} \qquad (4.20)$$

If there are more females, the denominator is $3(♀) + ♂$; if there are more males, the denominator is $♀ + 3(♂)$. Effective breeding number increases when you use pedigreed mating, because you artificially increase the genetic variance by ensuring that each brooder is represented in the next generation.

The only drawback to pedigreed mating is that you must be able to identify individual fish. Although several marking systems have been devised (Anon. 1956; Clemens and Sneed 1959; Moav et al. 1960A, 1960B; Groves and Novotny 1965; Monan 1966; Volz and Wheeler 1966; Everest and Edmundson 1967; Fujihara and Nakatani 1967; Hill et al. 1971; Brauhn and Hogan 1972; Thomas 1975; Rinne 1976; Joyce and El-Ibiary 1977; Welch and Mills 1981; Murray and Beacham 1990; Morris et al. 1990), most hatcheries are not able to mark fish or to isolate families until they can be marked.

When the breeding population cannot be replaced or increased in size, the only way to increase N_e is either to alter the sex ratio or to switch to pedigreed mating. The benefits produced by changes in the sex ratio or in the breeding program can be measured as the effective breeding efficiency of the population:

$$N_b = \frac{N_e}{N} \qquad (4.21)$$

where N_b is the effective breeding efficiency and N is the population size.

Within the constraints of a fixed population size N, N_b allows you to determine the efficiency that your sex ratio or breeding program has in maximizing N_e relative to other sex ratios or breeding programs. Figure 4.28 shows N_b for all possible sex ratios for both random and pedigreed mating. For example, if a trout farmer spawns 90 female and 10 male rainbow trout by random mating, he has an $N_b = 36\%$. By adjusting his sex ratio to 70 females and 30 males, he will increase N_b to 84%. The 70:30 sex ratio is 2.3 times more efficient in maximizing N_e, and inbreeding will be only 42% as great. This index shows the farmer that he can still use a skewed sex ratio for optimizing fingerling production, but

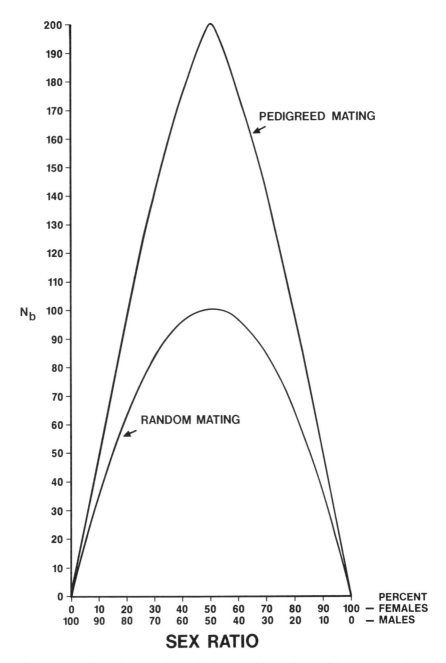

Figure 4.28 Effective breeding efficiency (N_b) at all possible sex ratios for both random and pedigreed mating.
Source: Tave (1984C)

if he moderates the ratio he can improve his N_e, and it quantifies the improvement.

Effective breeding efficiency can also inform a hatchery manager that he is doing a good job within the constraints of his hatchery. For example, if random mating is used, a 55:45 sex ratio has an N_b of 99%, so switching to a 50:50 sex ratio to achieve an N_b of 100% may not be worthwhile. The random mating parabola depicted in Fig. 4.28 shows that changes in sex ratios that are more skewed than 60:40 will produce large improvements, while changes in sex ratios less skewed than 60:40 will produce smaller improvements.

Effective breeding efficiency also provides an additional piece of information that is crucial in the decision making process. Effective breeding efficiency quantifies the effects that proposed changes in sex ratio will have on fingerling production. The index quantifies how many fewer females will be spawned, so it is possible to calculate how many fewer fingerlings will be produced. If the decrease is acceptable, then the proposed change in sex ratio will cause no problem. However, if the proposed change in sex ratio will lower fingerling production to an unacceptable level, either the change in sex ratio can be moderated or average female body weight can be increased so that the same number of eggs will be produced. Thus, you can both satisfy fingerling production quotas and also improve the genetics of the population.

Figure 4.28 also shows the advantage of spending the extra effort involved in switching to pedigreed mating. A 79:21 sex ratio with pedigreed mating has a larger N_b than a 50:50 sex ratio with random mating, within the constraint of a fixed population size: 102% vs 100%. Thus, within the constraint of a fixed population size, the use of pedigreed mating can produce an N_b that is larger than can possibly be produced by random mating. You can even produce an N_b of 200% with pedigreed mating (50:50 sex ratio).

A fourth approach is to alter certain management practices. When gametes are stripped, as is the case with salmonids, there is a tendency to pool eggs and/or sperm to facilitate this task. This shortcut could have disastrous consequences on a population's N_e. Gharrett and Shirley (1985) and Withler (1990) found that when sperm was added in a sequential manner to eggs, the males had unequal fertilizing success in pink salmon and chinook salmon, respectively. Withler (1988, 1990) found that when sperm was pooled before it was added to eggs, there was unequal fertilizing success in chinook salmon. Males that were unable to fertilize many eggs under these conditions did not have defective sperm. When sperm from only one male was added, all males achieved a satisfactory fertilization rate. Gile and Ferguson (1990) found that crossing methodology, as well as differential survival among families, altered parental contributions to the next generation of rainbow trout, and they felt that these factors could adversely affect N_e.

Differences in potency among males (fertilizing ability when there is competition among sperm) means that these practices produce unequal family size, which lowers N_e. To prevent this from occurring, sperm from one male should be used to fertilize eggs from one female. By using this procedure, N_e is maximized (and is easy to determine). Additionally, it enables equalization of family size, so variance of family size will not lower N_e.

Remember, all that adjustments in the sex ratio, breeding program, and spawning techniques do is prevent inbreeding and genetic drift from getting worse. They do not reconstruct a damaged gene pool. If you ensure that your N_e does not go through bottlenecks, these adjustments in broodstock management will help keep genetic problems in check. The only cure for the problems caused by bottlenecks is to acquire new broodstock and either replace the population or cross it with the new broodstock.

GENETIC–ENVIRONMENTAL INTERACTION VARIANCE

The final component of V_P that can be exploited by breeding to improve productivity is the interaction variance component between the genotype and the environment: V_{G-E}. This interaction occurs because some of the alleles responsible for the production of a phenotype are expressed differently in different environments. Several studies have shown that h^2s were different in different environments: Hagen (1973) found that the h^2 for the number of lateral plates in the threespine stickleback changed with temperature; McIntyre and Blanc (1973) found that the h^2 for hatching rate of steelhead trout eggs was different in different incubators; Tave (1984A) found that the h^2 for dorsal fin ray number in the guppy changed with temperature; Beacham (1988) found that h^2 for fry and alevin weight changed with temperature; and Kronert et al. (1989) found that h^2s for weight, gonad weight, and GSI in *T. nilotica* that were raised in Germany and Kenya were different. On a practical level, these studies show that selection to alter these phenotypes would be more efficient in one environment than in another, so the same effort would generate different responses.

Genetic–environmental interactions have been recognized as important production variables in agronomy and horticulture for years. That is why so many varieties of fruit, vegetables, and grains have been produced for specific soil types, hydrological budgets, and climates.

The importance of V_{G-E} as a factor that can influence productivity has also been demonstrated in fish culture, although detailed information is scarce. Genetic-environmental interactions have been found in rainbow trout (Gall 1969; Ayles et al. 1979; Ayles and Baker 1983; McKay et al. 1984; Iwamoto et al. 1986; Wangila and Dick 1988; Kindschi et al. 1991), steelhead trout (Reisenbichler and McIntyre 1977), common carp (Moav et al. 1975, 1976A; Hulata et al. 1982; Wohlfarth et al. 1983), catfishes (Ella 1984; Dunham et al. 1990), and the guppy (Vanelli et al. 1981).

For example, Wohlfarth et al. (1983) evaluated the performance of three different hybrid common carp under several different management programs in Israel. Some of their results are shown in Table 4.15. They found that the Chinese common carp hybrids did better than the European common carp hybrids when poultry manure was the sole nutrient input and stocking density was high, while the European common carp hybrids were better when the groups were fed supplemental rations. This makes sense because the stocks were developed in environments similar to the one in which they were superior. The growth rate of the Chinese × European common carp hybrid is interesting. The daily gain was better than either parent until the stocking density was lowered. When that happened, the European common carp hybrid, which is traditionally grown at low stocking densities, became the best hybrid. This study suggests that all environmental variables in fish culture can have an impact on productivity by interacting with the genome. In this experiment, nutrient input and stocking density were shown to cause variable expressivity of growth rate (V_{G-E}).

Although V_{G-E} can be a major influence in determining productivity, three studies found that V_{G-E} was of minor importance (Gunnes and Gjedrem 1978; Reinitz et al. 1978; Iwamoto et al. 1986). For example, Gunnes and Gjedrem (1978) examined Atlantic salmon from a number of rivers and found that V_{G-E} was a minor contributor to V_P. However, their experiment only suggests that V_{G-E} was of minor importance under the conditions that they examined. If the environments had been different, the importance of V_{G-E} might have been different.

On a practical level, V_{G-E} is variance that is best exploited by growing the best strain. Thus, exploitation of V_{G-E} is accomplished following yield trials of various strains. Many studies have shown that certain

Table 4.15 Daily Growth Rates (g/fish/day) of Three Hybrid Common Carp Raised under Different Management Programs in Israel

		Management program		
Group	Poultry manure; high stocking rate	Poultry manure and sorghum pellets; high stocking rate	Poultry manure and high protein pellets; high stocking rate	Poultry manure and high protein pellets; low stocking rate
European × European	1.69	3.48	5.60	8.43
European × Chinese	2.30	3.85	5.62	7.70
Chinese × Chinese	2.15	2.54	3.75	4.98

Source: After Wohlfarth *et al.* (1983).

strains are better than others for growth, food consumption, seinability, disease resistance, parasitic load, catchability by hook and line, heat tolerance, cold tolerance, developmental rate, body conformation, salinity tolerance, stress, brood size, viability, etc. Strain evaluations have been done with channel catfish (Smitherman *et al.* 1974; Burnside *et al.* 1975; Plumb *et al.* 1975; Shrestha 1977; Green *et al.* 1979; Chappell 1979; Youngblood 1980; Bice 1981; Broussard and Stickney 1981, 1984; Dunham and Smitherman 1981; Horn 1981; Al-Ahmad 1983; Dunham *et al.* 1986; Tomasso and Carmichael 1991), blue catfish (Ramboux and Dunham 1991), common carp (Kirpichnikov *et al.* 1967; Moav and Wohlfarth 1970; Kuzema 1971; Moav *et al.* 1974, 1975; Hines *et al.* 1974; Wohlfarth *et al.* 1975A, 1975B; Hulata *et al.* 1985B; Corti *et al.* 1988; Houghton *et al.* 1991), rainbow trout (Wales and Berrian 1937; Wales and Evins 1937; Wolf 1942; Gall 1969, 1972; Savost'yanova 1969; Cordone and Nicola 1970; Rawstron 1973, 1977; Gall and Gross 1978B; Klupp *et al.* 1978; Reinitz *et al.* 1978, 1979; Ayles *et al.* 1979; Brauhn and Kincaid 1982; Ayles and Baker 1983; Linder *et al.* 1983; Dwyer and Piper 1984; Morkramer *et al.* 1985; Ferguson *et al.* 1985; Hörstgen-Schwark *et al.* 1986; Siitonen 1986; Okamoto *et al.* 1987; R. R. Smith *et al.* 1988; Wangila and Dick 1988; Kindschi *et al.* 1991), Atlantic salmon (Gjedrem and Aulstad 1974; Refstie *et al.* 1977; Refstie and Steine 1978; Gunnes and Gjedrem 1978; Gunnes 1980), brook trout (Wolf 1941, 1954; Snieszko *et al.* 1959; Vincent 1960; Ehlinger 1964; Flick and Webster 1964; Wahl 1974; Robinson *et al.* 1976; Swarts *et al.* 1978; Webster and Flick 1981), brown trout (Wolf 1954; Ehlinger 1964), lake trout (Ihssen and Tait 1974), chinook salmon (Cheng *et al.* 1987), coho salmon (McGeer *et al.* 1991), Arctic charr (Delabbio *et al.* 1990), silver carp (S. Li *et al.* 1987A, 1987B), bighead carp (S. Li *et al.* 1987B), *T. nilotica* (Hulata *et al.* 1985A; Khater 1985; Abella and Palada 1986; Smitherman *et al.* 1988; Khater and Smitherman 1988; Capili *et al.* 1990), *Tilapia* spp. (Moreau *et al.* 1986; Pauly *et al.* 1988), tilapia hybrids (Hulata *et al.* 1988), guppy (Phang and Doyle 1989; Macaranas and Fujio 1990), and zebra danio (von Hertell *et al.* 1990).

Each study indicates the same fact: A certain strain was best under a particular set of conditions. It is not possible to generalize the findings of a study to say one strain is always better than another. Table 4.15 shows that such a generalization can be wrong. A hatchery manager who reads that strain A was the best strain in a yield trial should do several things before he sells his stock and replaces it with strain A. He must examine the conditions in which the fish were grown and then determine how similar those conditions are to the conditions that exist at his farm. For example, Chappell (1979) found that the Marion and Kansas strains of channel catfish had the greatest yields during a yield trial at Auburn University. Those channel catfish were raised in ponds that are constructed on Piedmont soil and that are filled with captured surface run-off water with an alkalinity of 12–40 ppm. Catfish farmers in Mississippi who raise channel catfish in ponds that are constructed on heavy clay aluvium soil and that are filled with well water with an alkalinity of 150–300 ppm should be careful about extrapolating Chappell's (1979) results to their farms. They should run yield trials in their locale before they decide to switch strains. This also applies to the use of hybrids. The superiority of the Marion × Kansas hybrid channel catfish that was demonstrated at Auburn University (Chappell 1979) may not exist at every catfish farm.

Yield trials are an important aspect of animal husbandry. If productivity is to be maximized, you must discover which strain or hybrid grows best in your area or on your farm.

Tave (1989) proposed that producer organizations or agricultural research stations develop yield trial centers where performance tests could be conducted on farm populations in order to determine the relative growth rates and other performance characteristics of cultured stocks of fish. Fish farmers can be scientific and precise about feeding, water quality management, disease control, etc. Paradoxically, these same farmers are culturing a crop of unknown quality. Because fish farmers do not know whether their strains are good or bad, they do not know whether good or poor yields are due to the biological (genetic) potential of their fish, to various aspects of management or mismanagement, or to climactic conditions.

Performance tests at a yield trial center would reveal the relative value of a farmer's strain and would give him a scientific basis for deciding that his fish are good or bad; additionally, the data would identify the strains that grow best in a given locale. The strains to be evaluated should be stocked communally (see page 262) and grown using standard management practices. Local farmers should help decide experimental protocol (stocking rates, feeding rates, etc.) so they will accept the results.

Each yield trial center should maintain a reference strain that is managed to minimize genetic changes. The reference strain is stocked communally in each pond and serves as an internal control or benchmark against which all strains will be compared. The reference strain is assigned a value of 100.0. Strains that are heavier or longer are given values that are greater than 100.0, while strains that are lighter or shorter receive values that are less than 100.0. For example, the following data are collected at harvest for three strains:

Strain	Mean weight	Value
Reference	400 g	(400/400)100 = 100.0
A	500 g	(500/400)100 = 125.0
B	380 g	(380/400)100 = 95.0

The ranking of the strains is determined by their numerical value.

The conversion of average weights to values is not necessary if such a comparison is going to be conducted only once and if it will be conducted in a single pond. However, this conversion is necessary if such comparisons are going to be made across years or across ponds. The culture of the reference strain in each pond will enable comparisons to be made across ponds and years, because the reference line will be used to standardize all values; once the values are standardized, they can be compared. For example:

Strain	Mean weight	Value
Pond 1		
Reference	500 g	(500/500)100 = 100.0
A	480 g	(480/500)100 = 96.0
B	450 g	(450/500)100 = 90.0
Pond 2		
Reference	400 g	(400/400)100 = 100.0
C	410 g	(410/400)100 = 102.5
D	380 g	(380/400)100 = 95.0

Strain C is the best because it has the largest value. Even though strain A averaged 70 g more at harvest, it was not as good as strain C because its value was only 96.0. Strain A was larger than strain C simply because it was raised in a better pond (higher dissolved oxygen values, etc.).

If yield trial centers are constructed, farmers who conduct selective breeding programs can use the center's reference strain as a control population. The ability to use the reference strain as a control for these breeding programs will make them easier to conduct, and they will be less expensive.

The development of strain registries which collate information about different strains, their performances in yield trials, their histories, and their phenotypic characteristics will help hatchery managers make more informed decisions about which strain they should raise. The trout (Kincaid 1981) and catfish (Dunham and Smitherman 1984) registries are important first steps in this process.

ENVIRONMENTAL VARIANCE

The final component of V_P is one that has no genetic basis: environmental variance (V_E). Environmental variance encompasses the obvious variables which, if made into a general statement, would read: If you improve the environment you will improve the phenotype. This should come as no surprise to anyone who has raised fish. If fish are fed properly, are raised under optimal growing temperatures with good water quality, and if health problems are minimized, they will grow faster than if you do not feed them, raise them in a marginal environment, and do nothing to keep them healthy.

For example, culture conditions have been found to affect percentage of sexually precocious Atlantic salmon (Saunders *et al.* 1983), to alter body and fin shape in coho salmon (Swain *et al.* 1991), to cause skull abnormalities in masu salmon (Romanov 1984), and to produce meristic changes in rainbow trout (Leary *et al.* 1991).

But V_E is far more than the obvious. Environmental variance can have a large and, if uncontrolled, a marked impact that can ruin a breeding program. In a breeding program you are trying to improve productivity by exploiting V_G, and you must be able to assess the gain due to genetic improvement in order to evaluate a program's success. If you confound V_G with V_E, you will not be able to quantify the improvement due to breeding. Since breeding programs are very costly, a program that confounds V_E with V_G is tantamount to a misappropriation of funds. If data cannot be properly analyzed, the breeding program should not be performed. Improperly designed breeding programs which fail to consider all sources of V_E will produce costly incorrect conclusions. Breeding programs must be designed so that influences due to V_E can be separated from those due to V_G.

Shooting

The first major effort to quantify the effects that V_E can have on V_P was done by N. Nakamura and Kasahara [1955, 1956, 1957, 1961; an English translation of these papers is found in Wohlfarth (1977)]. They noticed that total length in common carp populations did not assume a normal bell-shaped curve, but exhibited one that was skewed to the right. They wondered whether the larger individuals (tobi koi or shoot carp) which caused the skewed distribution were genetically superior. If they were, selection for increased growth could be done as soon as shooting occurred. (Shooting is the sudden and dramatic growth of the shoot carp.)

To answer this question, they performed a series of experiments to determine the cause of shooting.

The results of seven experiments that N. Nakamura and Kasahara (1955, 1956, 1957, 1961) performed are as follows: First, they discovered that shooting occurs when fry begin to feed on natural food such as cladocerans. Before that, the population exhibits a normal distribution (Fig. 4.29). Second, they discovered that food particle size had a marked effect on shooting. Common carp populations fed cladocera smaller than 400 μm did not have shoot carp, and total length distribution produced a normal curve. Common carp populations fed cladocera and other zooplankters larger than 400 μm had shoot carp, and total length

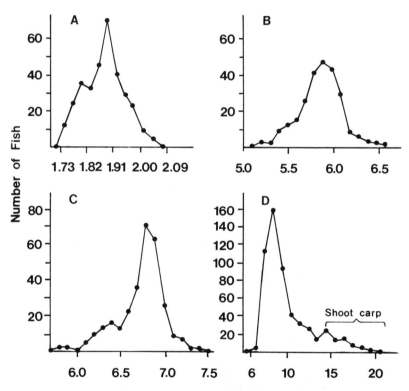

Figure 4.29 Frequency distribution of common carp egg diameter and fry body length: (A) egg diameter (mm); (B) total length at hatching (mm); (C) total length of 8-day-old fry that had not begun to feed (mm); (D) body length of 20-day-old fry that were feeding (mm). These graphs show that shooting occurred when common carp began to feed (D).
Source: After N. Nakamura and Kasahara (1955) and Kasahara (1985)

distribution was skewed (Fig. 4.30). Third, they found that the quantity of food had an effect on shooting. When food was abundant, no shoot carp were produced, but when food was scarce, they were; the more limited the food, the more skewed the population. Fourth, stocking density had a great influence on shooting; the denser the stocking rate, the more skewed the population. Fifth, they found that when they removed all competition by raising common carp in individual containers, no shooting occurred. Sixth, they found that when shoot carp were removed from the population, new shooters appeared to replace the old ones. Seventh, they discovered that the introduction of 20-mm goldfish or common carp which acted as artificial shoot carp prevented any fish from becoming shoot carp, and the population retained its normal distribution.

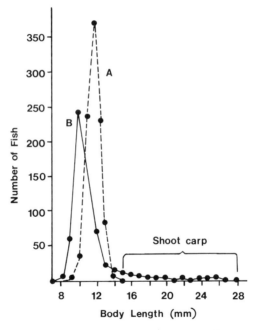

Figure 4.30 Body length curves of two groups of common carp fry that were fed food of different particle size: group A was fed food particles >400 μm; group B was fed food particles <400 μm. The curves show that shooting occurred in the group that was fed food particles >400 μm.
Source: After N. Nakamura and Kasahara (1956)

These seven experiments show that shooting is an environmentally controlled phenomenon. Shooting occurs when common carp begin to feed on food which has a particle size that not every fish can injest. Competition for food also produces shooting, so stocking density and feeding practices are major contributors to the production of shoot carp. These experiments demonstrated that shooting is caused by V_E and not V_G. Thus, if shooters were selected as broodstock in an effort to improve growth rate, no progress would be made because V_E is not transmitted from parents to offspring.

Two other studies also observed this phenomenon. Shooting was observed in channel catfish, although the degree of shooting was less pronounced than in common carp (McGinty 1980). W. M. Lewis and Heidinger (1971) found that stocking density and feeding practices caused skewness in sunfish populations, an observation which suggests that shooting occurred during their study.

These studies suggest that shooting may be a universal phenomenon. Because shooting is controlled by V_E, care must be taken to eliminate competition for food by increasing the amount of food and/or lowering the stocking density, and care must be taken to feed the fry with food particles which can be eaten by all fish. If these cultural practices are not adopted, these sources of V_E may confound and obscure the sources of V_G that you are trying to exploit.

Spawning Sequence

Another environmental variable that can be easily overlooked is spawning sequence. Older fish usually have a size advantage because they have had extra growing days. Wohlfarth and Moav (1970) found that a 24-hour age difference gave common carp a size advantage that was maintained or increased throughout the growing season. Thus, it is crucial that you do not mix or compare fish of different ages during breeding studies. If you do, you may choose older fish instead of genetically superior fish.

Age and Size of Parents

Age and size of the parents can influence certain phenotypes. Size of broodstock has been shown to affect spawning success and early growth of channel catfish fry (Bondari et al. 1985). They found that within 5-year-old broodstock there were negative correlations between: size of males and spawning success; female weight and egg weight; female weight and fry weight at 4 weeks. Siraj et al. (1983) found that maternal age had an effect on fecundity, egg length, egg weight, hatchability, and sac fry length in *T. nilotica*.

These studies indicate that broodstock age and size must be considered when planning a breeding program. Failure to control these factors may confound these sources of V_E with V_A when selecting for hatchability and early growth rates.

Egg Size

Egg size is another variable which can influence growth. A number of studies have shown that there is a positive correlation between egg size and growth rate. This has been found in common carp (Kirpichnikov 1970), brown trout (Bagenal 1969; Blanc *et al.* 1979), sockeye salmon (Bilton 1971), chinook salmon (Fowler 1972), Atlantic salmon (Naevdal *et al.* 1975), rainbow trout (Gall 1972, 1974; Chevassus 1976; Springate and Bromage 1985), *T. nilotica* (Siraj *et al.* 1983), and channel catfish (Reagan and Conley 1977). Most studies examined the influence of egg size on growth during the first year. Although there is a pronounced effect early in life, in many cases it dissipates over time, because there is some compensatory growth by fish that come from the smaller eggs. For example, Reagan and Conley (1977) found that the influence of egg size on growth rate became less pronounced as channel catfish grew to 180 days. Siraj *et al.* (1983) found that the effect was virtually gone by 20 days in *T. nilotica*. Even if the effect of egg size on body size disappears as the fish get older, it can be a problem if you select your fish before the effect disappears. If you select too early, you may confound a source of V_E with V_A.

Egg size is influenced by many environmental factors as well as genetics. Egg size is influenced by food supply, quality of the food (Meyer *et al.* 1973), and female size and age (Bilton 1971; Hulata *et al.* 1974; Gall 1972, 1974; Siraj *et al.* 1983). Thus, it is crucial to use females of the same age and comparable size and to ensure that they have been treated equally, or the differences that you detect in their progeny may be due to the mother's age, size, or diet, and not to genetic superiority.

Magnification

Initial size differences among groups are important because they become magnified (size differences become greater) as the fish grow (Fig. 4.31). Wohlfarth and Moav (1972) found that this was true in common carp, and Chappell (1979) found the same effect in channel catfish.

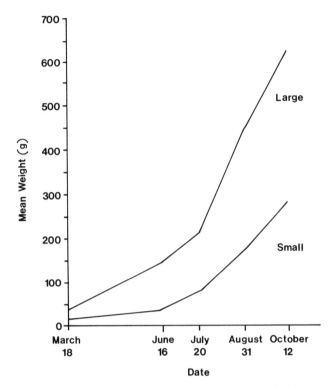

Figure 4.31   Growth rates of two size groups of channel catfish grown communally in earthen ponds. Initial mean weight of the large group was 30 g; initial mean weight of the small group was 12 g. The growth curves show the magnification effect. Weight difference continually increased (magnified) over the growing season from an initial difference of 18 g until it was over 300 g at harvest.
Source: After Chappell (1979)

Because initial size differences are magnified, it is important to compare groups that have the same initial size. If you do not, size differences that are observed when the fish are harvested may be due to initial size differences (V_E) and not to genetic differences.

Multiple Nursing

Moav and Wohlfarth (1968) proposed multiple nursing as a way to eliminate the problem of initial size differences. Multiple nursing is a technique in which the larger fish are "held back" by overstocking or under-feeding while the smaller fish are allowed to grow, so that the different groups are the same size when the experiment begins. Unfortunately, they found that common carp that were held back regained their super-iority when the constraints on growth were removed. Dunham *et al.* (1982), on the other hand, suggested that multiple nursing may work with catfish.

Communal Stocking

A final source of V_E that can confound V_G is experimental error due to pond-to-pond or raceway-to-raceway differences. Anyone who has managed two or more ponds knows all too well that ponds tend to have individual personalities and respond differently to management. To pre-vent this from obscuring the genetic differences that you wish to mea-sure, you must replicate the experiment so that the differences can be accurately assessed. The number of replicates that are needed for a particular experiment depends on many factors, and an excellent discus-sion about this problem is provided in "Fish Farming Research" by Shell (1983).

Replications can be very expensive, so Moav and Wohlfarth (1968, 1974) and Wohlfarth and Moav (1969) developed the communal pond concept to circumvent the problem of pond-to-pond variation. In the communal pond concept, the different groups that will be evaluated are marked and are then stocked together rather than in separate ponds. This technique will work only if the relative rankings of the groups are the same when stocked in communal and in separate ponds. If one group receives a competitive advantage in the communal pond, the rankings of the groups in the communal pond will be different than the rankings of the groups when they are stocked separately, and this tech-nique will not be a valid way of examining genetic differences. Moav and Wohlfarth (1968, 1974) and Wohlfarth and Moav (1969) found that communal stocking was a valid technique that could be used to compare different groups of common carp. In addition, this technique actually enhanced phenotypic differences among the groups so that it was easier to evaluate the genetic differences. This technique has also been shown to work with channel catfish (Chappell 1979; Dunham *et al.* 1982) and tilapia (McGinty 1984, 1987) and, as with common carp, communal stocking enhanced group differences. Wohlfarth and Moav (1991) found that this technique also worked when they evaluated groups of common carp in cages.

On the other hand, Williamson and Carmichael (1990) found that the relative rankings of strains of largemouth bass were not the same when they were stocked communally and separately. This suggests that one strain of largemouth bass received a competitive advantage in the communal setting, which means that this technique may not work with largemouth bass.

Regression equations have been developed to correct for the effect of initial size differences in communal ponds. Dunham *et al.* (1982) developed regression equations for catfish, and Wohlfarth and Milstein (1987) developed regression equations for common carp.

Communal stocking has tremendous implications in breeding research. It reduces costs by decreasing the number of ponds needed for the program, eliminates pond-to-pond variation so that it cannot confound V_G, and phenotypic differences are actually magnified so that it is easier to evaluate the genetic differences. Although the results with common carp, catfish, and tilapia suggest that communal stocking may be a valid technique for many species, the fact that it did not work with largemouth bass means that the technique should be evaluated with each species before it is used. Largemouth bass are piscivores, so these results may suggest that communal stocking does not work with carnivores.

It should be obvious that environmental variables are more than just "good" and "bad" growing conditions. It is important to quantify and to control as many environmental variables as possible during breeding programs.

If not controlled, these variables may prevent you from accurately assessing genetic differences, and this may hinder your program because it will be more difficult to exploit V_G.

CHAPTER 5

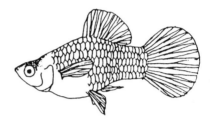

Biotechnology

5

Biotechnology

This chapter deals with subjects that would have been considered science fiction only 30 to 40 years ago. Although these technologies existed when the first edition of this book was written, relatively few farmers or hatchery managers were in a position to utilize them to manage a population or to manipulate its gene pool. Some of these technologies are becoming fairly routine and can be incorporated into routine management programs. Others are so sophisticated and technically demanding that it is unlikely they will ever be used except by research institutions or agribusinesses.

It is doubtful that these technologies will be universally used in the future, but it is likely that the fish that are raised at many hatcheries and farms will be affected by these technologies. Because of that, it is necessary to at least understand what these technologies are and how they can be used. It is not necessary to become an expert in any of them unless they will be incorporated into a breeding program. These technologies should become a standard part of some management programs but should be totally ignored by others. Their use or non-use should be determined only after a thorough examination of the goals that exist for a particular breeding program.

The chapter is divided into four sections: The first will discuss the use of exogenous hormones in order to create fish that are capable of producing monosex populations. The second section will address the manipulation of chromosome number or haploid chromosomal sets in order to produce triploid, tetraploid, haploid, gynogenetic, and androgenetic fish. The basic goal governing the research and technology of the first two sections is the control of sex and fertility. The ability to direct sex or create sterile fish is done to prevent early maturation, to prevent reproduction during grow-out, to produce larger fish, to improve carcass quality, to grow the sex that has the greater market value, and to utilize exotic species both in fish farming and in fisheries management. The third section will describe genetic engineering, a sophisticated, high-tech means of transferring a single gene from one fish to another or even of transferring a gene from one species to another. The last section will examine electrophoresis, a biochemical technique that reveals a fish's protein phenotypes and their genotypes.

SEX REVERSAL AND THE PRODUCTION OF MONOSEX POPULATIONS

As an order, fish are different from other animals that are raised on farms because, for the most part, fertilization occurs externally and embryological development takes place without the protection of an internal womb or a hard-shelled egg. This combination enables aquaculturists to manipulate fertilized eggs or embryos (fry) to an extent not possible with mammals or birds.

Because development occurs externally, hatchery managers can direct phenotypic sex by the addition of anabolic steroids in feed or water. The reason that it is possible to direct phenotypic sex is that while genotypic sex is determined at fertilization, phenotypic sex is not (see Chapter 2 for a discussion about sex determination). During early embryology, an embryo is, phenotypically, neither male nor female in that it does not possess ovaries, testes, or other characteristics associated with the reproductive systems. Instead, an embryo possesses embryological precursors of ovaries and testes (primordial germ cells), and at this stage an embryo is "totipotent" because it could develop into either a male or a female. At a specific time during embryological development (it is different for each species), a chemical signal originates from a gene or set of genes, and this signal "informs" the totipotent tissue which way to develop. Once this occurs and the tissue completes its development, the fish becomes either a phenotypic male or a phenotypic female. Additionally, once this occurs, it is not possible to alter phenotypic sex except by radical techniques, such as surgery, and it is unlikely that this will succeed.

There is a window of time, during which phenotypic sex can be altered, and this window is species specific. If a fish ingests or absorbs anabolic steroids during this period, the steroid can direct the development of the totipotent cells. This technology has been actively researched for the past 20 years and has become so routine that it is frequently incorporated in tilapia, salmon, trout and carp culture programs.

The direct alteration of phenotypic sex by the administration of hormones is neither genetics nor breeding; instead it is part of hatchery management. As such it will not be discussed in great detail. Hunter and Donaldson (1983) and Yamazaki (1983) provide excellent reviews of the types of hormones, dosages, and lengths of treatments that can be used to direct phenotypic sex in many species. To direct phenotypic sex, fry are typically fed diets containing either androgens (male hormones) or estrogens (female hormones) or are raised in water containing dilute concentrations of the hormones (Fig. 5.1). The exact dosage, duration of feeding, beginning and ending dates, and access to natural sources of food are major factors that determine the success of these endeavors.

Direct sex reversal has probably had its greatest impact in tilapia farming. (Note: some researchers prefer to use the term "sex inversion" rather than "sex reversal." I will use "sex reversal" because it is self-defining.) A major goal in tilapia farming is to produce all-male popula-

**DIRECT PRODUCTION OF MONOSEX POPULATIONS
BY ADMINISTRATION OF ANABOLIC STEROIDS**

Figure 5.1 General methods used to produce either monosex male or monosex female populations, either by feeding sexually undifferentiated fry rations that contain anabolic steroids or by raising fry in water that contains a dilute concentration of anabolic steroids.

tions in order to prevent reproduction and to eliminate females, which grow more slowly than males. To accomplish this, swim-up fry are typically fed rations containing 60 mg of 17 α-methyltestosterone/kg feed for 30 to 60 days. This usually changes about 90–100% of the females into males, which creates a population that is 95–100% male. The exact reasons why it is not always 100% successful are not fully understood.

In salmon farming, the goal is to produce all-female populations in order to eliminate precocious males who die after they mature. To accomplish this, when salmon fry first begin to feed, they are typically fed rations containing 20 mg of 17 β-estradiol/kg feed for 40 to 60 days.

The amount of hormone consumed by each fish is miniscule, and most is eliminated rapidly (Fagerlund and McBride 1978; Fagerlund and Dye 1979; Lone and Matty 1981; Johnstone *et al.* 1983; Goudie *et al.* 1986A, 1986B; Cravedi *et al.* 1989). For example Cravedi *et al.* (1989) found that rainbow trout excreted 67% of ingested 17 α-methyltestosterone within 24 hours, and Goudie *et al.* (1986A) found that only 0.9% of ingested 17 α-methyltestosterone remained in juvenile *T. aurea* 21 days after it was removed from the diet. This means that the hormones used to sex reverse fish might be detectable only in parts per billion when sex-reversed fish reach marketable size, so the use of these hormones should pose no health hazard.

However, even if it can be demonstrated that the consumption of hormone-treated fish poses no health problems, this approach is risky from a marketing aspect in a health-conscious society. Consumers are demanding food grown with little or no pesticides, fertilizers, antibiotics, or hormones. These demands are often without logic or merit but are based on fears and concerns about cancer or other health problems. As such, an industry that is based on the culture of hormone-treated fish could be destroyed by adverse publicity.

This problem can be circumvented by incorporating sex reversal into a breeding program where sex-reversed fish are created in order to become broodstock that are capable of producing monosex offspring. Such programs are sophisticated and require meticulous planning and adequate facilities, but they can work. The approach needed to create these broodfish depends on the sex-determining system of the species and whether the goal is to produce monosex male or monosex female populations.

For species that have the XY sex-determining system, and most important aquacultured species do (see Table 2.1), monosex male populations can be produced by the creation of supermales (males that are YY instead of XY). Figure 5.2 is a schematic diagram that outlines the procedure needed to produce supermales. The first step in this breeding

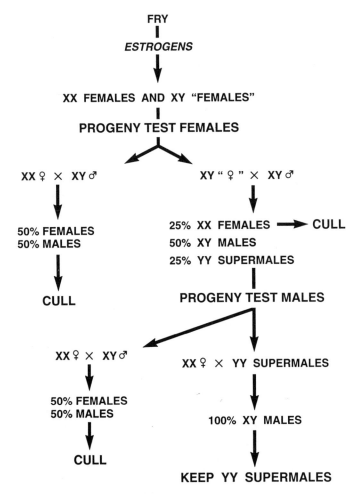

Figure 5.2   Schematic diagram of the protocol needed to produce YY supermales by creating sex-reversed broodstock. Supermales are capable of producing all-male populations for species with the XY sex-determining system.

program is to feed sexually undifferentiated fry a ration containing estrogens in order to produce sex-reversed females (females that are males genetically, in that they are XY, but that are phenotypic females). When the hormone-treated fish can be sexed, they will fall into three categories: normal XY males, sex-reversed XY "females" and normal XX females. The normal XY males should be culled because they are of no use in this breeding program.

Normal XX females are also of no use in this breeding program, but it is impossible to separate them from the XY "females" by examining external sexual characteristics. Consequently, the second step in the breeding project is to progeny test the females in order to identify and save the XY "females" and to identify and cull the XX females. This is accomplished by pairing each female with a normal male and by then raising each family in an individual hapa or pool until the offspring can be sexed. Females that produce offspring with a 50 : 50 sex ratio are XX and should be culled. Those that produce 75% sons are XY "females" and should be saved.

However, at this point the XY "females" are less important than one-third of their sons, which are the supermales. The mating of an XY "female" with a normal XY male produces the following offspring:

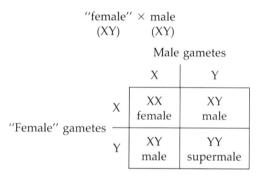

"female" × male
(XY)   (XY)

Male gametes

	X	Y
X	XX female	XY male
Y	XY male	YY supermale

"Female" gametes

Two-thirds of the sons from this mating are normal XY males, and one-third are YY supermales. The supermales are the fish that are the object of the breeding program, because supermales are capable of breeding true and of producing only male offspring:

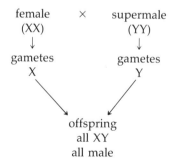

female   ×   supermale
(XX)   (YY)
↓   ↓
gametes   gametes
X   Y

offspring
all XY
all male

Unfortunately, it is impossible to separate males from supermales by an examination of external sexual characteristics, so a second round of progeny testing is necessary to identify and keep the supermales and to identify and cull the males. This is accomplished in the same manner as that used in the previous generation, except that males that produce offspring with a 50:50 sex ratio are culled, while those that produce 100% males are kept and pampered because they are supermales—fish that are capable of producing 100% male offspring without the use of hormones. The progeny test and means of differentiating XX females from XY "females" and XY males from YY supermales are identical to those described in Chapter 3 (pages 96–100).

Programs to produce supermales have been conducted with medaka (Yamamoto 1955, 1961, 1963, 1964B), channel catfish (K. B. Davis et al. 1990), T. mossambica (Varadaraj and Pandian 1989), T. nilotica (Mair et al. 1991A), rainbow trout (Chevassus et al. 1988), and goldfish (Yamamoto 1975C). An absolute prerequisite for this procedure to be successful in tilapia is to start with a pure species, because a mongrelized population that contains T. aurea or T. hornorum genes may contain more than one sex-determining system, which means that it will be difficult, if not impossible, to make this type of breeding program succeed.

Chevassus et al. (1988) produced YY supermale rainbow trout by cleverly utilizing individuals which developed abnormally and became hermaphrodites (fish that contain both testes and ovaries) during an experiment to produce sex-reversed XY "females." Because hermaphrodites contain both eggs and sperm, they can be self-fertilized: XY eggs were fertilized by XY sperm, which produced some YY supermale rainbow trout. The YY supermales were identified by progeny testing, and they produced all-male offspring.

Scott et al. (1989) found a presumed YY supermale T. nilotica that produced all-male offspring. This individual was produced by gynogenesis (this breeding technique will be described later in this chapter; Fig. 5.8) from what they hypothesized was a spontaneous sex-reversed XY "female."

It is easier to produce sex-reversed broodstock that are capable of producing monosex female populations in species with the XY sex-determining system (Fig. 5.3). In this case, androgens are used to produce XX "males." Depending on the species, the sexually undifferentiated fry are either fed a ration containing hormones or are raised in water containing hormones. The fish are raised until they can be sexed, and the fish are divided into two categories: females, which are culled, and males which are kept.

It is impossible to separate sex-reversed XX "males" from normal XY males by an examination of external sexual characteristics, so progeny testing is used to differentiate these males. Single-pair matings are made

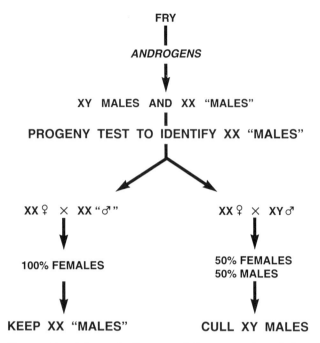

Figure 5.3   Schematic diagram of the protocol needed to create sex-reversed XX "male" broodstock that are capable of producing all-female populations for species with the XY sex-determining system.

between the males and normal females, and each family is raised in an individual hapa or pool until the offspring can be sexed. If the offspring exhibit a 50 : 50 sex ratio, the male is a normal XY male and should be culled. If the offspring are all females, then the male is an XX "male" and is kept for breeding purposes because these "males" are the fish that are capable of producing monosex female populations.

A number of these breeding programs have been conducted and have created sex-reversed broodstock that breed true and produce all-female populations. This has been accomplished with medaka (Yamamoto 1958), grass carp (Boney *et al.* 1984; W. L. Shelton 1986A), chinook salmon (Hunter *et al.* 1983; Solar *et al.* 1987, 1989), rainbow trout (Johnstone *et al.* 1979), Atlantic salmon (Johnstone and Youngson 1984), and *T. mossambica* (Varadaraj and Pandian 1990). Mair *et al.* (1991A) created sex-reversed XX "male" *T. nilotica,* and most of the sex-reversed broodstock that were evaluated produced all-female families. Wu *et al.* (1990) sex-reversed gynogenetic common carp (gynogenetic common carp are all XX females) and created broodstock that produced only daughters.

In rainbow trout, progeny testing is not necessary, because the sex-reversed males (XX "males") have incomplete sperm ducts which means that milt cannot be stripped from these fish (Johnstone *et al.* 1979; Bye and Lincoln 1986). Because of this, sperm from sex-reversed rainbow trout must be obtained by sacrificing the fish. Thus, the males can be categorized by an examination of the sperm ducts. Johnstone and Youngson (1984) found that this was also true in Atlantic salmon.

The breeding programs needed to produce monosex populations for species with the WZ sex-determining system are the opposite of those described for species with the XY sex-determining system: superfemales (WW females) must be created in order to produce monosex female populations, while a simple program is needed to produce monosex male populations. The first step in the production of superfemales is to feed sexually undifferentiated fry with feed that is laced with an androgen in order to produce sex-reversed males (Fig. 5.4). When the fish can be sexed, they are divided into females, which are culled, and males, which are kept. There are two types of males—WZ "males" and ZZ males—and it is impossible to separate them by an examination of external sexual characteristics.

To identify the sex-reversed males, the males are progeny-tested as described previously. Males that produce offspring with a 50:50 sex ratio are ZZ males and should be culled, while those that produce 75% females are WZ "males" and should be kept. One-third of the daughters produced by the WZ "males" are the superfemales that are capable of producing monosex female populations, while two-thirds are normal females that must be culled:

female × "male"
(WZ)   (WZ)

		"Male" gametes	
		W	Z
	W	WW superfemale	WZ female
Female gametes	Z	WZ female	ZZ male

The only way to differentiate the superfemales from the normal females is to progeny test the fish. This is done as described previously. If the sex ratio of the offspring is 50:50, the female is a normal WZ female and should be culled; on the other hand, if all offspring are

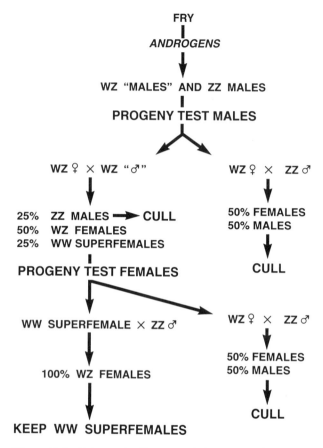

Figure 5.4 Schematic diagram of the protocol needed to create WW superfemales. Superfemales are capable of producing all-female populations for species with the WZ sex-determining system.

females, then the mother is a WW superfemale and should be saved because she is capable of producing only daughters:

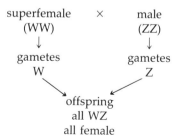

Few important aquacultured species have the WZ sex-determining system; the most important one that does is *T. aurea*, and it is unlikely that anyone would want to produce an all-female population. However, it is possible. Mair *et al.* (1991B) created a WW superfemale *T. aurea* by gynogenesis (half of gynogenetic *T. aurea* are WW superfemales and half are normal ZZ males), and she produced only daughters.

To produce sex-reversed broodstock that are capable of producing monosex male populations for species with the WZ sex-determining systems, fry are sex-reversed with estrogens (Fig. 5.5). This creates ZZ "females." As was the case earlier, the only way to differentiate ZZ "females" from normal WZ females is to progeny test the females. Normal WZ females produce offspring with a 50:50 sex ratio and should be culled. On the other hand, sex-reversed ZZ "females" will produce nothing but sons and are the sex-reversed broodstock that are capable of producing a monosex male population.

Hopkins (1979) and Mair *et al.* (1987A, 1991B) produced sex-reversed *T. aurea* broodstock that produced all-male progeny. Some females that were progeny-tested produced nearly all-male families or produced families that were composed of males and hermaphrodites. The reason for the incomplete success of these programs is unknown but might be due

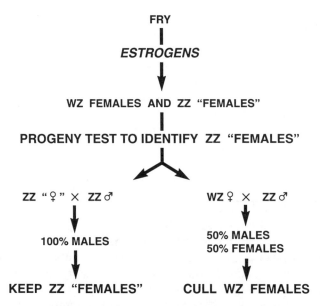

Figure 5.5 Schematic diagram of the protocol needed to create sex-reversed ZZ "female" broodstock that are capable of producing all-male populations for species with the WZ sex-determining system.

to contamination with sex chromosomes from another species, or it could be due to sex-influencing or sex-modifying genes. Females and males that produce 100% male offspring should be isolated, because they are capable of producing all-male populations.

Sex-reversal breeding programs have developed to the point where they supply fish for some large-scale commercial aquaculture industries. For example, over half the rainbow trout farmed in the United Kingdom (Bye and Lincoln 1986) and 40% of the chinook salmon farmed in British Columbia (Solar *et al.* 1989) are all-female progeny produced by sex-reversed broodstock.

CHROMOSOMAL MANIPULATION

Because fertilization occurs externally for most species of fish, it is relatively easy to manipulate chromosome number. A number of techniques have been developed to produce haploid, triploid, and tetraploid fish and even to produce fish whose chromosomes come solely either from their mothers (gynogens) or from their fathers (androgens). The lion's share of the work in this area has centered on the production of triploids.

The techniques that are used to alter the chromosome number are similar and can be divided into three categories: temperature, pressure, and chemical shocks. Most researchers use either pressure or temperature. These physical shocks are applied to eggs that are newly fertilized (Fig. 5.6). The exact timing and duration of the shock, relative to both meiosis (extrusion of the second polar body) and mitosis (first cleavage—when the zygote divides to become a 2-celled embryo), and the exact temperature or pressure determine not only the success rate but the type of chromosomal manipulation (Fig. 5.7).

Pressure chambers are used to apply pressure shocks. These devices range from cylinders attached to a mechanically operated screw-press to those controlled by computers. Pressures used to shock eggs range from 7,000 to 10,000 psi for several minutes.

The number of eggs that can be shocked at one time is constrained by the size of the cylinder. Additionally, the use of a pressure chamber poses a safety hazard. More than one pressure chamber has ruptured or "blown its top" and gone through a hatchery wall or roof when the air in the cylinder was not bled properly.

Because of this, as well as for other reasons, temperature baths are usually used at most hatcheries to shock the eggs. Both cold and warm temperatures can be used to alter ploidy number. Although both warm

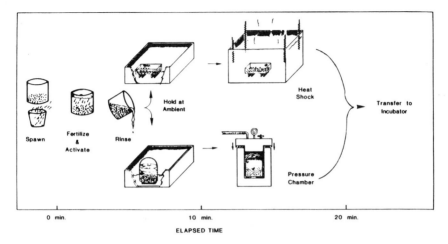

Figure 5.6 Schematic diagram of the techniques used to produce triploids.
Source: Thorgaard and Allen (1988)

and cold shocks can be used for either warm- or coldwater fishes, heat shocks seem to have produced better results for coldwater species, while cold shocks seem to have produced better results for warmwater species (Thorgaard and Allen 1988; Table 5.1). In general, the temperature needed to produce the best results is near the lethal limit for that species. The temperature needed to alter chromosome number is quite precise, so it is important that both water bath volume and temperature control mechanisms are sufficient to prevent temperature fluctuations while the eggs are being shocked.

There are several techniques that can be used to assess the success of chromosomal manipulations: One method is to count chromosomes. A second method is to examine the erythrocyte (red blood cell) size (Beck and Biggers 1983). A third method is to examine erythrocyte nuclear volume (Valenti 1975; Meriwether 1980; Wolters *et al.* 1982A); this is often done with a Coulter Counter (O. W. Johnson *et al.* 1984; Wattendorf 1986). A fourth method is to determine DNA content in erythrocyte nuclei by flow cytometry (Thorgaard *et al.* 1982; Allen 1983; Allen and Stanley 1983; O. W. Johnson *et al.* 1984; Pine and Anderson 1990; Ewing *et al.* 1991). A fifth method is to stain and count nucleoli (Phillips *et al.* 1986; May *et al.* 1988). A sixth technique is to use a Coulter Counter to analyze cell density (Cassani 1990). A seventh technique is to use electrophoresis (Dawley *et al.* 1985; Arai 1988; Seeb *et al.* 1988); this is done to examine triploid hybrids.

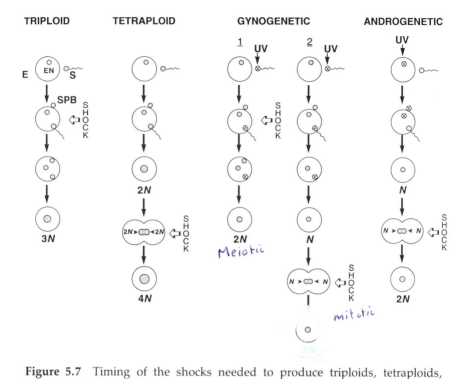

Handwritten annotations on figure: 2N **Meiotic** (gynogenetic 1); **mitotic** (gynogenetic 2)

Handwritten annotations at left margin: *& are heterozygous* (mitotic); *& are homozygous*

Figure 5.7 Timing of the shocks needed to produce triploids, tetraploids, gynogens, and androgens. Triploids are produced by shocking newly fertilized eggs (E) to prevent extrusion of the second polar body (SPB). The fertilized egg thus contains an egg haploid nucleus (EN), a sperm haploid nucleus (S), and a second polar body haploid nucleus which fuse to form a triploid nucleus and create a triploid zygote. Tetraploids are produced by shocking diploid zygotes as they undergo first cleavage. This prevents nuclear division, so the nucleus becomes 4N, and the zygote is a tetraploid. Gynogens are produced in one of two ways: 1) Eggs are activated by irradiated sperm (×) and are then shocked before the second polar body can be extruded. This creates an egg that contains a haploid egg nucleus and a haploid second polar body nucleus which fuse to form a diploid nucleus and a diploid zygote. Gynogens produced in this manner are called meiotic gynogens. 2) Eggs are activated with irradiated sperm, and a haploid zygote is produced because it contains only the haploid egg nucleus. The haploid zygote is shocked during first cleavage to prevent nuclear division. This produces a diploid nucleus and zygote. Gynogens produced in this manner are called meiotic gynogens. Androgens are produced by fertilizing irradiated eggs (×) with normal sperm. This produces a haploid zygote because it contains only the haploid sperm nucleus. The haploid zygote is shocked during first cleavage to prevent nuclear division, which produces a diploid nucleus and zygote.

Source: after Tave (1990C)

Many farmers in the U.S. use Coulter Counters to assess ploidy levels because, even though such devices are expensive, they are fast and accurate. The assessment of ploidy levels causes relatively little stress to the fish. All that is required is a drop of blood, which is usually taken by puncturing the isthmus or some other structure with a lancet, when the fish is lightly anesthetized.

Triploids

The creation of triploids is the type of chromosomal manipulation with which most aquaculturists are familiar. This technology has created fish farming industries, such as the triploid grass carp industry in the southeastern U.S.

Triploids are created by shocking newly fertilized eggs shortly after fertilization (Figs. 5.6 and 5.7). Fish eggs do not extrude the second polar body until they are fertilized. Because of this, if a newly fertilized egg is shocked, the shock prevents the second polar body from leaving the egg. Consequently, the fertilized egg will contain three haploid nuclei: one from the egg, one from the sperm, and one from the second polar body. The three haploid nuclei will fuse to form a triploid zygote nucleus which creates a triploid.

Triploids have been created in many species, including common carp (Gervai *et al.* 1980A; Meriwether 1980; Ueno 1984; Taniguchi *et al.* 1986B; Wu *et al.* 1986; Hollebecq *et al.* 1988; Cherfas *et al.* 1990), grass carp (Cassani and Caton 1985, 1986; B. Z. Thompson *et al.* 1987), bighead carp (Aldridge *et al.* 1990), rohu (Reddy *et al.* 1990), *T. aurea* (Valenti 1975; Don and Avtalion 1986, 1988B; Penman *et al.* 1987; Romana 1988; Mair *et al.* 1991B), *T. nilotica* (Penman *et al.* 1987; Don and Avtalion 1988B; Romana 1988; Hussain *et al.* 1991; Mair *et al.* 1991A), *T. mossambica* (Penman *et al.* 1987; Pandian and Varadaraj 1988A; Varadaraj and Pandian 1990), channel catfish (Wolters *et al.* 1981; Chrisman *et al.* 1983; Bidwell *et al.* 1985), Asian catfish (Manickam 1991), African catfish (Richter *et al.* 1987), European catfish (Krasznai and Márián 1986), Atlantic salmon (Benfey and Sutterlin 1984A; Johnstone 1985, 1987; Johnstone *et al.* 1989), chinook salmon (Utter *et al.* 1983), pink salmon (Utter *et al.* 1983), coho salmon (Utter *et al.* 1983; O. W. Johnson *et al.* 1986), chum salmon (Benfey *et al.* 1988), masu salmon (Arai 1988), rainbow trout (Chourrout 1980, 1984; Thorgaard *et al.* 1981; Chourrout and Quillet 1982; Lincoln and Scott 1983; Lou and Purdom 1984A; Solar *et al.* 1984; Kim *et al.* 1986, 1988; C. J. Shelton *et al.* 1986; Happe *et al.* 1988; Quillet *et al.* 1988A; Ueda *et al.* 1988A; Guo *et al.* 1990), brook trout (Arai 1988; Happe *et al.* 1988; Dubé *et al.* 1991), brown trout (Arai and Wilkins 1987),

Table 5.1 Biotypes of Some Species of Fish and the Type of Temperature Shock Which Has Been Effective in Producing Triploidy

	Biotype	Shock	Reference
Loach	warm	cold	Chao et al. (1986)
T. aurea	warm	cold	Suzuki et al. (1985B)
		cold	Valenti (1975)
		heat	Penman et al. (1987)
		heat	Don and Avtalion (1986)
		heat, cold	Don and Avtalion (1988B)
		heat	Romana (1988)
T. mossambica	warm	heat	Penman et al. (1987)
		heat	Pandian and Varadaraj (1988A)
		heat	Varadaraj and Pandian (1990)
T. nilotica	warm	heat	Penman et al. (1987)
		heat, cold	Don and Avtalion (1988B)
		heat	Romana (1988)
		heat, cold	Hussain et al. (1991)
Common carp	warm	cold	Nagy et al. (1978)
		cold	Gervai et al. (1980A)
		cold	Meriwether (1980)
		cold	Ueno (1984)
		cold	Taniguchi et al. (1986B)
		cold	Wu et al. (1986)
		heat	Hollebecq et al. (1988)
		cold	Cherfas et al. (1990)
Channel catfish	warm	cold	Wolters et al. (1981)
		heat	Chrisman et al. (1983)
		heat	Bidwell et al. (1985)

Species	Temperature	Shock	Reference
Grass carp	warm	cold	Cassani and Canton (1985)
		cold, heat better	B. Z. Thompson et al. (1987)
Rohu	warm	heat	Reddy et al. (1990)
African catfish	warm	cold	Richter et al. (1987)
Asian catfish	warm	cold	Manickam (1991)
Plaice	cool	cold	Purdom (1969, 1972)
White sturgeon	cool	heat	Kowtal (1987)
Brook trout	cold	heat	Arai (1988)
		heat	Dubé et al. (1991)
Brown trout	cold	heat	Arai and Wilkins (1987)
Rainbow trout	cold	cold, heat better	Chourrout (1980)
		heat	Thorgaard et al. (1981)
		heat	Chourrout and Quillet (1982)
		heat	Lincoln and Scott (1983)
		heat	Solar et al. (1984)
		heat	Happe et al. (1988)
		heat	Quillet et al. (1988A)
		heat	Kim et al. (1986, 1988)
Atlantic salmon	cold	heat	Benfey and Sutterlin (1984A)
		heat	Johnstone (1985, 1987)
Chinook salmon	cold	heat	Utter et al. (1983)
Pink salmon	cold	heat	Utter et al. (1983)
Coho salmon	cold	heat	Utter et al. (1983)
		heat	O. W. Johnson et al. (1986)
Masu salmon	cold	heat	Arai (1988)

Source: After Thorgaard and Allen (1988)

ayu (Taniguchi *et al.* 1986A; Ueno *et al.* 1986), white sturgeon (Kowtal 1987), loach (Suzuki *et al.* 1985B; Chao *et al.* 1986), plaice (Purdom 1972), threespine stickleback (Swarup 1959), white bass (Curtis *et al.* 1987), striped bass (Curtis *et al.* 1987), medaka (Naruse *et al.* 1985), and large-mouth bass (Garrett *et al.* 1992).

Triploids have been created by using temperature shocks (Swarup 1959; Purdom 1972; Valenti 1975; Chourrout 1980; Gervai *et al.* 1980A; Meriwether 1980; Thorgaard *et al.* 1981; Wolters *et al.* 1981; Chourrout and Quillet 1982; Chrisman *et al.* 1983; Lincoln and Scott 1983; Utter *et al.* 1983; Benfey and Sutterlin 1984A; Solar *et al.* 1984; Ueno 1984; Bidwell *et al.* 1985; Cassani and Caton 1985; Johnstone 1985, 1987; Naruse *et al.* 1985; Suzuki *et al.* 1985B; Chao *et al.* 1986; Don and Avtalion 1986, 1988B; O. W. Johnson *et al.* 1986; Kim *et al.* 1986, 1988; Krasznai and Márián 1986; Taniguchi *et al.* 1986A, 1986B; Ueno *et al.* 1986; Wu *et al.* 1986; Arai and Wilkins 1987; Kowtal 1987; Penman *et al.* 1987; Richter *et al.* 1987; B. Z. Thompson *et al.* 1987; Arai 1988; Happe *et al.* 1988; Hollebecq *et al.* 1988; Pandian and Varadaraj 1988A; Quillet *et al.* 1988A; Romana 1988; Cherfas *et al.* 1990; Guo *et al.* 1990; Reddy *et al.* 1990; Varadaraj and Pandian 1990; Dubé *et al.* 1991; Hussain *et al.* 1991; Manickam 1991), pressure shocks (Benfey and Sutterlin 1984A; Chourrout 1984; Lou and Purdom 1984A; Cassani and Caton 1986; Curtis *et al.* 1987; Johnstone 1987; Benfey *et al.* 1988; Aldridge *et al.* 1990; Hussain *et al.* 1991; Garrett *et al.* 1992), and chemical shocks (C. J. Shelton *et al.* 1986; Ueda *et al.* 1988A).

Most researchers and fish farmers use either temperature or pressure shocks. Chemical shocks are not used as often; they seem to produce a large number of mosaics (fish that have cells of various ploidy levels) (Allen and Stanley 1979; L. T. Smith and Lemoine 1979). Thermal shocks are the easiest, least expensive, and safest way to produce triploids. Consequently, this is the method used at most commercial hatcheries. The exact temperature and duration needed to produce triploids is determined by trial-and-error. The initial step is to determine the lethal temperatures. The optimal temperature is probably within a few degrees of the upper or lower lethal limit for a given species. The exact time that the shock should begin and end depends on how soon after fertilization the second polar body is extruded. The shock should begin before this event and continue until it is unlikely that the second polar body will be extruded.

✱ Triploids are created for two reasons: increased growth and sterility. In theory, triploids should be larger than their diploid counterparts because cell size is larger; the nucleus contains 33% more alleles for growth; and triploids do not divert energy for growth to gamete production, reproduction, or care of young. Unfortunately, improved growth has rarely been demonstrated; it has been found in *T. aurea* (Valenti 1975), channel catfish (Wolters *et al.* 1982B; Chrisman *et al.* 1983), rainbow trout (Kim *et al.* 1988; Quillet *et al.* 1988B; Guo *et al.* 1990), common carp (Taniguchi *et al.* 1986B), European catfish (Krasznai and Márián 1986), loach (Suzuki *et al.* 1985B), and striped bass ♀ × white bass ♂ hybrids (Kerby *et al.* 1991); however, the superior performance by some of these triploids was demonstrated for only a small part of their life histories. Benfey and Sutterlin (1984B) found that triploid Atlantic salmon were longer but that they weighed less than their diploid counterparts. Ueno *et al.* (1986) found that triploid ayu produced more flesh (had a greater dressing percentage) than their diploid counterparts. Even when growth was monitored after sexual maturity, which is when superior growth should be expressed, triploids did not produce the expected growth bonus. For example, Lincoln (1981A) found that even though triploid plaice × flounder hybrids did not divert energy to gamete development and reproduction and despite the fact that they grew when their diploid counterparts did not grow for the 4 months that it took for them to develop gametes and spawn, diploids exhibited compensatory growth after the spawning season and regained their superior size.

More importantly, triploids are created to produce sterile fish so that species that might cause adverse environmental impacts can be cultured on farms or used in natural resources management. A simplistic explanation for sterility is that gametic incompatibility occurs during meiosis because triploids have three sets of chromosomes (the chromosomes can't properly pair and align themselves, which makes it difficult to undergo meiosis, because the chromosomes can not be divided into two equal haploid sets). Gametes that are produced are usually aneuploid in that they contain unbalanced numbers of chromosomes. Aneuploid gametes usually produce abnormal and sub-viable offspring that do not live long.

A number of studies have examined the gonads of triploid fish to determine histologically if they are capable of producing viable offspring. The studies have shown that triploids have abnormal gonads, produce few gametes, and that the gametes that are produced are usually abnormal (Gervai *et al.* 1980A; Lincoln 1981C; Wolters *et al.* 1982B, 1991; Chrisman *et al.* 1983; Dawley *et al.* 1985; Lincoln and Scott 1983, 1984; Benfey and Sutterlin 1984B; Solar *et al.* 1984; Suzuki *et al.* 1985B; Allen *et al.* 1986; Benfey *et al.* 1986; O. W. Johnson *et al.* 1986; Taniguchi *et al.* 1986B; Ueno *et al.* 1986; Nagy 1987; Penman *et al.* 1987; Kim *et al.* 1988; Pandian and Varadaraj 1988B).

Even when offspring are produced by artificially inducing gameto-genesis and reproduction via hormonal injections, the offspring that are produced are abnormal and die during embryological development (Dawley *et al.* 1985; Nagy 1987; Penman *et al.* 1987). Lincoln (1981B) got triploid plaice to produce offspring, but none hatched. Although it is unlikely that triploids will produce viable, fertile offspring, it is possible. For example, Nagy (1987) mated sex-reversed triploid gynogenetic silver carp males with diploid common carp females and produced two indi-viduals that survived past first feeding; both of these fish were triploids, which means that even though the two offspring were viable, it is un-likely that they would be able to reproduce.

Chromosomal engineering can be combined with sex reversal to produce monosex sterile populations. The combination of these pro-grams creates fish that are both only the desired sex and incapable of reproduction. This has been accomplished for species with the XY sex-determining system. For example, Lincoln and Scott (1983) and Varadaraj and Pandian (1990) combined these breeding programs to create all female triploid *T. mossambica* and all-female rainbow trout, respectively.

The creation of triploids is an excellent way to utilize exotic fishes for fisheries management programs or for fish farming industries while minimizing adverse environmental impacts. For example, many states in the U.S. ban the importation of grass carp because they fear that these fish will escape from ponds, reproduce in the wild, become established, and displace native species. However, some of these states do allow the importation of triploid grass carp because they are sterile. Allen *et al.* (1986) induced gametogenesis and milt production in triploid grass carp by hormonal injections and found that virtually all sperm were abnor-mal; they calculated that the probability that a triploid male would pro-duce a fertile offspring was so remote the triploid grass carp could be considered to be sterile for management purposes.

Quality control is absolutely necessary in such an industry and adds to production costs. Direct production of triploids is seldom 100% suc-cessful. Because of this, it is often necessary to verify the ploidy level of each lot of fish. In some cases, the ploidy of each fish must be verified before it can be stocked. For example, in the U.S., to ensure that all triploid grass carp are triploids, the ploidy of each grass carp is deter-mined before it can be legally stocked in states where diploid grass carp are banned. This increases the cost of grass carp by as much as 400%.

A second place where triploids would be important is when fish are stocked in public waters as stop-gap management programs to help ailing fisheries. An excellent example of where such a program is needed is in the hybrid striped bass stocking programs in the U.S. Striped bass are considered to be one of the premier sport and commercial species in the U.S. Fishing pressures, pollution, and destruction of spawning grounds and nursery areas caused the dramatic depletions of many stocks of striped bass.

To help these stocks recover, fisheries management programs enlisted aquaculturists to produce millions of striped bass to stock public waters. Aquaculturists quickly discovered that striped bass were a difficult species to produce and began to look for ways to improve production. It was quickly discovered that striped bass ♀ × white bass ♂ hybrids were easier to produce. Striped bass ♀ × white bass ♂ hybrids are being stocked in rivers and lakes around the U.S. to help relieve the fishing pressure on striped bass and to help restore local fisheries. Unfortunately, this very practice could destroy the resource it is designed to save. If the hybrids are fertile, they could backcross with native striped bass and mongrelize the population. These programs could be the environmental disturbance that causes local stocks of striped bass to go extinct, because if the hybrids were to backcross with a stock of striped bass it would mean that the stock would no longer be striped bass but a bastard population.

The solution to this problem is to stock nothing but triploid hybrids. Because the hybrids would be sterile, they could not backcross with local striped bass populations, and their genetic integrity would be preserved. Preliminary research by Curtis et al. (1987) demonstrated that this approach is feasible.

The induction of triploidy is a technique that can be used to improve the results of interspecific hybridization when the cross produces subviable offspring (Chevassus et al. 1983; Scheerer and Thorgaard 1983; Seeb et al. 1986, 1988; Parsons et al. 1986; Arai 1988; Quillet et al. 1988B). The exact reason why this improves survival isn't known, but the induction of triploidy in these hybrids may increase survival because it produces hybrids with a normal diploid genome plus an extra set, rather than two mismatched sets, and this makes mitosis possible. Parsons et al. (1986) found that triploid rainbow trout × coho salmon triploids were more resistant to IHN, which means that chromosomal manipulation may be a technique that could be used to improve certain non-growth production traits.

Many researchers would like to produce triploids in some important cultured species, such as tilapia, to eliminate reproduction during growout. Unfortunately, the reproductive behavior of some species makes this technology impractical. A prime prerequisite for this technology to be effective and practical on a commercial scale is that spawning must be easily induced, and it must be easy to strip thousands of eggs. Some species, such as tilapia, are asynchronous spawners, are not very fecund, and are sequential spawners. This combination means that tens of thousands of broodfish would have to be handled annually to produce the eggs needed for one large-scale farm. Clearly, it is impractical to base a tilapia industry on the direct creation of triploids.

Triploids usually are not as viable as their diploid counterparts, in that early survival is lower. This has been attributed to the shock that is used to prevent extrusion of the second polar body (Chourrout and Quillet 1982; Scheerer and Thorgaard 1983; Solar *et al.* 1984; Johnstone 1985; Happe *et al.* 1988; Guo *et al.* 1990). The increased early mortality may be due to some sort of adverse effect that the shock has on egg cytoplasm. One way to prevent this is to create tetraploids and mate them with diploids in order to produce triploids. Chourrout *et al.* (1986A) and Blanc *et al.* (1987) found that this procedure improved early survival of triploids. Triploids produced in this manner are called "interploid triploids" to differentiate them from triploids produced by the prevention of meiosis (shocking fertilized eggs to prevent extrusion of the second polar body).

Tetraploids

Tetraploids are fish that have four sets of chromosomes instead of the normal two sets. Tetraploids can be created by shocking a diploid zygote when it is undergoing first cleavage (Fig. 5.7). The shock should be applied after the chromosomes have replicated and the zygote nucleus is about to divide into two. The shock prevents nuclear and cell division, so the resulting zygote nucleus will contain four sets of chromosomes instead of two. Tetraploid rainbow trout (Refstie 1981; Chourrout 1982A, 1984; Myers *et al.* 1986), channel catfish (Bidwell *et al.* 1985), *T. nilotica* (Myers 1986), *T. mossambica* (Myers 1986), *T. aurea* (Don and Avtalion 1988A), coho salmon (Myers *et al.* 1986), chinook salmon (Myers *et al.* 1986), grass carp (Cassani *et al.* 1990), rohu (Reddy *et al.* 1990), catla (Reddy *et al.* 1990), bighead carp (Aldridge *et al.* 1990), white sturgeon (Kowtal 1987), white bass (Curtis *et al.* 1987), striped bass (Curtis *et al.* 1987), and medaka (Naruse *et al.* 1985) have been produced.

The major interest in the creation of tetraploids is that they could be used to produce interploid triploids, which would eliminate the need to continually create triploids manually. The theory behind this is rather simple: Tetraploids are mated to diploids, and the ensuing progeny are interploid triploids. The logical mating would be to use tetraploid females because the micropyle of eggs from diploid females might not be large enough to admit sperm produced by tetraploid males (the sperm will be diploid):

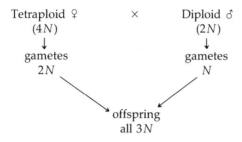

Once created, tetraploids would be mated to perpetuate themselves, so the manual creation of tetraploids would also not have to be done continually.

Tetraploid rainbow trout have produced interploid triploid offspring from this type of mating (Chourrout *et al.* 1986A; Blanc *et al.* 1987; Chourrout and Nakayama 1987; Diter *et al.* 1988; Quillet *et al.* 1988A; Myers and Hershberger 1991A, 1991B). Tetraploid rainbow trout have also been mated to produce second generation tetraploids (Chourrout *et al.* 1986A; Blanc *et al.* 1987). Although early results are somewhat encouraging, the fertilizing ability of tetraploid sperm is less than desirable.

Haploids

Haploids are individuals that have only a single set of chromosomes. Because they have only a single set of chromosomes, every detrimental allele will be expressed, since none will be hidden by a dominant allele in the heterozygous condition. As such, it is doubtful that many haploids will survive very long. Haploid medaka (Uwa 1965; Ijiri and Egami 1980), common carp (Gervai *et al.* 1980B; Wu *et al.* 1986), masu salmon (Ueda *et al.* 1988B), rainbow trout (Chourrout *et al.* 1980; Parsons and Thorgaard 1984; Bolla 1987; Ueda *et al.* 1988B), Atlantic salmon (Stanley 1983; Bolla 1987), chum salmon (Onozato 1982), and plaice (Purdom 1969) have been produced for experimental purposes, but none lived long.

There are two ways to create haploids, and both involve the destruction of genetic material from one of the parents. One way is to "fertilize" eggs with sperm that have had their genetic material destroyed by X-rays, gamma rays, chemicals, or UV irradiation. UV irradiation is probably the best method to destroy sperm DNA, because the other methods can leave viable chromosomal fragments which could produce partial diploids or reduce survival (Thorgaard *et al.* 1985; Thorgaard and Allen 1988). Additionally, UV irradiation is cheaper and safer. Sperm that have been treated this way do not actually fertilize eggs, because their genetic material has been destroyed; instead, they "activate" the egg. Once activated, an egg will begin embryological development, even though the sperm carries no functional genes. Because the sperm contributes no genetic material, the alleles contained in the egg haploid nucleus are the only genes that the embryo has, and the haploid egg nucleus becomes the zygote's nucleus. These haploid embryos are gynogenetic haploids, because all genetic material comes from the embryo's mother.

The second way to create haploids is to destroy the genetic material in eggs with X-rays or gamma rays and to fertilize these eggs with normal sperm. Because the sperm's haploid nucleus is the sole contributor to the zygote's genome, the embryo is a haploid. In this case, the haploid embryo is an androgenetic haploid, because all alleles come from the father.

Haploids are interesting because they enable breeders to study certain embryological genetic processes. On a practical level, it is important to know how to produce haploids, because that is the first step in the production of gynogenetic and androgenetic fish.

Gynogens

These are fish that are engineered so that both sets of chromosomes come from the mother. Gynogens are produced for several reasons. The first is to produce a monosex female population. If the species has the XY sex-determining system, females are XX, so all gynogens will be females.

A second reason is to discover a species' sex-determining system. As was discussed in Chapter 2, few fish have discernible sex chromosomes, so the sex-determining system usually has to be determined indirectly. The creation of gynogens can differentiate sex-determining systems where the female is homogametic from those where the female is heterogametic. When the female is homogametic, gynogenesis will produce only females; when the female is heterogametic, gynogenesis will produce equal numbers of males and females. These results do not prove which of the sex-determining systems exist, but because the XY and WZ sex-determining systems are the most common, the production of all-female gynogens suggests that the species has the XY sex-determining system, while a 50:50 sex ratio in the gynogens suggests that the species has the WZ sex-determining system.

Lastly, gynogenesis can be used to create highly inbred lines of fish. It can be used to create homozygous lines, particularly if it is coupled with sex reversal.

There are three ways to produce gynogens: The first is to activate eggs with sperm which have had their genetic material destroyed by X-rays, gamma rays, or UV irradiation (Fig. 5.7). Shortly after activation, the egg is shocked to prevent the second polar body from leaving. The egg now contains two haploid (N) nuclei: the egg nucleus and the second polar body nucleus. These will fuse to form a $2N$ nucleus, and both sets of chromosomes originated in the mother.

The second technique begins the same way: the first step is to activate eggs with sperm which have had their genetic material destroyed by irradiation. This will produce a gynogenetic haploid. When the haploid zygote undergoes first cleavage, the embryo is shocked, and this prevents cell division (Fig. 5.7). The two haploid (N) nuclei fuse to form a $2N$ zygote, and both sets of chromosomes come from the mother.

It is possible that some of the diploids produced during these processes will be normal diploids, because undamaged sperm could fertilize an egg. To preclude this from occurring, some researchers use sperm from another species that is known to produce non-viable hybrid embryos to activate the eggs. A second approach is to use genetic markers. For example, sperm from homozygous, normally pigmented rainbow trout can be used to activate eggs from albino females. All albino offspring will be gynogens, while normally pigmented offspring will be normal diploids.

Gynogens produced by suppressing first cleavage (mitotic gynogens) are homozygous at all loci and have inbreeding of 100%. Most of these gynogens are not likely to live very long, because if there are any detrimental recessive alleles, they will be expressed, and the fish will be sub-viable or will die. Gynogens produced by preventing the extrusion of the second polar body (meiotic gynogens) are inbred by definition (it is a form of self-fertilization), but they are quite heterozygous because of crossing over during meiosis. There is some disagreement about the exact level of inbreeding in gynogens (Nagy and Csányi 1982, 1984; Purdom 1983; D. Thompson 1983; Allendorf and Leary 1984; Guyomard 1984). Tsoi (1981) used gynogenesis and its concomitant inbreeding to reveal mutant recessive phenotypes produced by chemical reagents in common carp.

Gynogenetic silver carp (Mirza and Shelton 1988; Xia *et al.* 1990), grass carp (Stanley *et al.* 1975, 1976; Stanley 1976A, 1976B; Stanley and Jones 1976; Xia *et al.* 1990), ayu (Taniguchi *et al.* 1986A; 1990), zebra danio (Streisinger *et al.* 1981), goldfish (Stanley *et al.*1975), black buffalo (Stanley *et al.* 1975), bigmouth buffalo (Stanley *et al.* 1975), common carp (Stanley *et al.* 1975; Nagy *et al.* 1978, 1979; Gervai *et al.* 1980B; Tsoi 1981, 1987; Hollebecq *et al.* 1986; Linhart *et al.* 1986; Taniguchi *et al.* 1986B; Wu *et al.* 1986; Komen *et al.* 1988, 1991), *T. aurea* (Don and Avtalion 1988B; Romana 1988), *T. nilotica* (Mair *et al.* 1987B; Don and Avtalion 1988B; Romana 1988; Shah 1988), coho salmon (Refstie *et al.* 1982), rainbow trout (Chourrout 1980, 1982B, 1984, 1986; Chourrout and Quillet 1982; Guyomard 1984; Lou and Purdom 1984B; Onozato 1984; Purdom *et al.* 1985; Kaastrup and Hørlyck 1987; Quillet *et al.* 1988A), Atlantic salmon (Refstie 1983), chum salmon (Onozato 1984), masu salmon (Onozato 1984), catla (John *et al.* 1984), rohu (John *et al.* 1984), European catfish (Krasznai and Márián 1987), brown trout (Guyomard 1986), loach (Suzuki *et al.* 1985A; S.-J. Chen *et al.* 1986), plaice (Purdom 1969), flounder (Purdom 1969), medaka (Naruse *et al.* 1985), and red sea bream (Sugama *et al.* 1990) have been produced by using these techniques.

A third technique can be used to produce gynogens. This technique produces gynogens by fertilizing eggs from tetraploid females with sperm whose DNA has been destroyed by irradiation. Tetraploid females produce diploid eggs, so once they are activated they do not have to be shocked to produce a diploid chromosomal complement. Diter *et al.* (1988) and Chourrout and Nakayama (1987) have produced gynogenetic rainbow trout using this technique. Although these gynogens are inbred by definition, they should have considerably more heterozygosity than gynogens produced from diploid mothers. In fact, it is likely that many of these gynogens may exhibit little reduction in heterozygosity.

Gynogenesis can also be combined with sex reversal to produce broodstock that are capable of producing monosex populations. W. L. Shelton (1986A) combined gynogenesis and sex reversal to create grass carp that produced all-female offspring, and Wu *et al.* (1990) produced common carp broodstock that produced all-female progeny by combining these technologies. Varadaraj and Pandian (1989) proposed that sex reversal and gynogenesis could be combined to produce YY supermale *T. mossambica*. Scott *et al.* (1989) discovered a presumed YY supermale *T. nilotica* following gynogenesis. They hypothesized that the supermale's mother had been spontaneously sex reversed. The YY supermale *T. nilotica* produced only sons, which demonstrates that this technique could be used to produce all-male populations of tilapia. Figure 5.8 outlines the procedure needed to produce supermales by combining sex reversal and gynogenesis for species with the XY sex-determining system.

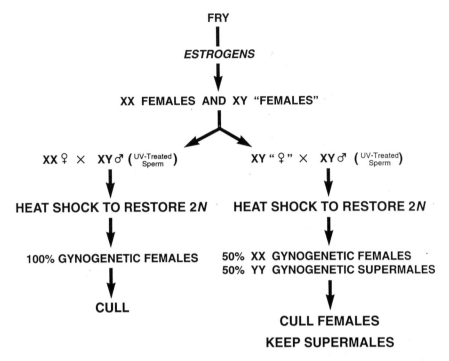

Figure 5.8 Schematic diagram of how sex reversal and gynogenesis can be combined to produce supermales.

Mair (1992) used gynogenesis to help decipher the inheritance of caudal deformity syndrome in *T. nilotica* (Table 3.1). Because the deformity is produced by a recessive lethal allele, homozygous recessive individuals cannot be spawned. To help prove that this deformity was caused by a recessive allele, Mair (1992) produced gynogens from suspected heterozygous females. If the females were heterozygotes, 50% of the gynogenetic offspring should have been deformed; they were, which confirmed the mode of inheritance.

Androgens

Androgens are the opposite of gynogens. These are fish whose chromosomes come solely from the father. Androgens can be produced by one of two methods: The most commonly used method is to fertilize eggs whose genetic material has been destroyed by gamma rays or X-rays with normal sperm. This produces a haploid androgen. When the haploid androgenetic zygote undergoes first cleavage, it is shocked to prevent cell division, and the two haploid (N) nuclei fuse to form a diploid ($2N$) nucleus (Fig. 5.7). All chromosomes in these fish come from the father. Androgenetic rainbow trout (Parsons and Thorgaard 1984, 1985; Scheerer et al. 1986; Thorgaard et al. 1990), brook trout (May et al. 1988), grass carp (Stanley 1976A; Stanley and Jones 1976; Stanley et al. 1976), and white sturgeon (Kowtal 1987) have been produced in this manner.

Androgens produced by this technique are highly inbred. Because the diploid complement of chromosomes comes from a single set of chromosomes that was allowed to replicate itself, it is 100% homozygous, and inbreeding is 100%. Such fish are likely to be sub-viable because if any deterimental recessive alleles are present, they will be expressed.

The second technique that can be used to produce androgens is to fertilize eggs whose genetic material has been destroyed by irradiation with sperm from tetraploid males. Tetraploid males produce diploid sperm; consequently, the sperm pronucleus is diploid rather than haploid, which means the ensuing zygote will also be diploid. Thorgaard et al. (1990) produced androgenetic rainbow trout using this technique. These androgens should be more viable, because they will be far less inbred.

If androgenesis can produce viable fish that survive to sexual maturity, it could be used to produce supermales for species that have the XY sex-determining system. In this sex-determining system, males are heterogametic, so half of the androgenetic fish produced from diploid fathers will be YY supermales.

Androgenesis has a very interesting application in the area of germ plasm maintenance and the conservation of endangered species (Thorgaard 1986; Thorgaard et al. 1990). Androgenesis could be used to recover genes from "stored" populations of cryopreserved sperm that exist in germ plasm resource centers. Ideally, both eggs and sperm would be cryopreserved and stored in these centers. Unfortunately, the technology to cryopreserve fish eggs has proven elusive, so the only practical way to stockpile genes and to recover them is to fertilize freshly stripped eggs with cryopreserved sperm. Using androgenesis, the genes from cryopreserved sperm could be recovered, even if the species goes extinct, because these sperm could be used to fertilize irradiated eggs from closely related species. This approach will not be perfect because the species' mitochondrial DNA will be lost (it is inherited maternally). Additionally, such fish will be hybrids in that the nuclear DNA comes from the father's species, while the mitochondrial DNA comes from the mother's.

GENETIC ENGINEERING

Genetic engineering is a molecular technique by which one or a few genes are transferred from one animal into another. In actuality, the transfer can be across species or even across kingdoms—bacterial or plant genes can be transferred to animals. The fruits of this technology will enable breeders to develop superior animals and plants and will enable physicians to cure or to prevent many inherited diseases.

This technology is so sophisticated that no ordinary farmer will conduct this type of breeding program. Additionally, there is a tangled web of federal, state, and local laws, rules, and regulations governing this research (Hallerman and Kapuscinski 1990A). Furthermore, this reseach is somewhat controversial in that some feel it is somehow immoral or unethical (Hallerman and Kapuscinski 1990B). Those who feel that genetic engineering is against the laws of God and Nature are the spiritual grandchildren of those who felt the same way about inbreeding some 200 years ago. These modern naysayers will, like their ancestral prophets of doom and dismay, be ignored because superstition and fear cannot thwart progress. Others fear this research because they feel that transgenic fish will cause adverse environmental impacts (Kapuscinski and Hallerman 1990). This issue is usually debated at an emotional rather than at an intellectual level; so, like most emotional issues, it will be with us for many years. Because of this, genetic engineering will be restricted to research institutions and to large agribusinesses that can afford million-dollar projects.

The basic premise behind this research is fairly simple, although it took years and hard work to make these technologies routine. The first step is to clone the desired gene. In order to accomplish this, the gene is removed from its chromosome in the donor species by restriction enzymes (enzymes that cut DNA at specific places). The gene is then inserted into bacterial plasmids (small rings of cytoplasmic DNA) by cutting the plasmid with restriction enzymes, attaching the gene to the plasmid, and then closing the plasmid with ligases (repair enzymes). The plasmid-gene complex is then transferred to bacteria via an osmotic shock, and the bacteria are then cultured. The bacteria replicate the plasmids as they grow and divide, and this produces millions of copies of the plasmid and the attached gene. The bacteria are cultured until there are enough for the project, at which point they are killed and ruptured to release the plasmids. The plasmids are then isolated and purified, and the gene is removed from the plasmids by restriction enzymes.

Once purified, thousands to millions of copies of the gene are in-

Figure 5.9 Microinjection of a medaka egg. The egg is on a grid that is attached to a microscope slide. The openings of the grid are 1 mm. The egg is being held in place by a micropipette. The diameter of the glass injection needle is a few microns.
Source: Penny Riggs and C. Larry Chrisman

jected into each egg (Fig. 5.9). Injection of genes into a zygote nucleus seems to produce better results than injection of the gene into the zygotic cytoplasm. However, fish zygote nuclei are small and virtually impossible to see in large eggs, so the genes are usually injected into the cytoplasm near the zygote nucleus. Once in the nucleus, the gene will, hopefully, be incorporated into a chromosome so that it can be transmitted from the transgenic fish (a fish that has a foreign gene in one of its chromosomes) to its offspring. The latter is the most crucial aspect of the process, because the goal of these programs is to produce superior fish that can transmit their superiority to their offspring. The gene could be replicated and transmitted to each cell in the host animal without being incorporated into a chromosome, but unless it is incorporated into a chromosome, genetically engineered offspring will not be produced.

Even if the gene is incorporated into a chromosome, the gene can be damaged, in that only part is incorporated, the gene can be incorrectly incorporated, or it can be incorporated in the wrong place. Even if the gene is properly incorporated, the message it carries may not be properly transcribed into messenger RNA, preventing production of the gene's phenotype.

The transfer of the gene does not guarantee success. Genes have regulators and promoters that turn them on and off at the proper times, and these must be transferred with the gene. Consequently, gene-regulator complexes are usually created and injected. The injection of a gene along with promoters and regulators is crucial.

To date, there have been a handful of experiments to produce transgenic fish; experiments to produce transgenic fish have been conducted with rainbow trout (Chourrout *et al.* 1986B; Maclean *et al.* 1987A, 1987B; Guyomard *et al.* 1989A, 1989B; Rokkones *et al.* 1989; Kolesnikov *et al.* 1990; Penman *et al.* 1990, 1992), Atlantic salmon (Fletcher *et al.* 1988; McEvoy *et al.* 1988; Rokkones *et al.* 1989, 1990), *T. nilotica* (Brem *et al.* 1988; Indig and Moav 1988), channel catfish (Dunham *et al.* 1987; Hayat *et al.* 1991), goldfish (Zhu *et al.* 1985; Yoon *et al.* 1989, 1990A, 1990B; Hallerman *et al.*1990), zebra danio (Stuart *et al.* 1988, 1989; Kolesnikov *et al.* 1990), madaka (Ozato *et al.* 1986), loach (Zhu *et al.* 1986; T. T. Chen *et al.* 1990; Kolesnikov *et al.* 1990), northern pike (Schneider *et al.* 1989), and common carp (T. T. Chen *et al.* 1990; P. Zhang *et al.* 1990; Hayat *et al.* 1991). The production of stable, genetically engineered germ plasm has been accomplished with rainbow trout (Guyomard *et al.* 1989A; Penman *et al.* 1992), loach (Maclean and Penman 1990; T. T. Chen *et al.* 1990), zebra danio (Stuart *et al.* 1988, 1989), and common carp (T. T. Chen *et al.* 1990).

Genetic engineering has caused many aquaculturists to daydream about fanciful fish that are perfect in every way. Unfortunately, genetic engineering is not a panacea that is going to produce fish that can tolerate 0 ppm dissolved oxygen, are resistant to all diseases, and have feed conversions of 1.0.

To date the types of genes that are available for transfer is somewhat limited. However, research is being conducted to produce genes that could be used to improve many traits for various species. T. T. Chen *et al.* (1987) and Maclean and Penman (1990) list the types of genes that are needed and some of the genes that are being developed.

Genetic engineering will produce fish that can grow faster, but these improvements will probably be no greater than those that can be accomplished by traditional breeding programs. It is unlikely that genetic engineering will produce disease-free, transgenic fish, because that would mean that resistance to all diseases is controlled by a single gene, something that is most unlikely. It is also unlikely that genetic engineering will produce transgenic tilapia that are capable of surviving winter weather in Iowa. I have heard several proposals to transfer the anti-freeze gene from winter flounder to tropical species, such as tilapia in order to improve cold tolerance. This will not improve cold tolerance. The antifreeze gene depresses the freezing temperature of body fluids. This is not going to keep tilapia alive when the temperature goes below 13°–15°C. Tilapia evolved in the tropics, and their enzyme systems have developed to function within a specific temperature range. If the temperature goes to the extremes of that range, the tilapia's biochemical reactions slow, and if the temperature goes outside of the range, its enzymes become denatured and death ensues. Furthermore, it is probably desirable that cold tolerance in tilapia cannot be improved to a great degree, because if this were possible, they would be banned in most temperate regions.

It is likely that genetic engineering will have the greatest impact in two areas of fish culture: The first is the production of vaccines that will enable aquaculturists to protect fish against many important diseases (Leong *et al.* 1990). This will be done by genetically engineering viruses to contain the genes from disease-causing bacteria that produce the components which elicit the immune reactions by fish. Prior to injection, the viruses will be attenuated so that they can not cause a disease. Such genetically engineered viruses will enable fish to develop immunity against many economically important diseases.

The second area where genetic engineering will create superior fish is in nutrition. One day, genes that enable fish to digest and utilize nutritionally poor feedstuffs will be transferred. Once this is accomplished, fish will be able to utilize starch as efficiently as fat, will be able to utilize chitin, and will be able to efficiently utilize poor sources of protein such as chicken feathers. This will lower feed and production costs.

ELECTROPHORESIS

Electrophoresis is a biochemical technique that enables geneticists to determine protein phenotypes and their genotypes and to study DNA and RNA. This technology is used extensively by population geneticists and by those who study evolution, because it enables them to compare gene frequencies across both time and space, and it also enables them to determine how different events affect gene frequencies. Before the advent of electrophoresis, these disciplines had to rely on easily measured gross phenotypes; since the development and refinement of electrophoresis, not only can dozens of phenotypes be studied simultaneously, but the genes that control these phenotypes can also be analyzed.

Hundreds of research papers have been written about electrophoretic studies of fish populations. Biochemical population genetics is probably the most extensively studied aspect of fish genetics. However, because this topic does not fall within the purview of this book, this will not be discussed. Kirpichnikov (1981) provides an excellent discussion of this topic in "Genetic Bases of Fish Selection." "Electrophoretic and Isoelectric Focusing Techniques in Fisheries Management" by Whitmore (1990) and "Population Genetics & Fishery Management" by Ryman and Utter (1988) are the best sources to discover how electrophoresis can be used to study and to manage fish populations. Utter (1991) wrote an historical account that traced the development and use of biochemical genetics in fishery management.

The principles that govern this technology are rather simple. Proteins, which are the phenotypes produced by genes, are composed of amino acids. All amino acids have the same backbone: an amine (NH_2) on one end and a carboxyl (COOH) on the other end. In between is a single carbon atom, to which is attached a so-called "radical." This radical is different for every amino acid, and the radical is what distinguishes one amino acid from another. The radical groups impart different charges to the amino acids. Some give an amino acid a negative charge, while others give it a positive charge. The sum of the different charges and their strengths give each protein a specific charge. That, coupled with protein size and shape, determines how fast and the direction in which a protein moves when it is placed in an electric current.

To separate and isolate the protein, a small sample of muscle, eye, liver, or blood, is placed in a medium such as starch or polyacrilamide gels. Then an electric current is run through the medium for a specific time. To determine how far the proteins have moved, which will differentiate and identify the phenotypes, specific stains are added to the media to stain the specific proteins which make them visible (Fig. 5.10).

Once visible, the pattern that each protein exhibits determines the fish's phenotype and also reveals its genotype. A discussion of every possible permutation that can be observed with this type of work is beyond the scope of this book; instead, a simple example of how electrophoresis can be used to determine phenotypes, genotypes, and gene frequencies will be presented: that of a monomeric enzyme controlled by a single gene with codominant gene action (this is the example used in Fig. 5.10).

In the diploid condition, each gene is composed of two alleles (for example, T and T'). Each allele produces a biochemical message that

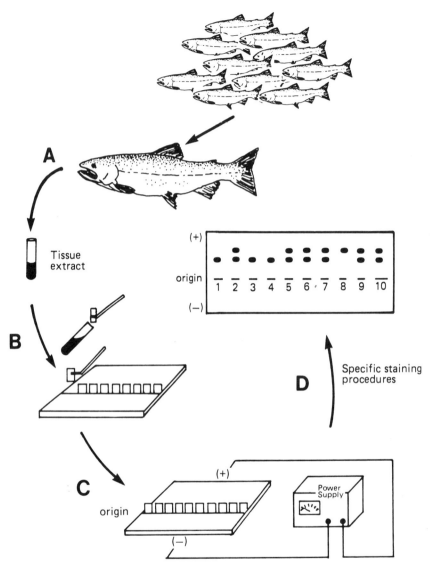

Figure 5.10 Schematic diagram of how protein phenotypes and genotypes are determined electrophoretically. First, a random sample of fish is obtained from a population, and protein is extracted from tissue such as muscle, liver, eye, or blood (A). Extracts from the tissue are then placed in a gel (B), and the gel is placed in a chamber where an electric current is run through the gel to move the proteins (C). After a specific time interval, the current is turned off, and the gel is stained to reveal the protein phenotypes (D). The gel in this figure contains protein extracts from 10 fish. The protein that is being examined is a monomeric enzyme controlled by a single gene with codominant gene action, which means that heterozygotes will produce two bands. Fish numbers

helps produce various protein phenotypes (enzymes are often the proteins that are examined). These phenotypes are expressed as bands on the media. A fish that is homozygous will produce only one band, while a fish that is heterozygous will produce two or more bands, depending on the mode of inheritance. Buth (1990) provides illustrations of the different types of heterozygous banding patterns that are caused by different modes of inheritance.

A monomeric enzyme controlled by a single gene with codominant gene action will produce two bands in the heterozygous condition. Consequently, when electrophoresis is used to examine the monomeric enzyme illustrated in Fig. 5.10, three phenotypes are observed: one is a single band called "T_1"; a second is a single band called "T_2"; the third is one where both the T_1 and T_2 bands are expressed simultaneously. The positions of the bands determine both the fish's phenotypes and their genotypes. For example, in Fig. 5.10, fish numbers 1, 2, and 8 have the following phenotypes and genotypes: fish number 1 has the T_1 phenotype, and its genotype is TT; fish number 2 has the heterozygous phenotype (both T_1 and T_2 bands are present), and its genotype is TT'; fish number 8 has the T_2 phenotype, and its genotype is $T'T'$.

By counting the bands that are produced by different fish in the population, it is possible to determine gene frequencies. For example, if the bands on the gel in Fig. 5.10 are examined, the gene frequencies for that population can be determined as follows: The bands produced by each fish are controlled by two alleles. If two bands are produced by a fish, a single allele is responsible for the production of each band. If only a single band exists, two alleles are responsible for the production of that band, because that fish is homozygous. Once the number of both alleles is determined, Eq. 3.1 can be used to determine gene frequencies.

For example, 20 alleles are depicted in the gel in Fig. 5.10 (there are 10 fish, and each has 2 alleles). There are 12 T alleles and 8 T' alleles (fish numbers 1, 3, and 4, each have 2 T alleles, and fish number 8 has 2 T' alleles); therefore:

$$f(T) = \frac{12}{20}$$

$$f(T) = 0.6$$

1, 3, and 4 have the T_1 phenotype and are TT: fish number 8 has the T_2 phenotype and is $T'T'$: fish numbers 2, 5, 6, 7, 9, and 10 are heterozygotes (because they have two bands) and are TT'.
Source: Utter et al. (1988)

and

$$f(T') = \frac{8}{20}$$

$$f(T') = 0.4$$

If this is done over time and space, changes in gene frequencies can be determined.

Electrophoresis can also be used to conduct taxonomic work. Species are traditionally separated by the presence of, absence of, shapes of, or counts of phenotypes—usually gross visible ones, such as dorsal fin ray number, presence or absence of an adipose fin, etc. Species can now be differentiated biochemically, and biochemical systematics has revolutionized the study of taxonomy. Examples of this research are studies by Crabtree and Buth (1987) on the systematics of *Catostomus*, by Kelsch and Hendricks (1986) on species differentiation in *Ictalurus*, and by Sodsuk and McAndrew (1991) on systematics in tilapia. Electrophoresis can also be used to identify hybrids (Buth *et al.* 1987; Harvey and Fries 1987; Bartley and Gall 1991; Hurrell and Price 1991; Jansson *et al.* 1991). Verspoor and Hammar (1991) provide an extensive review of this topic and show how electrophoresis has been used to uncover natural introgressive hybridization (the transfer of genes from one species to another). Electrophoresis can even be used as a forensics tool to differentiate flesh that comes from legally harvested animals from that which comes from endangered species and is thus illegally harvested and marketed (Harvey 1990).

Aquaculturists can also use these techniques to identify species and hybrids. For example, it is very difficult to identify tilapia, especially when they are young. The ability to correctly identify tilapia and to use only those that are a pure species is crucial for the success of crossbreeding programs used to produce monosex male populations. It is nearly impossible to select only fish that are, say, pure *T. aurea* or *T. nilotica* using gross, visible traits, but this is possible using electrophoresis. Biochemical phenotypes have been identified that can be used to differentiate species of tilapia and to identify hybrid populations (Avtalion *et al.* 1976; McAndrew and Majumdar 1983; Taniguchi *et al.* 1985; Macaranas *et al.* 1986; Brummett *et al.* 1988; Galman *et al.* 1988). Private companies now exist in the U.S. to perform this task, because some states allow one species of tilapia but not another, and the culture of the wrong species or of a hybrid population that contains genes from a banned species carries criminal penalties.

Biochemical phenotypes can be used as natural markers to assess growth or other characteristics of fish from different families, can be used to help establish family relationships, and can be used to help conduct mating programs to maximize growth or to help minimize inbreeding. An excellent series of papers on this topic was written by Moav *et al.* (1976B) and Brody *et al.* (1976, 1980, 1981).

Another use for electrophoresis is to detect genetic drift and inbreeding. Hatchery populations can be examined yearly to determine the frequencies of genes and genotypes to ascertain if they are changing over time. If inbreeding occurs, the percentage of homozygous genotypes should increase. If genetic drift occurs, gene frequencies should change, and some alleles should go extinct.

A number of biochemical genetic studies have found such changes in farm populations of cutthroat trout (Allendorf and Phelps 1980), brown trout (Ryman and Ståhl 1980), Atlantic salmon (Cross and King 1983), common carp (Paaver and Gross 1990; Sumantadinata and Taniguchi 1990), black sea bream (Taniguchi *et al.* 1983), *T. nilotica* (Brummett 1982), and channel catfish (Hallerman *et al.* 1986). Paaver and Gross (1990) examined several hatchery strains of common carp and concluded that differences among the populations were caused mainly by genetic drift. The mere fact that gene frequencies change over time should not be interpreted to say with absolute certainty that genetic drift was responsible for these changes, because selection and/or domestication can also change allelic frequencies. Hallerman *et al.* (1986) found that genetic drift and selection were both responsible for changes in gene frequencies in channel catfish.

If you decide to monitor your population electrophoretically to determine whether genetic drift has occurred, how many fish must be examined in order to provide an accurate estimate? If you sample too few fish, the sample itself is subject to genetic drift and may provide an inaccurate estimate. If you sample too many fish, you will waste a lot of effort. Although any change in gene frequency that is due to sampling error is genetic drift, the loss of rare alleles via genetic drift is far more important than small fluctuations from, say, 0.400 to 0.403. Thus, you need to know how many fish to sample in order to answer the question: Were any rare alleles ($f = 0.05$ or 0.01) lost as a result of genetic drift?

As with a progeny test, you need to know how many fish to examine before you decide that the allele is present or whether it was lost. If you find the allele right away, you do not need to examine any more fish because you know that the allele exists at $f \geq 0.01$ or 0.05. But what happens if you do not find the allele? How many fish must you examine before you can conclude that genetic drift eliminated the allele? That question is answered by using a slightly different form of Eq. (4.18):

$$P = (1.0 - q)^{2N}$$

where N is the number of fish that must be examined electrophoretically.

All you do is establish the desired probability (e.g., $P = 0.05$ produces a 95% chance of locating the allele if it exists) and the frequency of the alleles that you want to monitor (q) and solve for N (the sample size needed to establish your probability). You can use the P and N_e values in Table 4.11 to determine sample sizes for various probabilities at different gene frequencies.

If you examine several loci simultaneously, you will produce a more accurate picture about your population's gene pool and a more accurate estimate about the impact of genetic drift.

Electrophoresis can be used to determine the results of polyploidy experiments (K. R. Johnson and Wright 1986; Romana 1988; Crozier and Moffett 1989). Protein or DNA electrophoresis is how the results of genetic engineering are determined. Finally, electrophoresis can be used to help manage natural populations (Utter *et al.* 1976; Ryman and Utter 1988), and to assess the survival or reproductive success of different groups of fish or the contributions of stocked fish to local fisheries or local gene pools (Murphy *et al.* 1983; Taggart and Ferguson 1984; Pella and Milner 1988; Barbat-Leterrier *et al.* 1989; Garcia de Leániz *et al.* 1989; Leider *et al.* 1990; Morán *et al.* 1991; Pastene *et al.* 1991).

This section has discussed electrophoretic analysis of protein phenotypes produced by chromosomal genes, but this technique can also be used to examine protein phenotypes produced by mitochondrial genes. About 1% of a fish's DNA is mitochondrial DNA (mtDNA). mtDNA is a closed circular molecule of DNA that is located in the mitochondria (organelles in cells where cellular respiration occurs). This genetic material is transmitted maternally; i.e., offspring get their mtDNA solely from their mothers. mtDNA can be examined directly, or electrophoresis can be used to examine phenotypes produced by mtDNA genes to provide independent assessments of evolution (Billington *et al.* 1990) and population genetics (Hovey *et al.* 1989; Bernatchez and Dodson 1990; Shields *et al.* 1990; Birt *et al.* 1991). mtDNA can be used to identify species (Billington *et al.* 1991), subspecies (Seyoum and Kornfield 1992), and can be used to differentiate hatchery-produced and wild fish (Danzmann *et al.* 1991). Finally, because these genes are transmitted maternally, they can be used to identify hybrids (Herke *et al.* 1990; Gold *et al.* 1991; Knox and Verspoor 1991; Quattro *et al.* 1991). Ferris and Berg (1988) and Chapman and Brown (1990) provide good reviews of this topic.

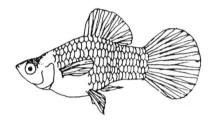

CHAPTER 6

Genetics of Broodstock Management

6

Genetics of Broodstock Management

The concepts that have been discussed in this book are of value only if they are used to manage a population. Genetic aspects of broodstock management are of the utmost importance because the genetics of a population determines its potential. A good hatchery environment is necessary to allow the expression of a stock's potential, but the population's alleles are its biological potential. Hatchery managers need to be aware of the techniques that can be used to manage, conserve, or exploit the alleles. They must know how these management practices affect productivity, because every management activity affects a population's gene pool: Every time fish are handled, some phenotypes and some alleles are culled.

First, a hatchery manager must make the decision: From what population will the fish be acquired? There is no right way to decide what population should be acquired; the answer to this question can be reached only through an examination of your goals, needs, previous research, and personal prejudice.

If the hatchery is to produce game fish that will be stocked in natural bodies of water, the obvious choice is to collect the fish from the existing populations in those bodies of water. It may be counterproductive to dilute a local gene pool with an exotic population. This is especially true if you are trying to restore a unique population. Garcia-Marin *et al.* (1991) found that exotic hatchery stocks of brown trout were genetically different from native populations in Spain and also found that stocking the exotic hatchery fish had had an adverse effect on native populations. They recommended that the exotic hatchery stocks be replaced with native stocks, in order to prevent further damage to the native stocks' gene pools.

Some current game fish management stocking programs may be great short-term solutions to certain pressing problems, but these programs can be a Pyrrhic victory. One project that could boomerang is the stocking of striped bass X white bass hybrids in the U.S. to restore the once great and famous striped bass fishery. The stocking of these hybrids has provided anglers with great fishing, but these fish could create some conservational headaches. First, their presence causes competition for food and other resources with striped bass, the very species this program is trying to save. Secondly, and perhaps more importantly, if the hybrids reproduce, they could mate with local populations of striped bass, which would mongrelize the striped bass and cause them to go extinct as a unique and identifiable biological entity. It would indeed be ironic if these programs caused the extinction of the very resource they were trying to save.

A solution to the second aspect of this problem is to stock sterile fish (W. L. Shelton 1986B, 1987). The creation of sterile triploid fish has become fairly routine and is the basis for certain industries. Curtis *et al.* (1987) demonstrated that triploid striped bass × white bass hybrids could be produced. A minimal amount of effort could make this technique simple and reliable and thus enable hatcheries to produce sterile fish that could be stocked in the wild. Even though the stocking of sterile fish would not negate the problem of competition with striped bass, it would eliminate the far more serious problem of introgressive hybridization and consequent contamination of the striped bass gene pool.

If the objective is to rescue an endangered or threatened population, there is an obvious decision: You must acquire the fish from the endangered population.

If you are going to produce a commodity, i.e., food fish or bait fish, you should try to discover which strain grows best in your area, which is the most disease resistant, which has the lowest feed conversion, etc. This information can come via personal communication, the scientific literature, extension agents, or from information that you generate from your own yield trials. Information produced during yield trials in other areas may not be valid in your locale because of V_E and V_{G-E}. Results from other studies should be used as intelligent first guesses, but the results should not be considered definitive in your area until they have been verified.

If you have a choice between wild or hatchery stocks, the hatchery stocks may produce better yields in fish farming situations because of domestication selection. Hatchery stocks generally outperform wild stocks in commercial yield trials.

You could even create your own strain by hybridizing two or more strains. A major liability of this plan is that the risk of introducing a disease is far greater than if the fish originate from a single location. A second drawback is the fact that selection should not begin until at least the F_2 generation.

Before you finalize a decision as to which strain to acquire, you must sleuth out the history of each population. You want to know its roots: Where did the population originate? How many hatchery transfers have there been? What kinds of yields have been attained with the population? What is the disease history of the population? This is done so that you do not waste money acquiring someone else's problems or so that you do not reacquire the problem that you just culled. The population's genealogy is an important piece of information.

An example of a strain's genealogy is the one that Dollar and Katz (1964) created for the McCloud River strain of rainbow trout (Fig. 6.1). Hatchery managers who know nothing about a strain's history could acquire what they think is a new strain when, in fact, they are simply replacing their own strain with another version of it. The genealogy presented in Fig. 6.1 shows that three hatcheries acquired the McCloud River strain of rainbow trout from different sources at different times.

A population's genealogy is not the only important historical information that must be collected in order to decide whether you should obtain a particular population. You need to know the population's N_e for every generation and every transfer. Obviously, this is not always possible, but information about a population's N_e will enable you to make a more intelligent decision about the suitability of the stock.

Most farmers do not calculate their N_es, so there are no historical records of what it was for each generation; however, if the farmer is a good manager, it might be possible to estimate N_es for many of the previous generations by examining his hatchery records. The number of spawns and the mating system can be used to calculate approximate N_es.

It does you little good to acquire your broodstock by sampling a population of 1,000,000 fish if that population was produced by only

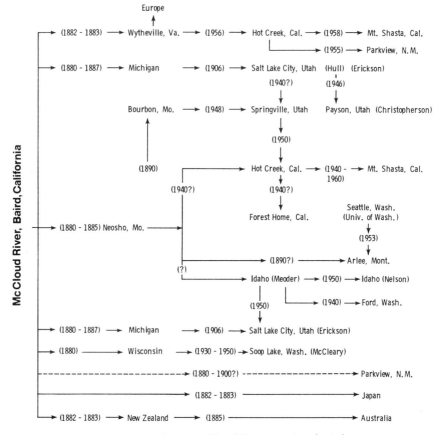

Figure 6.1   Pedigree of the McCloud River strain of rainbow trout.
Source: After Dollar and Katz (1964)

eight brooders. More than one hatchery stock has been started with a population of fish produced by a single mating—an N_e of only 2 (less if the parents were related and inbred). Such populations have little genetic variance, and future inbreeding and genetic drift will make it difficult to achieve production quotas.

Reductions in N_e can seriously affect a population's biological potential, can adversely affect productivity, and can ruin the opportunity to improve the stock via selection. What is the minimum N_e that is acceptable? To answer this question you have to determine: (1) how much inbreeding you will accept; (2) how rare are the alleles that you wish to save, and what guarantee do you want that they will be saved; (3) the number of generations that will be incorporated into your breeding program before inbreeding and the probability of losing rare alleles reach undesired levels.

The N_es listed in Table 4.13 (page 243) can be used as guidelines to prevent genetic drift and inbreeding from causing problems. The N_es recommended in this book are simply guidelines; they are not gospel. The methods that were developed to calculate N_es, the basic underlying assumptions behind the calculations, and the reasons for calculating desired N_es are what are truly important. Each hatchery manager must decide what N_es he needs for his populations, based on his goals, his budget, and the size of his hatchery.

When you collect the foundation population, make sure that you do not reduce N_e when you sample the fish. If you do, you will create what is called the founder effect, which means that you will eliminate a good deal of the genetic variance before you produce the F_1 generation. Most problems involving N_e probably occur when a population is acquired, so particular care must be taken when acquiring the population in order to prevent the founder effect.

If you acquire your foundation population from another hatchery, make sure that the generation that produced your fish had a large N_e. It will do you no good to collect a population of 500 full-sibs. If possible, take a few fish from each of as many spawns as practicable.

If you acquire your stock from a natural body of water, do whatever is necessary to collect a representative sample of the population—random sampling as well as nonrandom sampling. Use the life history of the species in that body of water to give you clues about how and where to collect fish. Collect fish from a wide geographic range. If you collect fish from only one spot you may collect a lot of fish that are related, and that will reduce N_e from what you think it is.

Collecting eggs may produce a population with more genetic variance than one produced by collecting fish, because you may be able to collect individuals from more families. Additionally, knowledge of the spawning behavior of the species should enable you to approximate the N_e of the generation that produced your foundation population. If you collect eggs, take small samples from many spawns.

You need to collect a larger population than your desired foundation N_e, because some fish will die and others will not spawn. For example, if mortality will be 50% before the fish reach sexual maturity, if spawning success is 60% at your hatchery, and if you want an $N_e = 344$, you need to collect

$$\frac{344}{(0.5)(0.6)} = 1147 \text{ fish}$$

The management of an endangered or threatened population can pose a problem because it may not be possible to acquire enough fish to produce an N_e of 344 or even 50. In some cases, efforts to collect this many fish in order to save the species could actually cause its extinction. When it is impossible to acquire a large foundation population, you simply have to work with the largest possible N_e.

The acquisition of broodstock is the most important step in the management of a population, because the foundation population determines how much genetic variation exists initially, and that determines the biological potential of the population. You cannot manage and exploit what does not exist.

Once you have acquired your population, how should you manage it? If you plan to propagate fish for ocean ranching, to produce game fish that will be stocked in natural bodies of water, or to establish and maintain standard reference strains of fish, N_e must be as large as possible. The major genetic goals in these programs are to conserve the population's genetic variance and prevent genetic drift and to keep inbreeding to a minimum. As such, you should try and keep alleles whose $f = 0.01$, and try to keep inbreeding $\leq 5\%$; minimum recommended N_es are listed in Table 4.13. Minimum N_es for these types of programs should be 344/generation; N_es greater than 1000 are not necessary. There is little point in producing populations for these purposes if you produce populations which have an edited gene pool. Populations with narrow gene pools are successful only in stable environments or in the absence of competition. Oceans, rivers, and lakes are dynamic, fluctuating, and diverse environments. To produce populations that survive and reproduce, you must provide them with broad genetic bases. There is little point in maintaining standard reference strains of fish if the populations' genomes change every generation.

Waples (1990A, 1990B), Waples and Teel (1990), and Waples *et al.* (1990) described ways to manage the N_e of Pacific salmon stocks to prevent the loss of genetic variance, and Smitherman and Tave (1987) outlined a breeding program to manage the N_e of hatchery populations to conserve genetic variance. Although management of N_e in order to conserve a population's genetic variance is of prime importance, Bentsen (1991) suggested that classic animal breeding programs should also be incorporated into management goals when managing hatchery stocks that are used to stock natural bodies of water.

If you are raising food fish or bait fish, minimum recommended N_es are between 61 and 250/generation (Table 4.13). Most farmers should have little trouble acquiring and maintaining N_es required to prevent inbreeding- and genetic drift-related problems. For example, if a catfish farmer spawns 25 males and 25 females per year, he can maintain an $N_e = 50$. That N_e will enable him to produce 5 generations of channel catfish (20 years) before F reaches 5%. At the end of 5 generations, that N_e will also produce a 97.1% guarantee of keeping an allele whose $f = 0.05$, which means that 97.1% of all alleles whose $f = 0.05$ will remain in the population, unless domestication selected against some of these alleles.

Farmers who manage small populations will not be able to prevent inbreeding from depressing productivity or prevent genetic drift from eroding their population's genetic variance if they use random mating. All they can do is realize the consequences of their small N_e, and when yields decline, import new broodstock to rejuvenate or replace the gene pool.

The best way to maintain a large N_e is to spawn as many fish as practicable, regardless of a station's production quota. If you need 1,000,000 fingerlings, it is better to produce 100 families of 10,000 than 10 families of 100,000. The best way to accomplish this is to produce more spawns than you need and keep only a sample from each spawn. Hatchery managers usually view such practices as a waste of effort, but genetically it is well worthwhile. A major goal in hatchery management is to spawn the fewest fish possible. While this may be a good idea in terms of personnel management and budgetary constraints, it is a bad idea for broodstock management. A hatchery can always recoup some of the additional expense by selling the excess fish.

If hatchery size or budgetary limitations require you to maintain a small spawning population or if you are working with an endangered or threatened population, you can increase N_e by switching from random to pedigreed mating. Another way to increase N_e without increasing the number of fish that will be spawned is to bring the spawning population closer to a 50:50 sex ratio. The improvement that is gained by altering the sex ratio and/or by switching to pedigreed mating can be measured as the N_b of the population (Fig. 4.28).

If you can mark your fish with individual marks, you can create individual pedigrees and prevent inbreeding, even in small populations. You still need to pay attention to the N_e because of genetic drift, but marking individual fish can remove one potential problem.

Spawning techniques can also be altered to improve N_e. If you strip gametes, as is done with salmonids, it may be counterproductive either to add sperm from several males to the eggs in a sequential manner or to mix the sperm from several males before fertilizing a batch of eggs. Gharrett and Shirley (1985) and Withler (1988, 1990) showed that both techniques produce N_es that are smaller than expected because of differences in potency (fertilizing ability) among the males. To maximize N_e, it may be advisable to subdivide the eggs and fertilize each sample with sperm from a single male.

If you have a choice between natural-type reproduction techniques, such as open pond spawning, or more controlled reproduction, such as pen spawning, the latter allows you to calculate your N_e, and you can prevent one male from mating with many females, which will lower N_e. For example, pond spawning of channel catfish is far more efficient in terms of manpower, and mortality of the broodstock is far lower, but pen spawning will produce a better population, genetically.

Cryopreservation of sperm is currently receiving a lot of attention. Its perfection will enable hatcheries to stockpile semen from thousands of males. Once the use of cryopreserved sperm becomes practical, this technique will enable hatchery managers to produce large N_es at minimal costs for certain species.

It is often difficult or impossible to maintain an N_e at the desired level, but this does not automatically condemn a population or mandate its destruction. The only reason a population should be scrapped is if its yield, fecundity, or viability declines. If the population is performing well, it should be kept. Managing the population's N_e may prevent depressions in production traits, and decreases in N_e can be used to explain certain problems. However, the important measurable characteristics of the population are growth, survival, fecundity, etc. The population's N_e is simply a parameter that can be used to manage the alleles that control these traits.

Broodstock management requires that information be collected from each generation. These data will provide a description of the population and will provide a base of information that will allow you to make intelligent decisions about how to manage the population. You need data on growth rates, survival, yield, epizootics, feed conversion, eggs/kg female, etc., in order to provide you with information about the quantitative characteristics of the population. You need means, SDs, CVs, and ranges so that you can quantify the variation that exists. The SDs and CVs will indicate whether enough variation exists to make selection a practical endeavor.

You also need information about the qualitative characteristics of the population: the number and frequencies of economically important phenotypes, and the number and frequencies of abnormalities. This information will let you know whether the frequencies of various phenotypes are increasing or decreasing over time. Like all fish stories, the number and types of abnormalities increase with each telling, so data will provide an accurate picture.

If the frequencies and types of abnormalities begin to increase, you may need to study them in order to determine their effects on productivity and to determine the underlying mechanisms that create them, so that you can control their frequencies. Most people assume that all abnormalities are genetic defects which occur as a result of inbreeding or mutation. Some are, but many abnormalities are caused by injury (Breder 1934, 1953; Gunter and Ward 1961), environmental disturbances (Garside 1959; Tomita and Matsuda 1961; Clemens and Inslee 1968; Komada 1977; Couch et al. 1979; Backiel et al. 1984; Parker and Klar 1987; Weigand et al. 1989), disease (Hoffman et al. 1962), nutritional deficiencies (Lim and Lovell 1978; Lovell and Lim 1978), culture techniques (Romanov 1984; Leary et al. 1991), or they can be nonheritable congenital defects (Tave et al. 1982; Dunham et al. 1991).

Because abnormalities have various underlying causes, the sudden occurrence of abnormalities should not be used as proof that inbreeding has occurred. The sudden occurrence of the stumpbody and the saddleback phenotypes in the Auburn University strain of *T. aurea* led many to conclude that inbreeding was beginning to uncover detrimental recessive alleles. However, analysis of these phenotypes showed that the stumpbody phenotype was a nonheritable congenital defect (Tave et al. 1982) and that the saddleback phenotype was controlled by a dominant allele (Tave et al. 1983). Neither phenotype was a by-product of inbreeding, because neither phenotype was produced by a recessive allele.

One way to monitor the population and to determine whether genetic drift has occurred is to collect electrophoretic data every generation. If you examine several loci simultaneously, you will produce a more accurate picture about your population's gene pool and a more accurate estimate about the impact of genetic drift.

If you cannot afford to examine your population electrophoretically, you can get a fairly accurate estimate of the impact of genetic drift on rare alleles by using the probability that your N_e generates. For example, if you have a guarantee of 99% of keeping an allele whose $f = 0.01$, you will lose 1% of all such alleles. You cannot quantify the number that you actually lose, because you need electrophoretic data to determine that, but you will have a relative indication about the likelihood of genetic drift and how much damage was done to the gene pool.

Leary et al. (1985A) suggested that another way to determine if genetic drift has robbed the population of genetic variance is to measure bilateral meristic phenotypes for 25–100 fish and check them for asymmetrical development. They felt that a decrease in heterozygosity created by genetic drift will reduce developmental stability and increase the frequency of asymmetrical development of bilateral phenotypes.

How you exploit the genetic variation that exists in your population depends on your goals. If you are raising fish that will be stocked in natural bodies of water, you must preserve as much of the variation as possible, including those fish that are small or that mature last, because you do not know which alleles are the most valuable in the wild. You need to monitor gene frequencies electrophoretically to ensure that genetic drift or unintentional selection does not change gene frequencies.

If you are raising food fish or bait fish, you may want to improve growth rate or other phenotypes. When managing these populations, you need to be aware of the effects of domestication selection on productivity. You need to know whether the effects of your management activities are positively or negatively correlated with production phenotypes. If they are positively correlated, you can increase productivity by increasing the intensity of domestication. But if they are negatively correlated, you need to alter domestication or it will depress productivity.

Before you initiate a breeding program, make sure that you are satisfied with the environment, i.e., make sure you are not going to make drastic changes in the way that you raise your fish. If you change the environment during or after a breeding program, you may find that you have wasted a lot of effort, because the population that was developed does better in an environment that you no longer use.

A good first step towards improving productivity via breeding is to evaluate a number of strains at your facility. If the communal pond concept has been shown to be a valid technique to evaluate groups for your species, it should be used to evaluate the strains, because it will save money during the yield trial and because genetic differences among the groups will be magnified.

When do you initiate a breeding program? A wise old breeder once told me: "Genetics begins at the beginning." I often hear farmers or researchers say, "We'll initiate a breeding program after we solve our other problems." This sentiment is misguided, and it is also naive. When are you going to solve all other problems? Farming, like many other ventures, is like the mythical Hydra—a multi-headed problem. If you chop off one head, the beast simply grows another. Once we figure out how to prevent and/or cure one disease, another supplants it. When are we ever going to concoct the perfect feed? Consequently, if you postpone the initiation of a breeding program until all other problems are solved, you will never begin, and since breeding is the aspect of management that manages the biological potential of the animal, yields will never be maximized.

A breeder on a commercial farm must understand and incorporate pathology, nutrition, production, marketing, economics, and everyday management into a breeding program. A scientist at a research facility usually does not have to worry about many of these other aspects of aquaculture when he conducts a breeding program. But a farmer must consider them, because he has to make a living and cannot sacrifice his livelihood for the sake of science.

What are some of the elements of a breeding program that a farmer must incorporate when he decides to conduct a breeding program? Much of this brief discussion comes from a paper written by Shultz (1986). Before initiating a breeding program you should conduct an industry assessment, determine your goals, quantify how the goals will be measured and assessed, conduct a literature review, determine the mating system that you will use, determine selection criteria, determine how progress will be monitored, examine the possibility of ancillary studies, and plan for periodic reassessments.

The first step is to conduct an industry assessment. This includes an analysis of the production system used to grow the fish, as well as the processing and marketing aspects. Most importantly, you must understand the consumers. The consumers are the people who keep you in business. You cannot sell them what they don't want to buy. Consumers' demands dictate what you need to do, not what you want to do. A clever advertising campaign can alter their perceptions and demands, but do not conduct your breeding program first and hope that you can change their minds second.

Part of the assessment involves determining what traits are truly important and what are not. Most of this boils down to economics. You need to improve traits that will increase profits, but you must ignore those that will not. To do this you must understand the costs of raising fish and determine how you are paid for the fish. This may sound silly, but it is not. For example, catfish farmers are paid by the pound (kilogram). Processing plants pay no bonus for albinos or for fish with large dressing percentages. Consequently, it makes little sense to incorporate these traits into a breeding program. Why spend effort to improve fillet yield or to decrease fat content when you are paid at a uniform rate? Granted, the processing plants would appreciate fish with better dressing percentages and those with less fat because they would make more money. Consumers would also appreciate fish with less fat and with greater dressing percentages because the price that they pay may decrease slightly and they will feel that they are purchasing a better quality product. But until processing plants or grocery stores pay premium prices for fish with improved dressing percentages, with lower fat content, or with cosmetic appeal (pretty skin color), why spend effort and money to improve these traits? Improve only those traits that can increase your income.

When making these decisions, future trends or desires should be anticipated, if that is possible. Just because cosmetic traits carry no premium in today's marketplace does not mean that they will not in the future. If marketing research suggests that they might, and if you are a gambler, you can incorporate these traits in a breeding program and be in a position to corner future markets.

A second aspect of the industry assessment is to determine your role in the industry: are you going to produce fingerlings for your use, or are you going to sell fingerlings? This decision determines how much money you should invest in the program. It also helps you define your goals.

If you are a small-scale farmer, you should not worry about initiating a breeding program. You should simply purchase your fingerlings from a reliable source. Large-scale farmers and those who sell fingerlings should conduct selective breeding programs. Fingerling farmers who conduct breeding programs and produce genetically improved fish should be able to increase their share of the business. Ads promoting genetically improved fish are becoming common in trade magazines.

The final facet of the industry assessment is financial: You must incorporate financing into your assessment. If you can not or will not invest the needed capital and effort into the program, it will not produce the desired results. Do not bother setting up a breeding program unless your financial backer or the company for which you work is dedicated to that program. If they are not 100% behind you, they may pull the plug on the program when other issues arise.

The second aspect of a breeding program that must be considered is the goal(s) and how they will be measured. You must establish precise, genetically feasible goals that are compatible with the conclusions from your industry assessment. When you establish your goals, carefully consider the traits that you wish to improve. You can conjure up an impressive wish list, and you should, because this will lessen the likelihood that you might forget to include an important trait. Once the list is created, carefully consider every goal and decide which one(s) is truly important. An excellent example of a breeding program's goals and the plans to achieve those goals is that outlined by Gall (1979) to improve egg size, number of eggs/kg female, hatchability, and weight at 1 year in rainbow trout.

The reason that you must pare the wish list to a bare-bones minimum is that you can apply only so much selection pressure to a population. The efficiency with which a trait is improved is inversely related to the number of traits that are incorporated into the program. Because of this, choose only traits that are truly important. For example, I have heard of several proposals to improve fecundity in channel catfish—the logic being that fewer females would be needed to reach production quotas. But this goal is not that important. Channel catfish are already quite fecund; they produce about 8300 eggs/kg, so a 4-kg female will produce over 33,000 eggs, 90+% of which will be fertile and hatch. Increasing fecundity is simply not important. If you want to produce more eggs, spawn more females. It is quicker, cheaper, and the results are instantaneous.

What about improving traits such as dressing percentage and feed conversion. Breeding programs have improved dressing percentages and carcass composition in livestock and poultry, and h^2s suggest that such traits could be improved in fish (Table 4.1). However, until processing plants pay premium prices for fish that have a certain dressing percentage, why incorporate this trait into a breeding program? You could succeed, but will you receive any return on your investment?

The idea of improving feed conversion is most appealing. Fish that have a lower feed conversion convert cheap feed into expensive flesh more efficiently, which increases profits. Since feed costs account for about 40% of annual operating costs in U.S. catfish farming, such a goal would seem like one which would be at the top of everyone's list. However, as desirable as this goal is, it is one which probably should be omitted at present. The major reason is the difficulty involved in measuring this trait, which means that it would be difficult to improve. It is relatively easy to measure overall feed conversion for the fish in a given pond. But how do you measure feed conversion for individual fish? Unless they are grown separately, you cannot measure exactly the feed consumed by each fish, so you cannot obtain an individual measure of feed conversion. If the improvement of this trait is of utmost importance, it might be possible to improve this trait via between-family selection.

Improving feed conversion is not as important as improving growth. That is the one trait that all food fish farmers can agree upon. After all, a farmer gets paid by the kilogram. As long as a kilogram of fish is worth more than the feed it took to produce that kilogram, lowering feed conversion will be less important than increasing growth rate. However, it is likely that feed conversion will be lowered via indirect selection during a program to improve growth rate. Burch (1986) found that improved growth rate of genetically improved channel catfish was due both to improved feed conversion and also to increased feed consumption.

All traits should be considered when compiling a wish list of traits that should be improved. For example, most farmers probably do not consider harvesting costs when they calculate their enterprise budgets.

This is a mistake, because it can cost between $0.11 and $0.22/kg to harvest channel catfish. Improvements in this area could increase profits significantly. Several studies have found that harvestability has a genetic component (Wohlfarth *et al.* 1975B; Chappell 1979; Tave *et al.* 1981; Dunham *et al.* 1986), and a breeding program to improve harvestability would be far simpler than one to improve feed conversion.

Some traits that make great sense should be avoided because it is easier to improve them by environmental manipulation than by breeding programs; two good examples of this type of trait are early viability and disease resistance. Improving early viability is a logical goal because it will result in more fingerlings, which will lower fingerling production costs. However, it is probably more efficient to accomplish this by manipulating the environment than by manipulating the gene pool. Early survival is strongly influenced by management—water quality, nutritionally complete feeds, feeding practices, control of predaceous insects, etc. If these aspects of husbandry are better controlled, early survival will improve dramatically and instantly, and the improvements will be less expensive and will be realized more quickly than can be achieved by breeding.

Improving resistance to diseases is a very logical goal and one that virtually everyone would include on a wish list, but it is a goal that should be removed from most lists. Disease resistance is not usually a qualitative phenotype that can be fixed. Disease resistance is expressed as an either/or phenotype in individual fish (a fish either gets sick and dies or it does not and lives), but disease resistance often seems to have a quantitative genetic basis. This means that improvements will be slow and gradual, so that after several generations of selection, the population will be more disease resistant, but the disease will still be able to cause trouble. If disease resistance has a quantitative genetic mode of inheritance it will be nearly impossible to produce true-breeding, disease-resistant strains. Furthermore, the disease resistance may be for only a single strain of the disease, so a mutant strain of one introduced from another locale will still cause severe problems.

On the other hand, if resistance to a particular disease is controlled by a single gene, then true-breeding, disease-resistant populations could be produced in only 1 to 2 generations. If this is the case for a disease, this goal should be included, unless there are negative pleiotropic effects associated with the disease-resistant genotype.

The best way to control diseases is by good management. However, this does not mean that genetic manipulation will not improve fish health. Genetic engineering will create vaccines that will enable fish farmers to inoculate their fish against virtually all important diseases within the next 10–15 years. Vaccines against some important diseases are already used, and others are being field-tested. One day, a single vaccine may inoculate fish against four or five diseases. That does not mean that disease problems will be a nightmare of the past. I guarantee that when the diseases that produce insomnia are controlled, another will rear its ugly head and become the bogeyman that keeps fish farmers awake at night.

The way goals will be measured must also be determined in advance. You need to determine the financial aspect of the measurements and must also determine the biological efficiency of the measurement. For example, if you want to improve growth, how will growth be measured? Growth could be improved by selecting for increased total length. It is rather easy to measure accurately hundreds of fish, so this is fairly simple and can be relatively error free. But will it accomplish the goal of improving growth? It will, because the fish will become longer; however, processing plants and customers buy their fish by the kilogram, not by the millimeter. Fortunately, length and weight are highly correlated (Table 4.5), so improvements in length will improve weight, but this will not maximize weight gain. Additionally, this program will produce longer fish, and this body shape may be less desirable. Finally, selection for total length includes head and tail lengths. Improving these are not going to improve profits; few eat fish tails.

An alternate way to measure this goal is to weigh fish to the nearest 0.1 g. This makes sense because fish are sold by weight. However, it is more difficult to obtain hundreds of accurate weights. Modern electronic balances have made this task easier, but it is still more costly than measuring the fish. Inaccuracies can also occur, because when a fish is weighed, you weigh feed in its stomach, feces in its intestine, and water in its mouth and gills. Excess weight caused by these variables are desirable when fish are sold to processing plants, but in a breeding program they are V_E, not V_A.

In addition, body weight also includes fat deposits, and few consumers dote on fried fish fat. There is an unfortunate positive correlation between growth rate and fat in animals. Swine breeders have had to devise breeding programs to select simultaneously for increased muscle gain and decreased fat deposits. Fish breeders will ultimately have to do this too.

If you do select for increased weight gain, head weight will also be included, and few want fish with larger heads. Some method of determining muscle mass will have to be devised. Such a program would select for increased muscle weight but would not increase fat or head weight.

One final aspect of selecting for increased growth must be determined. When will selection occur? Growth is not constant. Growth curves for fish look like a lazy S ($\int$); there is a lag phase when growth rate does not increase much, and there is a log phase when growth rate increases rapidly. Similar intensities of selection will produce different response, depending on when they occur; selection for increased growth rate will produce little gain at some ages, while the same intensity of selection will produce tremendous gains at other ages. Selection for increased growth rate may be most efficient if it is done more than once.

When compiling the wish list, avoid exotic, sexy traits—traits that are appealing but that have no economic value. Many people are fixated on cosmetics. This does not mean that they are unimportant. If consumers are willing to pay more for beauty, then this becomes a most important goal. For example, cosmetics is important for the U.S. poultry industry. All chickens and turkeys that are raised for meat have white feathers. Even though U.S. consumers do not purchase unplucked fowl, they prefer white-feathered chickens. This is because dark feathers produce spots of melanin in the skin, and when dark feathers are plucked, a dark spot remains in the skin. Even though this does not affect flavor or wholesomeness, it is unappealing, and the U.S. poultry industry is astute enough to realize that visual appearance is more important than taste when it comes to consumer appeal.

However, if the goal will not improve profits or if it is a goal that is simply a "fun project," you must veto it. Breeding programs must remain simple, because the fewer traits that are involved, the greater the rate of progress for all traits.

Before conducting a breeding program, you should conduct a review of the literature to discover what others have done. If you can acquire information this way, you will not have to re-invent it or repeat the mistakes that others have made. One bit of information that will be helpful is the h^2 of the trait. It is not mandatory to determine a trait's h^2 before embarking on a breeding program, but this information can help clarify the decision-making process. Occasionally, you will have to fly by the seat of your pants because nothing is known about your trait, but you may be able to transfer information from another program or from another species. A good guesstimate is better than nothing.

Before you begin, you must determine what mating system you will use. How will you achieve the desired results: crossbreeding, selection, or inbreeding? A trait's h^2 can provide guidelines that can help in this decision-making process. If selection will be used, will it be individual selection or family selection? Will this program combine selection and crossbreeding? Is crossbreeding the way to achieve the desired results? Or should you create inbred lines for crossbreeding?

You may need to utilize some of the new biotechnology breeding programs to achieve a desired goal. For example, a major goal may be to produce a monosex population or to produce sterile fish. If these goals are included in a breeding program, the creation of sex-reversed broodstock or supermales or the incorporation of chromosomal manipulation will be needed. Such programs can either stand alone or be utilized following crossbreeding or selection.

Most programs will probably involve selection. Selection occurs in all hatchery populations—even if it is not desired. Every aspect of management and the farm itself is selecting for or against certain alleles. This is called domestication, and it cannot be prevented. If domestication is analyzed, it can be used to improve the population by increasing growth rate, increasing disease resistance, improving docility, and improving reproductive success. This is the major reason why farm populations outperform wild populations in yield trials. On the other hand, if domestication is not done properly, it can decrease yields and ruin the population.

Understanding how domestication works can save lots of money and effort. Why conduct a selective breeding program to improve a trait if domestication will do it for free? For example, the breeders who conducted the selective breeding program with coho salmon (Hershberger 1988; Hershberger *et al.* 1990A, 1990B; described in Chapter 4) wanted to improve percent smoltification. However, an examination of the standard management practices revealed that they did not need to include this goal in a selective breeding program because domestication was already selecting for increased smoltification. The breeders realized that if smoltification were heritable, domestication was a selective breeding program and would achieve the desired results for free.

Can domestication selection ruin a population? The answer is a definite "yes." For example, the standard practice of selecting for the largest female catla and rohu in India decreased growth rate, because these fish were actually the oldest and slowest growing females (Eknath and Doyle 1985). This practice may be more widespread than is imagined. Many catfish farmers in the U.S. harvest their fish by partial harvesting and understocking ("topping") a pond for 5 to 7 years. After 5 to 7 years, the water is finally drained, and the remaining fish are harvested. At this point, a number of very large fish are removed, and they are often retained as broodfish. Unfortunately, this practice may select for decreased growth rate since these fish are likely to include slow growers. It probably selects for escape artists, since these fish have escaped a seine for 5–7 years.

When conducting a selective breeding program, you must be aware of all aspects of genetics and know how they can alter the outcome of the project.

Each time you select your fish you must consider how it will affect N_e. It does little good to spend years of effort to improve your stock if the selection program produces high levels of inbreeding and genetic drift loses valuable alleles. The N_e must be considered when you establish a cutoff value and select the fish. If the cutoff value is too high, you will automatically create a bottleneck. You also have to consider how many parents produced the select population. You may retroactively reduce N_e in the previous generation if your select population was produced by only a few matings.

Care must be taken to prevent V_E from confounding V_A during the selection program. Feeding rates, food particle size, age and size of brooders, and spawning date must not be allowed to obscure genetic differences. If these factors are not properly controlled, selection may not work.

Selective breeding programs are experiments—whether they are conducted on a farm or at a research station—and you must be able to assess the progress that has been made. Climactic conditions can become worse during a selective breeding program, and yields in the select line can decrease. Is the decrease due to the weather or due to a poorly run breeding program? Management skills can improve during a breeding program. Are increases in yields in the select line due to improvements in management or to the breeding program? There is no way to answer these questions unless there is benchmark population with which you can compare the genetically improved line.

This benchmark population can be a control that you maintain, or it can be a standard population that is maintained elsewhere. Before embarking on a long-term and expensive breeding program, the way in which progress will be monitored and measured must be determined. A breeding program is an investment of time and money. If you can not evaluate the return on your investment, do not begin.

Finally, you must conduct periodic reassessments of the program, because markets and consumer demands change. If you blindly stick to a particular program, you may find that you achieved your goal(s) but that no one wants what you produced. Swine and cattle breeders are changing some of their goals and programs to adjust to consumer demands for leaner meats.

The following is an example of a simple, bare-bones selective breeding program. This program is not free—there will be an investment in terms of land, labor, and capital. But then virtually anything worth having costs money. The costs involved in breeding programs should not be viewed as impediments; rather they should be viewed as another aspect of management. Good water quality management is not inexpensive. High quality, nutritionally complete rations are not cheap. Proper health management and disease control can be quite costly. Breeding programs are no different.

The program that I am going to outline is one to improve weight and harvestability in channel catfish. This program is simply an example of how a selective breeding program can be integrated into everyday management on a fish farm. This program is not complete, nor is it the only way such a program could be conducted.

Growth rate is the most important trait that can be improved in channel catfish, since it will produce the greatest benefits. Because of the way channel catfish are cultured in the U.S., harvestability is a second trait that would be worth improving. The water in typical catfish production ponds in the southeastern U.S. is not drained at harvest. The pond is filled, and the water is not drained for 5–7 years. The ponds are seined several times each year, and marketable fish are harvested. One of the reasons that the water is drained every 5–7 years is that too many fish escape the seine, and eventually there are many very large catfish that outcompete the smaller fish for feed. Consequently, it would improve management, as well as production costs, if harvestability could be improved. Other traits could be improved—egg production, body conformation, percent body fat, dressing percentage—but if only growth and harvestability are entered into the breeding program, the rate of improvement for both will be fairly rapid.

The most efficient way to improve weight gain is to select for weight gain and not for increased length. Catfish are sold twice: the first time is when they are fingerlings, and the second is when they go to the processing plants or to live haulers. Because of that, selection for increased weight should be done twice. Additionally, this would enable selection to be performed at two points on the growth curve, which should improve growth at a faster rate. Fingerlings are traditionally sold on an "inch" (length) basis, but fingerling producers now add an additional charge for weight, which means heavier fingerlings bring a premium price. Because of the large correlations between length and weight, selection for weight at the fingerling stage will also improve length.

To begin the program, a series of 0.04- to 0.1-ha earthen ponds should be constructed at the fish farm; the exact number of ponds depends on the magnitude of the project. The program could be conducted in the production ponds, but the construction of ponds for the breeding program "forces" you to invest time and effort in the breeding program, and once such an investment is made it also "forces" you to continue the project. Additionally, the production ponds are used to produce the crop, and it is probably advantageous not to have a conflict between farming and the breeding program, because the breeding program will lose every time.

To initiate the program, spawn the broodfish that will be used to produce your fingerlings, using normal hatchery management techniques. You may wish to acquire a new stock to initiate the breeding program, but that is not necessary unless problems with the existing stock have been detected. When the fish spawn, each egg mass should be segregated so that the genetic integrity of each family can be maintained.

It is important to minimize age differences during the breeding program so that age-related size differences do not obscure and confound genetically-related size differences. The optimal age cohort is 24 hours, and 48 hours should be used only as a last resort. To produce each age cohort, choose days during which at least 5 egg masses were collected; a minimum of 5 such cohorts should be created for the breeding program. If five cohorts, each containing 10 eggs masses, can be created, 100 broodfish will produce the offspring that are entered into the breeding program, and N_e will be approximately 100, which should suffice for most programs.

Each egg mass should be incubated so that when the eggs hatch, the fry are segregated from other families. When the fry reach swim-up, a random and equal sample of fry should be obtained from each family within each cohort. The exact number depends on the number of families, the size of the pond in which the fry will be stocked, and the stocking density that will be used to produce the fingerlings.

Fry from each cohort are stocked into a single pond. The stocking density and management should be identical to that normally used to produce fingerlings. At the end of the summer, obtain a random sample of 100–200 fingerlings (the exact number is not crucial) from each pond and weigh each fish to the nearest 0.1 g (the fingerlings from each pond are isolated and are not mixed with those from other ponds). This information is used to decide where the cut-off value is to be placed for each cohort. It is unlikely that the fish can be easily sexed at this time. If they can be sexed, selection will be more efficient, although it will be more laborious. If the fish are sexed, select the top 30% of the males and the top 30% of the females from each cohort for stocking in the genetic production ponds. If the fish can not be sexed, save the top 40% of the fingerlings from each cohort. The exact number of fish that needs to be saved depends on the size of the genetic production ponds and the stocking density that will be used to produce the fish. It may not be necessary to sample every fish in the pond to obtain the select fingerlings. Select fish from each age cohort are stocked in a single genetic production pond.

The fish are then raised using normal food fish production techniques, except that the water will be drained in the fall and all fish will be harvested. Just prior to harvest, obtain a random sample of 100–200 fish from each pond and weigh each fish to the nearest 0.1 g to determine where the cut-off value will be placed for each cohort. This time, sex each fish before it is weighed. At harvest, seine each pond before the water is drained. This simulates the typical "topping" harvest practice. The fish that are caught are those that are most seineable. If enough fish are caught, a second seining is not necessary. This harvest is the process that selects the most catchable fish while culling the least catchable. Each pond should be seined as many times as is needed, in order to create the desired size of the select population.

Sex and weigh the captured fish, and save the top 10–20% of the males and the top 10–20% of the females from each cohort for future broodfish; cull the remainder. You should save a total of at least 250 males and 250 females for the breeding project. Make sure there is equal representation by each cohort.

The culled fish can either be sold for food or can be kept and used for future broodfish to produce the farm fish. Even though these fish are culls, they were selected for fast growth as fingerlings. If they are saved for use as broodfish to produce the farm's fish, some of the progress will be transferred to the farm population. The farm population will improve, but it will be about one generation behind the select line. If you can afford to weigh every fish, "excess" select broodfish can be used as farm broodfish to produce fish that will be grown for food on the farm. This will transfer increased growth rate to the farm population immediately.

After the fish in each pond are selected, the water should be drained, and the remaining fish should be harvested. These fish can also be used as future broodfish for the farm. Any particularly large fish can be put with the select line, as was described for modified independent culling.

These select broodfish are then raised until they can be spawned and the process is repeated. This program can be repeated each year, so it is possible to have 2–4 select lines, or it can be done once every 3–4 years. The number of lines that are produced depends on the financial resources that can be committed to the program.

Several options are available to determine the progress that has been made: The first is to maintain a randomly spawned control line and to compare the growth of the two lines each generation. The second is to use someone else's fish as your control. You can purchase the control fingerlings from a nearby farmer and compare the growth of your select fish to his. Just make sure that this person is not selecting his fish too. If a reliable tagging system can be developed, the control fish can be stocked in each pond and can serve as an internal control for each age cohort. This will eliminate the need to raise several replicate ponds of control fish.

The program should be reassessed every year to evaluate the program and to determine if it needs to be modified. The program is not a quick fix that will produce the superfish that every catfish farmer dreams about. However, if it is done correctly, it is likely that over a 20-year period (4 generations), yields could be increased by 20–40%, and that is significant.

Crossbreeding is the other major breeding program that can be used to improve productivity. This is not as orderly and as predictable as selection, but the results can be almost as dramatic. The best way to begin to use a crossbreeding program is to find two strains that have desirable qualities and cross them to try and produce F_1 hybrids that possess all desirable qualities. There is no guarantee that you will obtain outstanding F_1 hybrids by mating two outstanding strains, but the results that have been observed with channel catfish suggests that this approach improves the odds of producing good hybrids.

The major problem with using crossbreeding is that the mating combinations that you must evaluate can become unwieldy. For example, if you wish to examine the hybrids from the four top strains, you must produce 12 different F_1 hybrids. There is no way to predict which of the hybrids will be best, which means that a crossbreeding program that does not include all reciprocal combinations may never uncover the best hybrid. If previous research has demonstrated that a particular hybrid outperforms all others, the prudent farmer should evaluate this hybrid first, and if the results are encouraging, this simple breeding program could suffice for 10 or 15 years.

Although the production of superior F_1 hybrids is basically fortuitous, an examination of the fish's biology, etc., coupled with proper planning, can increase the likelihood of success. Research suggests that crossbreeding hatchery strains produces better hybrids. In general, intraspecific hybrids are better than interspecific hybrids; this is not always true, but the likelihood of successful matings and of producing viable F_1 hybrids increases at the species level. Inter-family crosses are generally unlikely to succeed.

Crossbreeding can be used to produce fish other than F_1 hybrids. Classic genetic theory states that F_1 hybrids should exhibit maximum heterosis. However, recent research by Jayaprakas *et al.* (1988) and Tave *et al.* (1990D) suggested that heterosis can be greater in F_2 and backcross hybrids because of maternal heterosis. If maternal heterosis exists in the species you are culturing, yields can be increased by using crossbred dams. Although it takes two generations before the F_2 or backcross hybrids are produced, once this occurs, the use of crossbred dams is a fairly simple breeding program.

If one strain has a most desirable trait, you can use introgressive hybridization (repeated backcrossing) to transfer that trait to your strain and thus improve its value. This technique has been used with great success in other forms of agriculture. Behrends and Smitherman (1984) used it to transfer the alleles for red body color from one line of tilapia to another.

Crossbreeding can be used to create synthetic lines. If interspecific hybrids are fertile, it can even be used to create synthetic two-species hybrids. This type of breeding program will be common in 50 years, because it can create truly outstanding lines; this technique is the breeding program that has been used to produce many new breeds of livestock and pets. The only unfortunate aspect of this breeding program is that it takes three generations until you produce your first crop of "synthetic" fish.

Crossbreeding is a good technique that can be used to prevent inbreeding from depressing growth rate during grow-out. If F_1 hybrids are cultured, $F = 0\%$ as long as the two strains or species are not related.

Fingerling producers may want to incorporate crossbreeding as the final step of a selective breeding program. Selective breeding programs take time and money, so it is counter-productive to sell your genetically improved stock to your competitors. If you sell only F_1 hybrids, you make it more difficult for your competitors to use your fish as broodfish. Generally, they will have to sell the fish because F_1 hybrids usually do not breed as expected. This is a common practice in other forms of farming. The only exception to this would be that if the F_1 hybrid dams exhibited maternal heterosis, you would sell valuable broodstock to your competitors.

If the facility is large enough, you may wish to select simultaneously in two strains or species that have been shown to produce outstanding F_1 hybrids. If this approach is used, you can exploit V_A via selection and can then hybridize the strains to exploit V_D. As a bonus, the hybrids will have F = 0%, so inbreeding depression will not be a problem.

The relatively new techniques in biotechnology that are being developed and perfected will enable the larger and wealthier farmers to produce unique fish for grow-out or for other purposes. Farmers are now raising monosex populations of female trout and salmon to eliminate the occurrence of precocious males. Some farmers routinely produce triploid grass carp because of legal restrictions against diploid grass carp. Triploids are more expensive, but since the choice is either to use the more expensive triploids or to use nothing, these genetically engineered sterile fish have a definite niche in the fish farming community.

The creation of triploids should also be a priority for public game and fish agencies. The practice of stocking hybrids into public waters may be counterproductive if these hybrids backcross with the stocks that are the object of conservation management. If this occurs, it could lead to the destruction of the very resource that the project is designed to protect. The only hybrids that should be stocked are triploids, because they will not be able to reproduce with the native stocks. Curtis *et al.* (1987) showed that this approach is possible.

The use of anabolic steroids to produce monosex populations is a program that is fairly routine, but it is also an emotional issue because consumers are continually frightened by stories about how everything that they eat and breathe causes cancer, heart disease, or the heartbreak of psoriasis. Pressure by consumer groups to prohibit the use of hormones and other such compounds in farming is mounting, so an industry that depends on this practice may collapse.

Anabolic steroids can be used to produce sex-reversed broodstock that are then capable of producing monosex populations. Such work will undoubtedly continue in the future, and if this approach is integrated into a breeding program as the final step, farmers can not only exploit an additional source of increased production, but the breeder will not have to sell his prize broodfish to his competitors.

Genetic engineering is an aspect of genetic improvement that is simply too sophisticated for the average farmer. The laws and regulations that govern this type of research are staggering. Additionally, there is no guarantee that the genetically altered fish will be able to transfer the desired gene to their offspring, and even if they do, there is no guarantee that it will improve growth rate or other production phenotypes. This type of research should be left to research organizations and to agribusinesses. In the future, farm populations of fish will be genetically engineered so that they can more efficiently digest and utilize nutritionally poor feedstuffs. The insertion of growth hormone genes or other such genes will enable these fish to grow faster, but such developments are still in the future.

One area that will benefit from genetic engineering is fish health management. It is likely that within the next 10 years, genetic engineering will create vaccines that inoculate fish against many of the major viral and bacterial diseases. It might even be possible to create immunity against parasitic diseases. But once again, this type of research should be done only by research institutions and by agribusinesses.

A schematic representation of how breeding programs, or the lack of them, can alter productivity is shown in Fig. 6.2.

Unfortunately, there is no one right way to manage a population. No simple formula or recipe exists that will produce better populations. Each hatchery and each population requires a program tailormade to a specific set of circumstances, so a hatchery manager should seek advice about possible options and then make realistic decisions based on hatchery size, budget, species, and production goals. The ideas that have been outlined in this book will provide a good starting point, but they are not absolute guidelines.

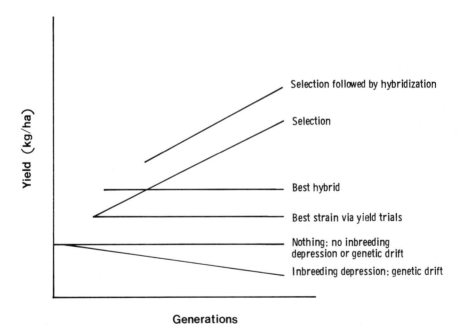

Figure 6.2 Schematic diagram of yield over time using various breeding programs.

Whether your budget is generous enough to enable you to hire a geneticist or whether it will only enable you to acquire new brooders every few years, the genetic aspects of broodstock management must become an integral part of hatchery management. The alleles in a population determine that population's potential. If you damage the potential, you will make it difficult to achieve production quotas. But if you manage and exploit it properly, you will exceed production quotas. Poor quality fish perform poorly; good quality fish give good results.

APPENDIX A

Scientific Names of Fishes Cited in Text and Tables

African catfish, *Clarias gariepinus*
Amago salmon, *Oncorhynchus rhodurus*
Angelfish, *Pterophyllum scalare*
Arctic charr, *Salvelinus alpinus*
Asian catfish, *Clarias batrachus*
Atlantic salmon, *Salmo salar*
Ayu, *Plecoglossus altivelis*
Bighead carp, *Hypophthalmichthys nobilis*
Bigmouth buffalo, *Ictiobus cyprinellus*
Black buffalo, *Ictiobus niger*
Black sea bream, *Acanthopagrus schlegeli*
Blue catfish, *Ictalurus furcatus*
Blue poecilia (also called steel-blue limia), *Limia caudofasciata*
Brook trout, *Salvelinus fontinalis*
Brown trout, *Salmo trutta*
Catarina pupfish, *Megupsilon aporus*
Catla, *Catla catla*
Channel catfish, *Ictalurus punctatus*
Chinook salmon, *Oncorhynchus tshawytscha*

Chum salmon, *Oncorhynchus keta*
Coho salmon, *Oncorhynchus kisutch*
Comanche Springs pupfish, *Cyprinodon elegans*
Common carp, *Cyprinus carpio*
Conger eel, *Astroconger myriaster*
Convict cichlid, *Cichlasoma nigrofasciatum*
Crimson sea bream, *Evynnis japonica*
Cutthroat trout, *Oncorhynchus clarki* (formerly *Salmo clarki*)
Dollar hatchetfish, *Sternoptyx diaphana*
Dwarf gourami, *Colisa lalia*
Eastern mosquitofish, *Gambusia holbrooki*
European catfish, *Silurus glanis*
Eye-spot rasbora, *Rasbora dorsiocellata*
Fathead minnow, *Pimephales promelas*
Filefish, *Stephanolepis cirrhifer*
Flounder, *Platichthys flesus*
Fourspine stickleback, *Apeltes quadracus*
Freshwater goby, *Gobionellus shufeldti*
Golden shiner, *Notemigonus crysoleucas*
Goldfish, *Carassius auratus*
Grass carp, *Ctenopharyngodon idella*
Guatopote culiche, *Poecilia prolifica*
Guatopote del Mocorito, *Poecilia lucida*
Guppy, *Poecilia reticulata*
Half-banded barb, *Barbus semifasciolatus*
Japanese eel, *Anguilla japonica*
Jewel tetra, *Hyphessobrycon callistus*
Lake trout, *Salvelinus namaycush*
Largemouth bass, *Micropterus salmoides*
Leon Springs pupfish, *Cyprinodon bovinus*
Loach, *Misgurnus anguillicaudatus*
Masu (cherry) salmon, *Oncorhynchus masou*
Medaka, *Oryzias latipes*
Mexican cave characin, *Astyanax fasciatus* (synonymous with *Astyanax mexicanus* and *Anoptichthys antrobius*)
Molly, *Poecilia sphenops*
Montezuma swordtail, *Xiphophorus montezumae cortezi*
Mosquitofish, *Gambusia affinis*
Nibe-croaker, *Nibea mitsukurii*
Northern pike, *Esox lucius*
Pecos gambusia, *Gambusia nobilis*
Pink salmon, *Oncorhynchus gorbuscha*

Plaice, *Pleuronectes platessa*
Platyfish, *Xiphophorus maculatus*
Rainbow trout, *Oncorhynchus mykiss* (formerly *Salmo gairdneri*)
Red sea bream, *Pagrus major*
Rohu, *Labeo rohita*
Rudd, *Scardinus erythrophthalmus*
Sailfin molly, *Poecilia latipinna*
Siamese fighting fish, *Betta splendens*
Siberian whitefish, *Coregonus peled*
Silver carp, *Hypophthalmichthys molitrix*
Sockeye salmon, *Oncorhynchus nerka*
Steelhead trout, *Oncorhynchus mykiss* (formerly *Salmo gairdneri*)
Striped bass, *Morone saxatilis*
Sumatran tiger barb, *Barbus tetrazona*
Swordtail, *Xiphophorus helleri*
Threespine stickleback, *Gasterosteus aculeatus*
Velvet belly shark, *Etmopterus spinax*
Virolito (common name in Argentina), *Apareiodon affinis*
Viviparous eelpout, *Zoarces viviparus*
White bass, *Morone chrysops*
White catfish, *Ictalurus catus*
White sturgeon, *Acipenser transmontanus*
Winter flounder, *Pseudopleuronectes americanus*
Zebra danio, *Brachydanio rerio*

Note: I have used the scientific names when referring to tilapia, because most aquaculturists use the scientific names. Additionally, many culturists give their tilapia exotic sounding names for marketing purposes, and many do not know what the proper common names are. Unfortunately, using the scientific names is also confusing because the scientific names have changed several times in recent years, and they may change once again. I use *Tilapia* because that is what is given in "Common and Scientific Names of Fishes from the United States and Canada" (Robins *et al.* 1991). The following are the scientific names I used, scientific names accepted by others, and their common names:

Tilapia aurea (= *Oreochromis aureus*), blue tilapia
Tilapia hornorum (= *T. urolepis*) (= *Oreochromis urolepis hornorum*), Wami tilapia
Tilapia mossambica (= *Oreochromis mossambicus*), Mozambique tilapia
Tilapia nilotica (= *Oreochromis niloticus*), Nile tilapia

APPENDIX B

How to Describe a Quantitative Phenotype

A quantitative phenotype is described by its central tendency and the distribution around that tendency: mean, variance, standard deviation, coefficient of variation, and range. These values enumerate a population's average phenotypic value, the variability that exists in a population, the distribution that individual phenotypes have relative to a population's average, and the difference between the extreme values. To describe a quantitative phenotype accurately and to exploit it in a scientifically sound breeding program, it is necessary to calculate the above values.

What these values mean and how they are calculated will be described by an example. Say a catfish farmer randomly samples 10 channel catfish fingerlings from a pond in order to estimate fingerling length in the population in that pond. (To produce accurate estimates of length, he should sample between 30 and 60 fish, depending on the variance of the sample, but a sample that big would be too unwieldy for this discussion.) He measures the fish to the nearest centimeter, and his data are

12, 12, 13, 14, 14, 16, 16, 17, 17, 19

What is the mean of this sample? The mean is the arithmetic average. It is determined by adding the values of the observations and then by dividing the total by the number of observations:

$$\overline{X} = \frac{\Sigma X}{N}$$ (A.1)

where $\overline{X}$ is the mean, ΣX is the sum or total of all the measurements (Σ is the symbol for add), and N is the number of measurements.
  In this example

$$\Sigma X = 150 \text{ cm}$$

$$N = 10$$

$$\overline{X} = \frac{150 \text{ cm}}{10}$$

$$\overline{X} = 15 \text{ cm}$$

The range is the difference between the extreme values that are observed, i.e.,

$$\text{range} = \text{high value} - \text{low value}$$ (A.2)

In this example, lengths ranged from 12 to 19 cm so

$$\text{range} = 19 \text{ cm} - 12 \text{ cm}$$

$$\text{range} = 7 \text{ cm}$$

The range quantifies the difference between the extreme values, but it gives no indication about the way the population is dispersed relative to the mean. For example, the following series of numbers have the same range (20), but they are dispersed differently about the mean (10):

Observation	Population 1	Population 2	Population 3
1	0	0	0
2	2	0	10
3	4	0	10
4	6	0	10
5	8	0	10
6	10	10	10
7	12	20	10
8	14	20	10
9	16	20	10
10	18	20	10
11	20	20	20

When working with a population, it is important to obtain a measure of phenotypic variability, because it provides a concise description of the phenotype in that population. As was demonstrated by the previous example, the range does not completely describe the distribution. The distribution of a phenotype is based on the mean and the variation. The variation is described by the variance and the standard deviation. It is important to quantify the variation, because it is the raw material with which breeders work and because it helps determine where cutoff values are placed during selection programs.

The variance (s^2) is the average squared deviation from the mean, and the standard deviation (s) is the square root of the variance. The symbols s^2 and s are used to designate the variance and standard deviation of a sample, because they are estimates of the variance (σ^2) and standard deviation (σ) of the population. If every individual in a population is measured, then the symbols σ^2 and σ should be used.

The variance (s^2) or average squared deviation from the mean can be estimated by using this formula:

$$s^2 = \frac{\sum x^2}{N - 1} \tag{A.3}$$

where s^2 is the variance, N is the number of measurements, and x is the deviation of each score from the mean ($X - \overline{X}$).

The deviation scores are squared for two reasons: (1) If they were not, the sum of the deviation scores would always be zero; squaring each score removes the negative values. (2) Squaring the deviation scores magnifies the extreme measurements, and that helps to differentiate distributions that are closely bunched around the mean from those that are widely scattered. In this example, the deviation scores are derived in Table A.1. Using the values derived in Table A.1, the variance is computed as follows:

$$s^2 = \frac{50}{10 - 1}$$

$$s^2 = 5.6$$

While Eq. (A.3) clearly demonstrates that the variance is the average squared deviation from the mean, it is a cumbersome formula. A more usable formula, which is algebraically equivalent to Eq. (A.3), can also be used to compute the variance:

$$s^2 = \frac{\sum X^2 - \dfrac{(\sum X)^2}{N}}{N - 1} \tag{A.4}$$

Table A.1   Calculation of Deviation Scores[a]

X	x or $(X - \bar{X})$	x^2
12	$(12 - 15) = -3$	9
12	$(12 - 15) = -3$	9
13	$(13 - 15) = -2$	4
14	$(14 - 15) = -1$	1
14	$(14 - 15) = -1$	1
16	$(16 - 15) = 1$	1
16	$(16 - 15) = 1$	1
17	$(17 - 15) = 2$	4
17	$(17 - 15) = 2$	4
19	$(19 - 15) = 4$	16
	$\sum x = 0$[b]	$\sum x^2 = 50$

[a] The mean in this example is 15.
[b] $\sum x$ must be zero. If it is not, an error has been made.

where $\sum X^2$ is the sum of each measurement squared, $(\sum X)^2$ is the square of the sum of measurements, and N is the number of measurements. The summation $\sum X^2$ is determined by squaring each measurement and adding the squared numbers; $(\sum X)^2$ is determined by adding the measurements and then squaring the total.

In this example, the values for $\sum X^2$ and $(\sum X)^2$ are derived in Table

Table A.2   Calculation of $\sum X^2$ and $(\sum X)^2$

X	X^2
12	144
12	144
13	169
14	196
14	196
16	256
16	256
17	289
17	289
19	361
$\sum X = 150$	$\sum X^2 = 2,300$
$(\sum X)^2 = 150^2$	
$(\sum X)^2 = 22,500$	

A.2. Using the information from Table A.2, s^2 is calculated as follows:

$$s^2 = \frac{2,300 - \frac{(150)^2}{10}}{10 - 1}$$

$$s^2 = \frac{2,300 - \frac{22,500}{10}}{9}$$

$$s^2 = \frac{2,300 - 2,250}{9}$$

$$s^2 = \frac{50}{9}$$

$$s^2 = 5.6$$

The standard deviation, when used in conjunction with the mean, is the best way to describe a population. In a normal distribution: $\overline{X} \pm \sigma = 68\%$ of the population; $\overline{X} \pm 2\sigma = 95\%$ of the population; and $\overline{X} \pm 3\sigma = 99.7\%$ of the population (Fig. 4.1). The standard deviation is the square root of the variance and can be calculated by taking the square root of the values derived by formulas (A.3) or (A.4):

$$s = \sqrt{\frac{\sum x^2}{N - 1}} \tag{A.5}$$

or

$$s = \sqrt{\frac{\sum X^2 - \frac{(\sum x)^2}{N}}{N - 1}} \tag{A.6}$$

In this example, the standard deviation is

$$s = \sqrt{5.6}$$

$$s = 2.4$$

The standard deviation can be standardized so that it can be compared to those in other populations, to those at different ages, to those for different year-classes, or to those for other phenotypes. The relative standard deviation is derived by dividing the standard deviation by the mean and by multiplying that value by 100. This measure is called the coefficient of variation (CV):

$$CV = \left(\frac{s}{X}\right) 100 \qquad\qquad (A.7)$$

In this example, the CV is

$$CV = \left(\frac{2.4}{15}\right) 100$$

$$CV = 16\%$$

It is not very difficult to calculate the variance and standard deviation by hand when sample size is small, but when it is large the calculations can become tedious and unmanageable. Fortunately, these values can be calculated easily and quickly by personal computers and even by inexpensive pocket calculators.

Glossary

A

Additive gene action The type of gene action where each allele at a given locus has an equal unidirectional phenotypic effect.

Additive genetic variance (V_A) The portion of total phenotypic variance that depends on the additive effects of the genes. The proportionate amount of additive genetic variance is heritability.

Allele An alternate form of a gene.

Anabolic steroids Androgens and estrogens.

Aneuploid A cell that has an unbalanced number of chromosomes (not an even multiple of N). It could have an extra chromosome (or part of a chromosome) so that it is either $N + 1$ or $2N + 1$, or it could be missing a chromosome so that it is either $N - 1$ or $2N - 1$.

Androgens 1) Anabolic steroids that result in the production of masculine traits. 2) Fish whose chromosomes come solely from the father.

Autosomal gene A gene that is located on an autosome.

Autosomal phenotype A phenotype that is controlled by an autosomal gene.

Autosome A chromosome that is not a sex chromosome. Autosomes are the pairs of chromosomes that males and females share in common.

B

Backcross   The mating of an F_1 hybrid to one of its parents or parental groups.

Between-family selection   A selective breeding program where whole families (or random samples of the families) are either saved or are culled.

Bottleneck   A severe restriction in a population's effective breeding number.

Breed   A group of fish which have a common origin and which share certain distinguishing features that separate them from other such groups.

Breeding value   The genetic value of a fish or a population, in terms of its or their ability to transmit certain desired phenotypes. This can be for only one phenotype or for the fish as a whole.

C

Chromosomes   The structures on which genes are located. There are two types of chromosomes: autosomes and sex chromosomes.

Chromosomal manipulation   The alteration of ploidy levels to N, $3N$, or $4N$ by temperature, pressure or chemical shocks.

Codominance   The type of gene action where more than one allele at a given locus is dominant.

Coefficient of variation (CV)   The standard deviation (SD) expressed as a percentage of the mean [$CV = (SD/\overline{X})100$].

Common ancestor   A fish that is an ancestor of both parents. Parents that share a common ancestor produce inbred offspring.

Communal stocking   A stocking technique where more than one strain, race, or closely related species are stocked in the same pond to assess genetic differences.

Compensatory growth   A dramatic increase in a fish's growth rate that enables it to compensate for the lack of growth during a period of deprivation caused by factors such as overcrowding and/or under-feeding.

Confound   The inability to separate treatment effects due to a faulty experimental design.

Crossing-over   The exchange of chromosomal segments between homologous chromosomes of a chromosome pair during meiosis.

Crossbreeding   A breeding program in which fish from two different breeds are mated to produce hybrids. Although the term refers specifically to the mating of two breeds, it is also used to designate all forms of hybridization.

Cull   The removal or weeding out of individuals during selection. Fish that are culled are not allowed to reproduce.

D

Deoxyribonucleic acid (DNA)   The molecule that contains the genetic code.

Diploid (2N)   A cell or fish in which chromosomes occur in pairs.

Dominance genetic variance (V_D)   The portion of total phenotypic variance that is produced by the interaction of alleles at each locus.

Dominant allele   An allele whose phenotype is expressed when an individual is either homozygous or heterozygous at that locus.

Dominant gene action   The type of gene action where one allele is expressed more strongly than another. One allele is dominant, and the other is recessive.

Dominant phenotype   A phenotype controlled by a dominant allele.

E

Electrophoresis   A biochemical technique that deciphers protein genotypes.

Effective breeding efficiency (N_b)   The ratio of the effective breeding number (N_e) to the breeding population (N): $N_b = N_e/N$.

Effective breeding number (N_e)   The breeding (genetic) size of a finite population. It is determined by the number of breeding individuals, the sex ratio, the breeding program, the variance of family size, and previous inbreeding.

Environment   All the external factors that influence phenotypic expression.

Environmental variance (V_E)   The portion of total phenotypic variance that is due to the influence of environmental variables.

Epistasis   The interaction of two or more genes for the production of a phenotype.

Epistatic genetic variance (V_I)   The portion of total phenotypic variance that is due to the interaction of alleles between and among loci.

Estrogens   Anabolic steroids that result in the production of feminine traits.

Expressivity   The degree to which a phenotype is expressed.

F

F_1   The first filial generation. This is the first generation that is produced in a breeding program.

F_2   The second filial generation. This is the second generation that is produced in a breeding program.

F_n   The nth filial generation.

Family   A group of fish that share either one or two parents in common.

Family selection   A selective breeding program where selection is based on family means rather than on individual values.

Fitness   The viability, in terms of survival and reproduction, that one fish has relative to others in a particular environment.

Fix   An allele is fixed when its frequency is 100%.

Founder effect   The loss of genetic variance that occurs when a population is started with a small number of fish.

Full-sibs   Brothers and sisters that have two parents in common.

G

Gamete   An egg or a sperm.

Gene   The basic unit of inheritance.

Gene frequency   The frequencies of the alleles at a given locus in a population.

Genetic engineering   Molecular technique used to transfer a gene from one organism to another.

Genetic-environmental interaction variance (V_{G-E})   The portion of total phenotypic variance that is due to interactions between the genes and the environment.

Genetic drift   Random changes in gene frequency that occur as a result of sampling error (broodstock selection or shipment of fish to another hatchery).

Genetic variance (V_G)   The proportion of total variance that is due to genetics; V_G is the sum of the additive genetic variance (V_A), dominance genetic variance (V_D), and epistatic genetic variance (V_I).

Genome   A fish's genetic makeup.

Genotype   A fish's genetic makeup.

Gynogens   Fish whose chromosomes come solely from the mother.

H

Half-sibs   Brothers and sisters that have only one parent in common.

Haploid (N)   A cell or individual that has only one homologue of each chromosome pair.

Hardy-Weinberg equilibrium When a population is in Hardy-Weinberg equilibrium, gene and genotypic frequencies remain the same from one generation to the next.

Heritability (h^2) The proportionate amount of additive genetic variance: $h^2 = V_A/V_P$.

Heterogametic sex The sex that is able to produce gametes with different sex chromosomes. Males are the heterogametic sex in fish with the XY sex-determining system, and females are the heterogametic sex in fish that have the WZ sex-determining system.

Heterosis The superiority or inferiority of hybrids over the parents.

Heterozygote A fish that has two different alleles at a given locus.

Homogametic sex The sex that is able to produce gametes with only one type of sex chromosome. Males are the homogametic sex in fish that have the WZ sex-determining system, and females are the homogametic sex in fish with the XY sex-determining system.

Homologues The two chromosomes of a chromosome pair.

Homozygote A fish that has two identical alleles at a given locus.

Hybrid A fish that was produced by mating two different breeds, strains, or species.

Hybrid vigor Positive heterosis.

I

Inbred lines A strain or line of fish that has been inbred to fix certain alleles. Because the fish are homozygous at certain loci, they will breed true for those phenotypes.

Inbreeding The mating of relatives.

Independent assortment The simultaneous and uncorrelated segregation of allelic and chromosomal pairs during meiosis.

Individual selection A selective breeding program where selection is based on individual merit, without regard to family relationships.

Interploid triploid A triploid produced by mating a tetraploid with a diploid.

Interspecific hybrid A hybrid produced by mating fish from two different species.

Intraspecific hybrid A hybrid produced by mating fish from two strains, breeds, or races within a species.

Introgressive hybridization Backcrossing to one of the parental lines.

K

Karyotype A picture that shows an individual's or species' chromosomes. If the fish is a diploid, the chromosomes are paired.

L

Linebreeding An inbreeding program in which an outstanding fish is mated to its descendants to increase its percentage of an individual's or a population's genome.

Linkage Genes are linked when they are transmitted as pairs or sets because they are located closely together on the same chromosome.

Locus (plural: loci) The location of a particular gene on a chromosome. Colloquially, the terms locus and gene are used interchangeably.

M

Magnification The continuous increase in size differences among groups that have initial size differences when the groups are stocked communally.

Mass selection A synonym for individual selection.

Mean The population average.

Meiosis The process that produces gametes.

Mendelian genetics The genetics that is normally seen for qualitative phenotypes.

Migration In nature, migration is the movement of fish from one population into another. In aquaculture, it is the importation of new broodstock.

Mitochondrial DNA (mtDNA) DNA located in the mitochondria; mtDNA is inherited maternally.

Mitosis Cell division.

Monomorphic When only one distinct phenotype (e.g., wild-type coloration) or one allele at a given locus exists in a population.

Monosex population A population that has only one sex.

Multiple nursing A technique where groups of fish are treated differently (different stocking and feeding rates) prior to an experiment so that all groups will have the same average size at the beginning of the experiment.

Mutation A mistake that occurs during the replication of a gene.

N

Nick The production of superior hybrids.

Normal curve The theoretical bell-shaped distribution that describes the distribution of quantitative phenotypes in a population.

Normal distribution A synonym for normal curve.

Null hypothesis The working hypothesis in an experiment. Data gathered from the experiment allow you to either accept or reject the null hypothesis.

O

Oogenesis The production of an egg from the primary oocyte.

P

P₁ The parental generation; the first generation. The P_1 generation will produce the F_1 generation.

Penetrance The proportion of individuals that express a given phenotype when they have a particular genotype.

Pedigree A fish's family tree.

Pedigreed mating A system of mating in which each male leaves one son and each female leaves one daughter for the following generation's broodstock. The choice of the sons and daughters must be made in a random manner.

Phenotype The chemical or physical expression of a gene.

Phenotypic variance The variance that is observed or measured for a phenotype in a population.

Pleiotropy The production of additional or secondary phenotypes that occur because a gene influences two or more phenotypes.

Polar body A nonfunctional cell produced during oogenesis.

Polymorphic When two or more distinct phenotypes (e.g., albino and wild-type pigmentation) or two or more alleles at a given locus exist in a population, and the frequency of at least two of the phenotypes or alleles is greater than or equal to 0.01.

Polyploid A state where a cell or an individual contains three or more sets of chromosomes: e.g., triploid ($3N$) and tetraploid ($4N$).

Population Group of fish.

Probability The mathematical likelihood that a particular event will occur. It ranges from 0 (0%) to 1.0 (100%).

Progeny Offspring.

Progeny test A breeding technique that deciphers a fish's genotype by examining the phenotypes of progeny produced when it is mated to a test animal (a fish whose genotype is known).

Q

Qualitative phenotype A phenotype that is described, e.g., albinism.

Quantitative phenotype A phenotype that is measured, e.g., weight.

R

Range The numerical difference between the extreme phenotypes that are measured in a population.

Random mating A system of mating where fish are mated without regard to phenotype.

Reciprocal hybrids The two hybrids that can be produced by mating fish from two different groups: Group A ♀ × Group B ♂ and Group B ♀ × Group A ♂.

Recessive allele An allele whose phenotype is expressed only when an individual is homozygous at that locus.

Recessive phenotype A phenotype controlled by a recessive allele.

Recurrent selection A selection program that can be used to improve the results of hybridization. If selection is done with one of the parental groups, it is called recurrent selection. If it is done with both parental groups it is called reciprocal recurrent selection.

Reduction division The separation of homologous chromosomes of each chromosome pair during meiosis.

Ribonucleic acid (RNA) There are several types of RNA: messenger RNA (mRNA) which carries a gene's message to the ribosomes (organelles in the cytoplasm) where the protein that the gene encodes will be produced; transfer RNA (tRNA) which attach to amino acids and brings them to the ribosomes for protein synthesis; ribosomal RNA (rRNA) which is found in the ribosomes (site of protein synthesis).

S

Segregation The separation of alleles and homologous chromosomes during meiosis.

Selection A breeding program in which the breeder chooses which fish will be the next generation's broodstock, based on some predetermined criteria.

Sex-limited phenotype A phenotype that is expressed only in one sex.

Sex-linked gene A gene that is located on one of the sex chromosomes.

Sex-linked phenotype A phenotype produced by a gene that is located on one of the sex chromosomes.

Sex chromosomes The chromosomes that determine sex.

Sex reversal The process of altering phenotypic sex by exposing fry to anabolic steroids.

Sex-reversed fish Fish that are one sex genotypically but the opposite sex phenotypically.

Shooters Fish that were able to exploit environmental variables (e.g., unequal food particle size) and grew more quickly than the rest of the fish in the population. These fish change a normal curve into one which is skewed to the right.

Shooting   The sudden and dramatic growth of a few fish in a population. Their growth is caused by environmental factors such as food particle size, stocking density, and feeding rate. This phenomenon changes a normal curve into one which is skewed to the right.

Sibs   Brothers and sisters.

Sib selection   A form of between-family selection.

Spermatogenesis   The production of sperm.

Standard deviation (SD)   The square root of the variance.

Strain   A group of fish that comes from a particular location or that are produced by a particular breeding program.

Superfemale   WW female (a female with two W and no Z chromosomes in species with the WZ sex-determining system); superfemales produce only daughters.

Supermale   YY male (a male with two Y and no X chromosomes in species with the XY sex-determining system); supermales produce only sons.

T

Test animal   A fish whose genotype is known. These fish are used in progeny tests to decipher other fish's genotypes.

Tetraploid (4N)   A cell or individual that has four sets of chromosomes.

Transgenic fish   A fish that contains a "foreign" gene as a result of genetic engineering.

Triploid (3N)   A cell or individual that has three sets of chromosomes.

V

Variance   The average squared deviation from the mean.

Variation   The measurable or observable differences among individuals.

W

Within-family selection   A selective breeding program where selection is conducted within each family, and individuals within each family are saved or are culled based on their relationship to their family mean.

Wild-type allele   The allele that produces the common phenotype. The symbol + is often used to designate the wild-type allele.

Wild-type phenotype   The common phenotype.

X

X-linked gene A gene located on the X chromosome.
X-linked phenotype A phenotype controlled by a gene located on the
  X chromosome.

Y

Y-linked gene A gene located on the Y chromosome.
Y-linked phenotype A phenotype controlled by a gene located on the
  Y chromosome.

Z

Zygote The cell that is produced by the fusion of an egg and a sperm.

References

A

Abella, T. A. 1987. Improved tilapia strains through broodstock development in the Philippines. *In* Tilapia Farming, R. D. Guerrero, III, D. L. Guzman, and C. M. Lantican (Editors). Philipine Council for Agriculture, Forestry and Natural Resources Research and Development, Bureau of Fisheries and Aquatic Resources, and Southeast Asian Fisheries Development Center Aquaculture Department, PCARRD Book Series No. 48/1987, Los Baños, Laguna, Philippines.

Abella, T. A. and Palada, M. S. 1986. Evaluation of tilapia strains and hybrids. CLSU Scientific J. *6(2)*, 39–42.

Abella, T. A., Palada, M. S., and Newkirk, G. F. 1990. Within family selection for growth rate with rotation mating in *Oreochromis niloticus. In* The Second Asian Fisheries Forum, R. Hirano and I. Hanyu (Editors). Asian Fisheries Society, Manila, Philippines.

Aida, T. 1921. On the inheritance of color in a fresh-water fish, *Aplocheilus latipes* Temmick and Schlegel, with special reference to sex-linked inheritance. Genetics *6*, 554–573.

Aida, T. 1930. Further genetical studies of *Aplocheilus latipes.* Genetics *15*, 1–16.

Al-Ahmad, T. A. 1983. Relative effects of feed consumption and feed efficiency on growth of catfish from different genetic backgrounds. Doctoral Dissertation, Auburn Univ., AL.

Aldridge, F. J., Marston, R. Q., and Shireman, J. V. 1990. Induced triploids and

353

tetraploids in bighead carp, *Hypophthalmichthys nobilis*, verified by multi-embryo cytofluorometric analysis. Aquaculture *87*, 121–131.

Allard, R. W. 1960. Principles of Plant Breeding. Wiley, New York.

Allen, S. K., Jr. 1983. Flow cytometry: assaying experimental polyploid fish and shellfish. Aquaculture *33*, 317–328.

Allen, S. K., Jr. and Stanley, J. G. 1979. Polyploid mosaics induced by cytochalasin B in landlocked Atlantic salmon *Salmo salar*. Trans. Am. Fish. Soc. *108*, 462–466.

Allen, S. K., Jr. and Stanley, J. G. 1983. Ploidy of hybrid grass carp × bighead carp determined by flow cytometry. Trans. Am. Fish. Soc. *112*, 431–435.

Allen, S. K., Jr., Thiery, R. G., and Hagstrom N. T. 1986. Cytological evaluation of the likelihood that triploid grass carp will reproduce. Trans. Am. Fish. Soc. *115*, 841–848.

Allendorf, F. W. and Leary, R. F. 1984. Heterozygosity in gynogenetic diploids and triploids estimated by gene-centromere recombination rates. Aquaculture *43*, 413–420.

Allendorf, F. W. and Phelps, S. R. 1980. Loss of genetic variation in a hatchery stock of cutthroat trout. Trans. Am. Fish. Soc. *109*, 537–543.

Allendorf, F. W. and Ryman, N. 1988. Genetic management of hatchery stocks. *In* Population Genetics & Fishery Management, N. Ryman and F. Utter (Editors). Washington Sea Grant Program, University of Washington Press, Seattle, WA.

Allendorf, F. W. and Thorgaard, G. H. 1984. Tetraploidy and the evolution of salmonid fishes. *In* Evolutionary Genetics of Fishes, B. J. Turner (Editor). Plenum Press, New York.

Allendorf, F. W. and Utter, F. M. 1979. Population genetics. *In* Fish Physiology, Vol. VIII. Bioenergetics and Growth, W. S. Hoar, D. J. Randall, and J. R. Brett (Editors). Academic Press, New York.

Andrijasheva, M. A. 1981. Methods and results of peled whitefish (*Coregonus peled*) selection. Selection for productivity and some biological traits. Biology and Selection of Fishes. Proc. Gos-NIORCH, Leningrad *174*, 59–79. (in Russian; cited in Kirpichnikov [1987]).

Andrijasheva, M. A., Mantelman, I. I., Kaidanova, T. I., Chernyaeva, E. V., Lokshina, A. B., Efanov, G. V., and Poljakova, L. A. 1983. Selection and genetic studies of some whitefish species. *In* Biological Bases of Fish Breeding: Problems of Genetics and Selection. Nauka, Leningrad. (in Russian; cited in Kirpichnikov [1987]).

Angus, R. A. 1983. Genetic analysis of melanistic spotting in sailfin mollies. J. Hered. *74*, 81–84.

Angus, R. A. 1989. Inheritance of melanistic pigmentation in the eastern mosquitofish. J. Hered. *80*, 387–392.

Angus, R. A. 1991. Personal communication. Department of Biology, University of Alabama at Birmingham, Birmingham, AL.

Angus, R. A. and Blanchard, P. D. 1991. Genetic basis of the gold phenotype in sailfin mollies. J. Hered. *82*, 425–428.

Ankorion, Y. 1966. Studies on the heredity of some morphological characteristics in the common carp (*Cyrpinus carpio* L.). Master's Thesis, The Hebrew University, Jerusalem, Israel. (in Hebrew; cited in Wohlfarth *et al.* [1987]).

Anon. 1956. Yippee! It's branding time. Penn. Angler *25(3)*, 2–5.

Arai, K. 1988. Viability of allotriploids in salmonids. Bull. Jpn. Soc. Sci. Fish. *54*, 1695–1701.

Arai, K. and Wilkins, N. P. 1987. Triploidization of brown trout (*Salmo trutta*) by heat shocks. Aquaculture *64*, 97–103.

Aulstad, D. and Kittelsen, A. 1971. Abnormal body curvatures of rainbow trout (*Salmo gairdneri*) inbred fry. J. Fish. Res. Board Can. *28*, 1918–1920.

Aulstad, D., Gjedrem, T., and Skjervold, H. 1972. Genetic and environmental sources of variation in length and weight of rainbow trout (*Salmo gairdneri*). J. Fish. Res. Board Can. *29*, 237–241.

Avtalion, R. R. and Don, J. 1990. Sex-determining genes in tilapia: a model of genetic recombination emerging from sex ratio results of three generations of diploid gynogenetic *Oreochromis aureus*. J. Fish Biol. *37*, 167–173.

Avtalion, R. R. and Hammerman, I. S. 1978. Sex determination in Sarotherodon (Tilapia). I. Introduction to a theory of autosomal influence. Bamidgeh *30*, 110–115.

Avtalion, R. R., Duczyminer, M., Wojdani, A., and Pruginin, Y. 1976. Determination of allogeneic and xenogeneic markers in the genus of *Tilipia*. II. Identification of *T. aurea*, *T. vulcani* and *T. nilotica* by electrophoretic analysis of their serum proteins. Aquaculture *7*, 255–265.

Ayles, G. B. and Baker, R. F. 1983. Genetic differences in growth and survival between strains and hybrids of rainbow trout (*Salmo gairdneri*) stocked in aquaculture lakes in the Canadian prairies. Aquaculture *33*, 269–280.

Ayles, G. B., Bernard, D., and Hendzel, M. 1979. Genetic differences in lipid and dry matter content between strains of rainbow trout (*Salmo gairdneri*) and their hybrids. Aquaculture *18*, 253–262.

B

Babouchkine, Y. O. 1987. La selection d'une carpe resistant a l'hiver. [Selection for winter resistant carp.] *In* Selection, Hybridization, and Genetic Engineering in Aquaculture, Vol. 1, K. Thiews (Editor). H. Heenemann GmbH and Co., Berlin, Germany.

Backiel, T., Kokurewicz, B., and Ogorzałek, A. 1984. High incidence of skeletal anomalies in carp, *Cyprinus carpio*, reared in cages in flowing water. Aquaculture *43*, 369–380.

Bagenal, T. B. 1969. Relationship between egg size and fry survival in brown trout *Salmo trutta* L. J. Fish. Biol. *1*, 349–353.

Bailey, J. K. and Friars, G. W. 1989. Genetic and environmental components of growth in cage cultured Atlantic salmon (*Salmo salar*). J. World Aquacul. Soc. *20*, 14A.

Baily, J. K. and Friars, G. W. 1990. Inheritance of age at smolting in hatchery-reared Atlantic salmon (*Salmo salar*). Aquaculture *85*, 317.

Bailey, J. K. and Loudenslager, E. J. 1986. Genetic and environmental components of variation for growth of juvenile Atlantic salmon (*Salmo salar*). Aquaculture *57*, 125–132.

Bakker, T. C. M., Feuth-De Bruijn, E., and Sevenster, P. 1988. Albinism in the threespine stickleback, *Gasterosteus aculeatus*. Copeia *1988*, 236–238.

Barbat-Leterrier, A., Guyomard, R., and Krieg, F. 1989. Introgression between introduced domesticated strains and mediterranean [sic] native populations of brown trout (*Salmo trutta* L.). Aquat. Living Resour. *2*, 215–223.

Bartley, D. M. and Gall, G. A. E. 1991. Genetic identification of native cutthroat trout (*Oncorhynchus clarki*) and introgressive hybridization with introduced rainbow trout (*O. mykiss*) in streams associated with the Alvord Basin, Oregon and Nevada. Copeia *1991*, 854–859.

Beacham, T. D. 1988. A genetic analysis of early development in pink (*Oncorhynchus gorbuscha*) and chum salmon (*Oncorhynchus keta*) at three different temperatures. Genome *30*, 89–96.

Beacham, T. D. 1989. Genetic variation in body weight of pink salmon (*Oncorhynchus gorbuscha*). Genome *32*, 227–231.

Beacham, T. D. and Murray, C. B. 1988A. A genetic analysis of body size in pink salmon (*Oncorhynchus gorbuscha*). Genome *30*, 31–35.

Beacham, T. D. and Murray, C. B. 1988B. Genetic analysis of growth and maturity in pink salmon (*Oncorhynchus gorbuscha*). Genome *30*, 529–535.

Beacham, T. D. and Withler, R. E. 1985A. Heterozygosity and morphological variability of pink salmon (*Oncorhynchus gorbuscha*) from southern British Columbia and Puget Sound. Can. J. Genet. Cytol. *27*, 571–579.

Beacham, T. D. and Withler, R. E. 1985B. Heterozygosity and morphological variability of chum salmon (*Oncorhynchus keta*) in southern British Columbia. Heredity *54*, 313–322.

Beardmore, J. A. and Shami, S. A. 1976. Parental age, genetic variation and selection. *In* Population Genetics and Ecology, S. Karlin and E. Nevo (Editors). Academic Press, New York.

Beck, M. L. and Biggers, C. J. 1983. Erythrocyte measurements of diploid and triploid *Ctenopharyngodon idella* × *Hypophthalmichthys nobilis* hybrids. J. Fish. Biol. *22*, 497–502.

Becker, W. A. 1985. Manual of Quantitative Genetics, 4th edition. Academic Enterprises, Pullman, WA.

Behrends, L. L. and Smitherman, R. O. 1984. Development of a cold-tolerant population of red tilapia through introgressive hybridization. J World Maricul. Soc. *15*, 172–178.

Behrends, L. L., Kingsley, J. B., and Price, A. H., III. 1988. Bidirectional-backcross selection for body weight in a red tilapia. *In* The Second International Symposium on Tilapia in Aquaculture, R. S. V. Pullin, T. Bhukaswan, K. Tonguthai, and J. L. Maclean (Editors). ICLARM Conference Proceedings 15, Department of Fisheries, Bangkok, Thailand and International Center for Living Aquatic Resources Management, Manila, Philippines.

Bellamy, A. W. and Queal, M. L. 1951. Heterosomal inheritance and sex determination in Platypoecilus maculatus. Genetics *36*, 93–107.

Benfey, T. J. and Sutterlin, A. M. 1984A. Triploidy induced by heat shock and hydrostatic pressure in landlocked Atlantic salmon (*Salmo salar* L.). Aquaculture *36*, 359–367.

Benfey, T. J. and Sutterlin, A. M. 1984B. Growth and gonadal development in triploid landlocked Atlantic salmon (*Salmo salar*). Can. J. Fish. Aquat. Sci. *41*, 1387–1392.

Benfey, T. J., Solar, I. I., de Jong, G., and Donaldson, E. M. 1986. Flow-cytometric confirmation of aneuploidy in sperm from triploid rainbow trout. Trans. Am. Fish. Soc. *115*, 838–840.

Benfey, T. J., Bosa, P. G., Richardson, N. L. and Donaldson, E. M. 1988. Effectiveness of a commercial-scale pressure shocking device for producing triploid salmonids. Aquacul. Engin. *7*, 147–154.

Bentsen, H. B. 1991. Quantitative genetics and management of wild populations. Aquaculture *98*, 263–266.

Bergot, P., Chevassus, B., and Blanc, J.-M. 1976. Déterminisme génétique du nombre de caeca pyloriques chez la Truite fario (*Salmo trutta* Linné) et la Truite arc-en-ciel (*Salmo gairdneri* Richardson). I.—Distribution du caractère et variabilité phénotypique intra et interfamilles. [Genetic analysis of the number of pyloric caeca in brown trout (*Salmo trutta* Linnaeus) and rainbow trout (*Salmo gairdneri* Richardson). I.—Character distribution and phenotypic variability within and between families.] Ann. Hydrobiol. *7*, 105–114.

Bergot, P., Blanc, J. M., and Escaffre, A. M. 1981A. Relationship between number of pyloric caeca and growth in rainbow trout (*Salmo gairdneri* Richardson). Aquaculture *22*, 81–96.

Bergot, P., Blanc, J. M., Escaffre, A. M., and Poisson, H. 1981B. Effect of selecting sires according to their number of pyloric caeca upon the growth of offspring in rainbow trout (*Salmo gairdneri* Richardson). Aquaculture *25*, 207–215.

Bernatchez, L. and Dodson, J. J. 1990. Mitochondrial DNA variation among anadromous populations of cisco (*Coregonus artedii*) as revealed by restriction analysis. Can. J. Fish. Aquat. Sci. *47*, 533–543.

Bertollo, L. A. C., Takahashi, C. S., and Filho, O. M. 1983. Multiple sex chromosomes in the genus *Hoplias* (Pisces: Erythrinidae). Cytologia *48*, 1–12.

Bice, T. O. 1981. Spawning success, fecundity, hatchability, and fry survival in strain and reciprocal pairings of Marion and Kansas channel catfish. Master's Thesis, Auburn Univ., AL.

Bidwell, C. A., Chrisman, C. L., and Libey, G. S. 1985. Polyploidy induced by heat shock in channel catfish. Aquaculture *51*, 25–32.

Billington, N., Hebert, P. D. N., and Ward, R. D. 1990. Allozyme and mitochondrial DNA variation among three species of *Stizostedion* (Percidae): Phylogenetic and zoogeographical implications. Can. J. Fish. Aquat. Sci. *47*, 1093–1102.

Billington, N., Danzmann, R. G., Hebert, P. D. N., and Ward, R. D. 1991. Phylogenetic relationships among four members of *Stizostedion* (Percidae) determined by mitochondrial DNA and allozyme analyses. J. Fish Biol. *39* (*Suppl. A*), 251–258.

Bilton, H. T. 1971. A hypothesis of alternation of age of return in successive generations of Skeena River sockeye salmon (*Oncorhynchus nerka*). J. Fish. Res. Board Can. *28*, 513–516.

Birt, T. P., Green, J. M., and Davidson, W. S. 1991. Mitochondrial DNA variation reveals genetically distinct sympatric populations of anadromous and nonanadromous Atlantic salmon, *Salmo salar*. Can. J. Fish. Aquat. Sci. *48*, 577–582.

Blanc, J.-M. 1973. Genetic aspects of resistance to mercury poisoning in steelhead trout (*Salmo gairdneri*). Master's Thesis, Oregon State Univ., Corvallis, OR.

Blanc, J. M., Chevassus, B., and Bergot, P. 1979. Déterminisme génétique du nombre de caeca pyloriques chez la Truite fario (*Salmo trutta*, Linné) et la Truite arc-en-ciel (*Salmo Gairdneri*, Richardson). III.—Effet du génotype et de la taille des oeufs sur la réalisation du caractère chez la Truite fario. [Genetic analysis of the number of pyloric caeca in brown trout (*Salmo trutta* Linnaeus) and rainbow trout (*Salmo gairdneri* Richardson). III.—Influence of the genotype and egg size on the realization of the character in brown trout.] Ann. Genet. Sel. Anim. *11*, 93–103.

Blanc, J.-M., Chourrout, D., and Krieg, F. 1987. Evaluation of juvenile rainbow trout survival and growth in half-sib families from diploid and tetraploid sires. Aquaculture *65*, 215–220.

Blanco, G., Sánchez, J. A., Vazquez, E., García, E., and Rubio, J. 1990. Superior developmental stability of heterozygotes at enzyme loci in *Salmo salar* L. Aquaculture *84*, 199–209.

Bolla, S. 1987. Cytogenetic studies in Atlantic salmon and rainbow trout embryos. Hereditas *106*, 11–17.

Bondari, K. 1982. Cage performance and quality comparisons of tilapia and divergently selected channel catfish. Proc. Southeast. Assoc. Fish Wild. Agen. *34*(1980), 88–98.

Bondari, K. 1983A. Response to bidirectional selection for body weight in channel catfish. Aquaculture *33*, 73–81.

Bondari, K. 1983B. Efficiency of male reproduction in channel catfish. Aquaculture *35*, 79–82.

Bondari, K. 1984A. A study of abnormal characteristics of channel catfish and blue tilapia. Proc. Southeast. Assoc. Fish Wild. Agen., *35*(1981), 566–578.

Bondari, K. 1984B. Growth comparison of inbred and randombred catfish at different temperatures. Proc. Southeast. Assoc. Fish Wild. Agen. *35*(1981), 547–553.

Bondari, K. 1984C. Comparative performance of albino and normally pigmented channel catfish in tanks, cages, and ponds. Aquaculture *37*, 293–301.

Bondari, K. 1986. Response of channel catfish to multi-factor and divergent selection of economic traits. Aquaculture *57*, 163–170.

Bondari, K. and Dunham, R. A. 1987. Effects of inbreeding on economic traits of channel catfish. Theor. Appl. Genet. *74*, 1–9.

Bondari, K., Dunham, R. A., Smitherman, R. O., Joyce, J. A., and Castillo, S. 1983. Response to bidirectional selection for body weight in blue tilapia. *In*

International Symposium on Tilapia in Aquaculture, L. Fishelson and Z. Yaron (Editors). Tel Aviv Univ., Tel Aviv, Israel.

Bondari, K., Ware, G. O., Mullinix, B. G., Jr., and Joyce, J. A. 1985. Influence of brood fish size on the breeding performance of channel catfish. Prog. Fish-Cult. *47*, 21–26.

Boney, S. E., Shelton, W. L., Yang, S.-L., and Wilken, L. O. 1984. Sex reversal and breeding of grass carp. Trans. Am. Fish. Soc. *113*, 348–353.

Brauhn, J. L. and Hogan, J. W. 1972. Use of cold brands on channel catfish. Prog. Fish-Cult. *34*, 112.

Brauhn, J. L. and Kincaid, H. 1982. Survival, growth, and catchability of rainbow trout of four strains. N. Am. J. Fish. Manage. *2*, 1–10.

Breder, C. M., Jr. 1934. The ultimate in tailless fish. Bull. N. Y. Zool. Soc. *37*, 141–145.

Breder, C. M., Jr. 1953. A case of survival of a goldfish following the loss of its tail. Zoologica *38*, 49–52.

Brem, G., Brenig, B., Hörstgen-Schwark, G., and Winnacker, E.-L. 1988. Gene transfer in tilapia (*Oreochromis niloticus*). Aquaculture *68*, 209–219.

Bridges, W. R. 1973. Rainbow trout breeding projects. Progress in Sport Fishery Research 1971. Div. Fish. Res. Bur. Sport Fish. Wild. Res. Pub. *121*, 60–63.

Bridges, W. R. 1974. Fish genetics. *In* Sport Fishery and Wildlife Research 1972, V. T. Harris and P. H. Eschmeyer (Editors). Bureau of Sport Fisheries and Wildlife, U. S. Government Printing Office, Washington, DC.

Bridges, W. R., and von Limbach, B. 1972. Inheritance of albinism in rainbow trout. J. Hered. *63*, 152–153.

Brody, T., Moav, R., Abramson, Z. V., Hulata, G., and Wohlfarth, G. 1976. Applications of electrophoretic genetic markers to fish breeding. II. Genetic variation within maternal half-sibs in carp. Aquaculture *9*, 351–365.

Brody, T., Storch, N., Kirsht, D., Hulata, G., Wohlfarth, G., and Moav, R. 1980. Application of electrophoretic genetic markers to fish breeding III. Diallel [sic] analysis of growth rate in carp. Aquaculture *20*, 371–379.

Brody, T., Wohlfarth, G., Hulata, G., and Moav, R. 1981. Application of electrophoretic genetic markers to fish breeding. IV. Assessment of breeding value of full-sib families. Aquaculture *24*, 175–186.

Brooks, M. J. 1977. A study of length variation in blue *Ictalurus furcatus* (Lesueur), white *I. catus* (Linnaeus), and four groups of channel *I. punctatus* (Rafinesque) catfish. Master's Thesis, Auburn Univ., AL.

Brooks, M. J., Smitherman, R. O., Chappell, J. A., and Dunham, R. A. 1982. Sex-weight relations in blue, channel, and white catfishes: Implications for brood stock selection. Prog. Fish-Cult. *44*, 105–107.

Broussard, M. C., Jr. and Stickney, R. R. 1981. Evaluation of reproductive characters for four strains of channel catfish. Trans. Am. Fish. Soc. *110*, 502–506.

Broussard, M. C., Jr. and Stickney, R. R. 1984. Growth of four strains of channel catfish in communal ponds. Proc. Southeast. Assoc. Fish Wild. Agen. *35* (1981), 541–546.

Brummett, R. E. 1982. Isozymic variability within and among populations of

Tilapia aurea, T. hornorum, T. mossambica and *T. nilotica*. Master's Thesis, Auburn Univ., AL.

Brummett, R. E. 1986. Effects of genotype × environment interactions on growth, variability and survival of improved catfish. Doctoral dissertation, Auburn University, AL.

Brummett, R. E., Halstrom, M. L., Dunham, R. A., and Smitherman, R. O. 1988. Development of biochemical dichotomous keys for identification of American populations of *Oreochromis aureus, O. mossambicus, O. niloticus, O. urolepis hornorum* and red tilapia. *In* The Second International Symposium on Tilapia in Aquaculture, R. S. V. Pullin, T. Bhukaswan, K. Tonguthai, and J. L. Maclean (Editors). ICLARM Conference Proceedings 15, Department of Fisheries, Bangkok, Thailand and International Center for Living Aquatic Resources Management, Manila, Philippines.

Burch, E. P. 1986. Heritabilities for body weight, feed consumption and feed conversion and the correlations among these traits in channel catfish, *Ictalurus punctatus*. Master's Thesis, Auburn Univ., AL.

Burger, C. V. 1974. Genetic aspects of lead toxicity in laboratory populations of guppies (*Poecilia reticulata*). Master's Thesis, Oregon State Univ., Corvallis, OR.

Burger, G. and Chevassus, B. 1987. Étude des possibilités de sélection de la précocité sexuelle chez la truite arc-en-ciel (*Salmo gairdneri* R.). [Potential of selection for age at first sexual maturation in rainbow trout (*Salmo gairdneri* R.).] Bull. Français Pêche Piscicul. *307*, 102–117.

Burnside, M. C., Avault, J. W., Jr., and Perry, W. G., Jr. 1975. Comparison of a wild and a domestic strain of channel catfish grown in brackish water. Prog. Fish-Cult. *37*, 52–54.

Busack, C. A. 1983. Four generations of selection for high 56-day weight in the mosquitofish (*Gambusia affinis*). Aquaculture *33*, 83–87.

Busack, C. A. and Gall, G. A. E. 1983. An initial description of the quantitative genetics of growth and reproduction in the mosquitofish, *Gambusia affinis*. Aquaculture *32*, 123–140.

Buth, D. G. 1990. Genetic principles and the interpretation of electrophoretic data. *In* Electrophoretic and Isoelectric Focusing Techniques in Fisheries Management, D. H. Whitmore (Editor). CRC Press, Boca Raton, FL.

Buth, D. G., Murphy, R. W., and Ulmer, L. 1987. Population differentiation and introgressive hybridization of the flannelmouth sucker and of hatchery and native stocks of the razorback sucker. Trans. Am. Fish. Soc. *116*, 103–110.

Bye, V. J. and Lincoln, R. F. 1986. Commercial methods for the control of sexual maturation in rainbow trout (*Salmo gairdneri* R.). Aquaculture *57*, 299–309.

C

Calaprice, J. R. 1969. Production and genetic factors in managed salmonid populations. *In* Symposium on Salmon and Trout in Streams, T. G. Northcote (Editor). Institute of Fisheries, The Univ. of British Columbia, Vancouver, B.C., Canada.

Campton, D. E. 1992. Heritability of body size of green swordtails, *Xiphophorus helleri:* I. Sib analyses of males reared individually and in groups. J. Hered. *83*, 43–48.

Campton, D. E. and Gall, G. A. E. 1988. Responses to selection for body size and age at sexual maturity in the mosquitofish, *Gambusia affinis*. Aquaculture *68*, 221–241.

Capili, J. B., Luna, S. M., and Palomares, M. L. D. 1990. A multivariate analysis of the growth of three strains of tank-reared tilapia *Oreochromis niloticus*. In The Second Asian Fisheries Forum, R. Hirano and I. Hanyu (Editors). Asian Fisheries Society, Manila, Philippines.

Cassani, J. R. 1990. A new method for early ploidy evaluation of grass carp larvae. Prog. Fish-Cult. *52*, 207–210.

Cassani, J. R. and Caton, W. E. 1985. Induced triploidy in grass carp, *Ctenopharyngodon idella* Val. Aquaculture *46*, 37–43.

Cassani, J. R. and Caton, W. E. 1986. Efficient production of triploid grass carp (*Ctenopharyngodon idella*) utilizing hydrostatic pressure. Aquaculture *55*, 43–50.

Cassani, J. R., Maloney, D. R., Allaire, H. P., and Kerby, J. H. 1990. Problems associated with tetraploid induction and survival in grass carp, *Ctenopharyngodon idella*. Aquaculture *88*, 273–284.

Chan, S. T. H. and Yeung, W. S. B. 1983. Sex control and sex reversal in fish under natural conditions. In Fish Physiology, Vol. 9, Reproduction; Part B, Behavior and Fertility Control, W. S. Hoar, D. J. Randall, and E. M. Donaldson (Editors). Academic Press, New York.

Ch'ang, M. T. 1971A. Influence of inbreeding on tilapia (*Tilapia mossambica* Peters). Sov. Genet. *7*, 1277–1282.

Ch'ang, M. T. 1971B. Determination of realized weight heritability in tilapia (*Tilapia mossambica* Peters.). Sov. Genet. *7*, 1550–1554.

Chao, N.-H., Chen, S.-J., and Liao, I.-C. 1986. Triploidy induced by cold shock in cyprinid loach, *Misgurnus anguillicaudatus*. In The First Asian Fisheries Forum, J. L. Maclean, L. B. Dizon, and L. V. Hosillos (Editors). Asian Fisheries Society, Manila, Philippines.

Chapman, R. W., and Brown, B. L. 1990. Mitochondrial DNA isolation methods. In Electrophoretic and Isoelectric Focusing Techniques in Fisheries Management, D. H. Whitmore (Editor). CRC Press Inc., Boca Raton, FL.

Chappell, J. A. 1979. An evaluation of twelve genetic groups of catfish for suitability in commercial production. Doctoral Dissertation, Auburn Univ., AL.

Chen, F. Y. 1969. Preliminary studies on the sex-determining mechanism of Tilapia mossambica Peters and T. hornorum Trewavas. Int. Assoc. Theor. Appl. Limnol. Proc. *17*, 719–724.

Chen, S. C. 1928. Transparency and mottling, a case of Mendelian inheritance in the goldfish *Carassius auratus*. Genetics *13*, 434–452.

Chen, S. C. 1934. The inheritance of blue and brown colours in the goldfish, *Carassius auratus*. J. Genet. *29*, 61–74.

Chen, S.-J., Chao, N.-H., and Liao, I.-C. 1986. Diploid gynogenesis induced by cold shock in cyprinid loach, *Misgurnus anguillicaudatus*. *In* The First Asian Fisheries Forum, J. L. Maclean, L. B. Dizon, and L. V. Hosillos (Editors). Asian Fisheries Society, Manila, Philippines.

Chen, T. R. 1969. Karyological heterogamety of deep-sea fishes. Postilla No. 130.

Chen, T. R. and Ebeling, A. W. 1968. Karyological evidence of female hetero-gamety in the mosquitofish, *Gambusia affinis*. Copeia *1968*, 70–75.

Chen, T. T., Agellon, L. B., and van Beneden, R. J. 1987. Genetic engineering of fish. *In* Selection, Hybridization, and Genetic Engineering in Aquaculture, Vol. 2, K. Thiews (Editor). H. Heenemann GmbH and Co., Berlin, Germany.

Chen, T. T., Lin, C. M., Zhu, Z., Gonzalez-Villasenor, L. I., Dunham, R. A., and Powers, D. A. 1990. Gene transfer, expression and inheritance of rain-bow trout and human growth hormone genes in carp and loach. *In* Transgenic Models in Medicine and Agriculture, R. B. Church (Editor). Wiley-Liss, Inc., New York, NY.

Cheng, K. M., McCallum, I. M., McKay, R. I., and March, B. E. 1987. A compar-ison of survival and growth of two strains of chinook salmon (*Oncorhynchus tshawytscha*) and their crosses reared in confinement. Aquaculture *67*, 301–311.

Cherfas, N. B., Kozinsky, O., Rothbard, S., and Hulata, G. 1990. Induced dip-loid gynogenesis and triploidy in ornamental (koi) carp, *Cyprinus carpio* L. Bamidgeh *42*, 3–9.

Chevassus, B. 1976. Variabilité et héritabilité des performances de croissance chez la Truite arc-en-ciel (*Salmo gairdnerii* Richardson). [Variability and heri-tability of growth in rainbow trout (*Salmo gairdnerii* Richardson).] Ann. Genet. Sel. Anim. *8*, 273–283.

Chevassus, B. 1979. Hybridization in salmonids: Results and perspectives. Aquaculture *17*, 113–128.

Chevassus, B., Blanc, J. M., and Bergot, P. 1979. Déterminisme génétique du nombre de caeca pyloriques chez la Truite fario (*Salmo trutta*, Linné) et la Truite arc-en-ciel (*Salmo gairdneri*, Richardson). II.—Effet du génotype du milieu d'élevage et de l'alimentation sur la réalisation du caractère chez la Truite arc-en-ciel. [Genetic analysis of the number of pyloric caeca in brown trout (*Salmo trutta* Linnaeus) and rainbow trout (*Salmo gairdneri* Richard-son). II.—Effect of the genotype, rearing environment, and feeding on the realization of the character in rainbow trout.] Ann. Genet. Sel. Anim. *11*, 79–92.

Chevassus, B., Guyomard, R., Chourrout, D., and Quillet, E. 1983. Production of viable hybrids in salmonids by triploidization. Genet. Sel. Evol. *15*, 519–531.

Chevassus, B., Devaux, A., Chourrout, D., and Jalabert, B. 1988. Production of YY rainbow trout males by self-fertilization of induced hermaphrodites. J. Hered. *79*, 89–92.

Childers, W. F. 1967. Hybridization of four species of sunfishes (Centrarchidae). Ill. Nat. Hist. Surv. Bull. *29*, 159–214.

Chourrout, D. 1980. Thermal induction of diploid gynogenesis and triploidy in the eggs of the rainbow trout (*Salmo gairdneri* Richardson). Reprod. Nutr. Develop. *20*, 727–733.

Chourrout, D. 1982A. Tetraploidy induced by heat shocks in the rainbow trout (*Salmo gairdneri* R.). Reprod. Nutr. Develop. *22*, 569–574.

Chourrout, D. 1982B. Gynogenesis caused by ultraviolet irradiation of salmonid sperm. J. Exp. Zool. *223*, 175–181.

Chourrout, D. 1984. Pressure-induced retention of second polar body and suppression of first cleavage in rainbow trout: production of all-triploids, all-tetraploids, and heterozygous and homozygous diploid gynogenetics. Aquaculture *36*, 111–126.

Chourrout, D. 1986. Use of grayling sperm (*Thymallus thymallus*) as a marker for the production of gynogenetic rainbow trout (*Salmo gairdneri*). Theor. Appl. Genet. *72*, 633–636.

Chourrout, D. and Nakayama, I. 1987. Chromosome studies of progenies of tetraploid female rainbow trout. Theor. Appl. Genet. *74*, 687–692.

Chourrout, D. and Quillet, E. 1982. Induced gynogenesis in the rainbow trout: Sex and survival of progenies production of all-triploid populations. Theor. Appl. Genet. *63*, 201–205.

Chourrout, D., Chevassus, B., and Herioux, F. 1980. Analysis of an Hertwig effect in the rainbow trout (*Salmo gairdneri* Richardson) after fertilization with γ-irradiated sperm. Reprod. Nutr. Develop. *20*, 719–726.

Chourrout, D., Chevassus, B., Krieg, F., Happe, A., Burger, G., and Renard, P. 1986A. Production of second generation triploid and tetraploid rainbow trout by mating tetraploid males and diploid females—Potential of tetraploid fish. Theor. Appl. Genet. *72*, 193–206.

Chourrout, D., Guyomard, R., and Houdebine, L.-M. 1986B. High efficiency gene transfer in rainbow trout (*Salmo gairdneri* Rich.) by microinjection into egg cytoplasm. Aquaculture *51*, 143–150.

Chrisman, C. L., Wolters, W. R., and Libey, G. S. 1983. Triploidy in channel catfish. J. World. Maricul. Soc. *14*, 279–293.

Clemens, H. P. and Inslee, T. 1968. The production of unisexual broods by *Tilapia mossambica* sex-reversed with methyl testosterone. Trans. Am. Fish. Soc. *97*, 18–21.

Clemens, H. P. and Sneed, K. E. 1959. Tattooing as a method of marking channel catfish. Prog. Fish-Cult. *21*, 29.

Cooper, E. L. 1961. Growth of wild and hatchery strains of brook trout. Trans. Am. Fish. Soc. *90*, 424–438.

Cordone, A. J. and Nicola, S. J. 1970. Harvest of four strains of rainbow trout, *Salmo gairdnerii*, from Beardsley Reservoir, California. Calif. Fish Game *56*, 271–287.

Corti, M., Thorpe, R. S., Sola, L., Sbordoni, V., and Cataudella, S. 1988. Multivariate morphometrics in aquaculture: a case study of six stocks of the common carp (*Cyprinus carpio*) from Italy. Can. J. Fish. Aquat. Sci. *45*, 1548–1554.

Couch, J. A., Winstead, J. T., Hansen, D. J., and Goodman, L. R. 1979. Vertebral dysplasia in young fish exposed to the herbicide trifluralin. J. Fish Dis. *2*, 35–42.

Crabtree, C. B. and Buth, D. G. 1987. Biochemical systematics of the catostomid genus *Catostomus:* Assessment of *C. clarki, C. plebeius* and *C. discobolus* including the Zuni sucker, *C.d. yarrowi.* Copeia *1987*, 843–854.

Crandell, P. A. and Gall, G. A. E. 1992. Body weight analysis for individually tagged rainbow trout: males, females, and precocious males. Aquaculture *100*, 99.

Cravedi, J. P., Delous, G., and Rao, D. 1989. Disposition and elimination routes of 17 α-methyltestosterone in rainbow trout (*Salmo gairdneri*). Can. J. Fish. Aquat. Sci. *46*, 159–165.

Cross, T. F. and Challanain, D. N. 1991. Genetic characterisation of Atlantic salmon (*Salmo salar*) lines farmed in Ireland. Aquaculture *98*, 209–216.

Cross, T. F. and King J. 1983. Genetic effects of hatchery rearing in Atlantic salmon. Aquaculture *33*, 33–40.

Crozier, W. W. and Moffett, I. J. J. 1989. Application of an electrophoretically detectable genetic marker to ploidy testing in brown trout (*Salmo trutta* L.) triploidised by heat shock. Aquaculture *80*, 231–239.

Curtis, T. A., Sessions, F. W., Bury, D., Rezk, M., and Dunham, R. A. 1987. Induction of ployploidy with hydrostatic pressure in striped bass, white bass, and their hybrids. Proc. Southeast. Assoc. Fish Wild. Agen. *41*, 63–69.

D

Danzmann, R. G., Ferguson, M. M., and Allendorf, F. W. 1988. Heterozygosity and components of fitness in a strain of rainbow trout. Biol. J. Linnean Soc. *33*, 285–304.

Danzmann, R. G., Ferguson, M. M., and Allendorf, F. W. 1989. Genetic variability and components of fitness in hatchery strains of rainbow trout. J. Fish. Biol. *35(Suppl. A)*, 313–319.

Danzmann, R. G., Ihssen, P. E., and Hebert, P. D. N. 1991. Genetic discrimination of wild and hatchery populations of brook charr, *Salvelinus fontinalis* (Mitchill), in Ontario using mitochondrial DNA analysis. J. Fish. Biol. *39 (Suppl. A)*, 69–77.

Davis, K. B., Simco, B. A., Goudie, C. A., Parker, N. C., Cauldwell, W., and Snellgrove, R. 1990. Hormonal sex manipulation and evidence of female homogamety in channel catfish. Gen. Comp. Endocrin. *78*, 218–223.

Davis, R. H., Jr. 1976. Evaluation of growth of inbred lines and their F₁ hybrids in brook trout, *Salvelinus fontinalis*, brown trout, *Salmo trutta*, and rainbow trout, *Salmo gairdneri*. Doctoral Dissertation, Pennsylvania State Univ., University Park, PA.

Dawley, R. M., Graham, J. H., and Schultz, R. J. 1985. Triploid progeny of pumpkinseed × green sunfish hybrids. J. Hered. *76*, 251–257.

Delabbio, J. L., Glebe, B. D., and Sreedharan. 1990. Variation in growth and survival between two anadromous strains of Canadian Arctic charr (*Salvelinus alpinus*) during long-term saltwater rearing. Aquaculture *85*, 259–270.

Denton, T. E. 1973. Fish Chromosome Methodology. Charles C Thomas, Springfield, IL.

Diter, A., Guyomard, R., and Chourrout, D. 1988. Gene segregation in induced tetraploid rainbow trout: genetic evidence of preferential pairing of homologous chromosomes. Genome *30*, 547–553.

Dollar, A. M. and Katz, M. 1964. Rainbow trout brood stocks and strains in American hatcheries as factors in the occurrence of hepatoma. Prog. Fish-Cult. *26*, 167–174.

Don, J. and Avtalion, R. R. 1986. The induction of triploidy in *Oreochromis aureus* by heat shock. Theor. Appl. Genet. *72*, 186–192.

Don, J. and Avtalion, R. R. 1988A. Production of viable tetraploid tilapias using the cold shock technique. Bamidgeh *40*, 17–21.

Don, J. and Avtalion R. R. 1988B. Ploidy and gynogenesis in tilapias. *In* Reproduction in Fish Basic and Applied Aspects in Endocrinology and Genetics, Y. Zohar and B. Breton (Editors). INRA Publications, Paris, France.

Donaldson, L. R. and Menasveta, D. 1961. Selective breeding of chinook salmon. Trans. Am. Fish. Soc. *90*, 160–164.

Donaldson, L. R. and Olson, P. R. 1957. Development of rainbow trout brood stock by selective breeding. Trans. Am. Fish. Soc. *85*(1955), 93–101.

Donaldson, L. R., Hansler, D. D., and Buckridge, T. N. 1957. Interracial hybridization of cutthroat trout, *Salmo clarkii*, and its use in fisheries management. Trans. Am. Fish. Soc. *86*(1956), 350–360.

Doyle, R. W. 1983. An approach to the quantitative analysis of domestication selection in aquaculture. Aquaculture *33*, 167–185.

Doyle, R. W. and Talbot, A. J. 1986. Effective population size and selection in variable aquaculture stocks. Aquaculture *57*, 27–35.

Dubé, P., Blanc, J.-M., Chouinard, M., and de la Noüe, J. 1991. Triploidy induced by heat shock in brook trout (*Salvelinus fontinalis*). Aquaculture *92*, 305–311.

Dunham, R. A. 1981. Response to selection and realized heritability for body weight in three strains of channel catfish grown in earthen ponds. Doctoral Dissertation, Auburn Univ., AL.

Dunham, R. A. and Smitherman, R. O. 1981. Growth in response to winter feeding of blue, channel, white, and hybrid catfishes. Prog. Fish-Cult. *43*, 63–66.

Dunham, R. A. and Smitherman, R. O. 1983. Response to selection and realized heritability for body weight in three strains of channel catfish, *Ictalurus punctatus*, grown in earthen ponds. Aquaculture *33*, 89–96.

Dunham, R. A. and Smitherman, R. O. 1984. Ancestry and breeding of catfish in the United States. Circular 273. Alabama Agricultural Experiment Station, Auburn Univ., AL.

Dunham, R. A. and Smitherman, R. O. 1985. Improved growth rate, reproductive performance, and disease resistance of crossbred and selected catfish from AU-M and AU-K lines. Circular 279. Alabama Agricultural Experiment Station, Auburn Univ., AL.

Dunham, R. A., Smitherman, R. O., Chappell, J. A., Youngblood, P. N., and Bice, T. O. 1982. Communal stocking and multiple rearing technique for catfish genetics research. J. World Maricul. Soc. *13*, 261–267.

Dunham, R. A., Smitherman, R. O., Horn, J. L., and Bice, T. O. 1983. Reproductive performances of crossbred and pure-strain channel catfish brood stocks. Trans. Am. Fish. Soc. *112*, 436–440.

Dunham, R. A., Joyce, J. A., Bondari, K., and Malvestuto, S. P. 1985. Evaluation of body conformation, composition, and density as traits for indirect selection for dress-out percentage of channel catfish. Prog. Fish-Cult. *47*, 169–175.

Dunham, R. A., Smitherman, R. O., Goodman, R. K., and Kemp, P. 1986. Comparison of strains, crossbreeds and hybrids of channel catfish for vulnerability to angling. Aquaculture *57*, 193–201.

Dunham, R. A., Eash, J., Askins, J., and Townes, T. M. 1987. Transfer of the metallothionein-human growth hormone fusion gene into channel catfish. Trans. Am. Fish. Soc. *116*, 87–91.

Dunham, R. A., Brummett, R. E., Ella, M. O., and Smitherman, R. O. 1990. Genotype-environment interactions for growth of blue, channel and hybrid catfish in ponds and cages at varying densities. Aquaculture *85*, 143–151.

Dunham, R. A., Smitherman, R. O., and Bondari, K. 1991. Lack of inheritance of stumpbody and taillessness in channel catfish. Prog. Fish-Cult. *53*, 101–105.

Durborow, R. M., Avault, J. W., Jr., Johnson, W. A., and Koonce, K. L. 1985. Differences in mortality among full-sib channel catfish families at low dissolved oxygen. Prog. Fish-Cult. *47*, 14–20.

Dwyer, W. P., and Piper, R. G. 1984. Three-year hatchery and field evaluation of four strains of rainbow trout. N. Am. J. Fish. Manage. *4*, 216–221.

Dzwillo, M. 1959. Genetische Untersuchungen an domestizierten Stämmen von Lebistes reticulatus (Peters). [Genetic studies in domesticated stocks of Lebistes reticulatus (Peters).] Mitt. Hamb. Zool. Mus. Inst. *57*, 143–186.

E

Edds, D. R. and Echelle, A. A. 1989. Genetic comparisons of hatchery and natural stocks of small endangered fishes: Leon Springs pupfish, Comanche Springs pupfish, and Pecos gambusia. Trans. Am. Fish. Soc. *118*, 441–446.

Edwards, D. and Gjedrem, T. 1979. Genetic variation in survival of brown trout eggs, fry and fingerlings in acidic water. Acid Precipitation—Effects on Forest and Fish Project. Norwegian Forest Research Institute SNSF Research Report FR 16/79. Oslo-Ås, Norway.

Ehlinger, N. F. 1964. Selective breeding of trout for resistance to furunculosis. N.Y. Fish Game J. *11*, 78–90.

Ehlinger, N. F. 1977. Selective breeding of trout for resistance to furunculosis. N.Y. Fish Game J. *24*, 25–36.

Eknath, A. E. and Doyle, R. W. 1985. Indirect selection for growth and life-history traits in Indian carp aquaculture. I. Effects of broodstock management. Aquaculture *49*, 73–84.

El Gamal, A. R. A. L. 1987. Reproductive performance, sex ratios, gonadal development, cold tolerance, viability and growth of red and normally pigmented hybrids of *Tilapia aurea* and *T. nilotica*. Doctoral Dissertation, Auburn Univ., AL.

El Gamal, A. A., Smitherman, R. O., and Behrends, L. L. 1988. Viability of red and normal-colored *Oreochromis aureus* and *O. niloticus* hybrids. *In* The Second International Symposium on Tilapia in Aquaculture, R. S. V. Pullin., T. Bhukaswan, K. Tonguthai, and J. L. Maclean (Editors). ICLARM Conference Proceedings 15, Department of Fisheries, Bangkok, Thailand and International Center for Living Aquatic Resources Management, Manila, Philippines.

El-Ibiary, H. M. and Joyce, J. A. 1978. Heritability of body size traits, dressing weight and lipid content in channel catfish. J. Anim. Sci. *47*, 82–88.

El-Ibiary, H. M., Andrews, J. W., Joyce, J. A., Page, J. W., and DeLoach, H. L. 1976. Sources of variations in body size traits, dress-out weight, and lipid content and their correlations in channel catfish, *Ictalurus punctatus*. Trans. Am. Fish. Soc. *105*, 267–272.

El-Ibiary, H. M., Hill, T. K., Joyce, J. A., and Andrews, J. W. 1979. Phenotypic correlations between commercial characters in channel catfish. Proc. Southeast. Assoc. Fish Wild. Agen. *32*(1978), 420–425.

Ella, M. O. 1984. Genotype-environment interactions for growth rate of blue, channel and hybrid catfish grown at varying stocking densities. Master's Thesis, Auburn Univ., AL.

Embody, G. C. and Hayford, C. O. 1925. The advantage of rearing brook trout fingerlings from selected breeders. Trans. Am. Fish. Soc. *55*, 135–148.

Emmens, C. W. 1970. Guppy Handbook. T.F.H. Publications, Neptune City, NJ.

Everest, F. H. and Edmundson, E. H. 1967. Cold branding for field use in marking juvenile salmonids. Prog. Fish-Cult. *29*, 175–176.

Ewing, R. R., Scalet, C. G., and Evenson, D. P. 1991. Flow cytometric identification of larval triploid walleyes. Prog. Fish-Cult. *53*, 177–180.

F

Fagerlund, U. H. M. and Dye, H. M. 1979. Depletion of radioactivity from yearling coho salmon (*Oncorhynchus kisutch*) after extended ingestion of anabolically effective doses of 17 α-methyltestosterone-1, 2-^{3}H. Aquaculture *18*, 303–315.

Fagerlund, U. H. M. and McBride, J. R. 1978. Distribution and disappearance of radioactivity in blood and tissues of coho salmon (*Oncorhynchus kisutch*) after oral administration of ^{3}H-testosterone. J. Fish. Res. Board Can. *35*, 893–900.

Falconer, D. S. 1957. Breeding methods—I Genetic considerations. *In* The UFAW Handbook on the Care and Management of Laboratory Animals, second ed., A. N. Worden and W. Lane-Petter (Editors). The Universities Federation for Animal Welfare, London, England.

Falconer, D. S. 1981. Introduction to Quantitative Genetics, 2nd edition. Longman, New York.

FAO/UNEP. 1981. Conservation of the genetic resources of fish: Problems and recommendations. Report of the expert consultation on the genetic resources of fish, Rome, 9–13 June 1980. FAO Fisheries Technical Paper No. 217.

Ferguson, M. M., Danzmann, R. G., and Allendorf, F. W. 1985. Developmental divergence among hatchery strains of rainbow trout (*Salmo gairdneri*). I. Pure strains. Can. J. Genet. Cytol. *27*, 289–297.

Fernando, A. A. and Phang, V. P. E. 1989. X-linked inheritance of red and blue tail colourations of domesticated varieties of guppy, *Poecilia reticulata* and its implications to the farmer. Singapore J. Pri. Ind. *17*, 10–18.

Ferris, S. D. 1984. Tetraploidy and the evolution of the catostomid fishes, *In* Evolutionary Genetics of Fishes, B. J. Turner (Editor). Plenum Press, New York.

Ferris, S. D. and Berg, W. J. 1988. The utility of mitochondrial DNA in fish genetics and fishery management. *In* Population Genetics & Fishery Management, N. Ryman and F. Utter (Editors). Washington Sea Grant Program, University of Washington Press, Seattle, WA.

Fevolden, S. E., Refstie, T., and Røed, K. H. 1991. Selection for high and low cortisol stress response in Atlantic salmon (*Salmo salar*) and rainbow trout (*Oncorhynchus mykiss*). Aquaculture 95:53–65.

Filho, O. M., Bertollo, L. A. C., and Junior, P. M. G. 1980. Evidences for a multiple sex chromosome system with female heterogamety in *Apareiodon affinis* (Pisces, Parodontidae). Caryologia *33*, 83–91.

Fletcher, G. L., Shears, M. A., King, M. J., Davies, P. L., and Hew, C. L. 1988. Evidence for antifreeze protein gene transfer in Atlantic salmon (*Salmo salar*). Can. J. Fish. Aquat. Sci. *45*, 352–357.

Flick, W. A. and Webster, D. A. 1964. Comparative first year survival and production in wild and domestic strains of brook trout, *Salvelinus fontinalis*. Trans. Am. Fish. Soc. *93*, 58–69.

Fowler, L. G. 1972. Growth and mortality of fingerling chinook salmon as affected by egg size. Prog. Fish-Cult. *34*, 66–69.

Frankel, J. S. 1982. Inheritance of shoulder spotting in the jewel tetra, *Hyphessobrycon callistus*. J. Hered. *73*, 310.

Frankel, J. S. 1985. Inheritance of trunk striping in the Sumatran tiger barb, *Barbus tetrazona*. J. Hered. *76*, 478–479.

Frankel, J. S. 1987. Inheritance of trunk coloration in the eye-spot rasbora. J. Hered. *78*, 112.

Frankel, J. S. 1991. Inheritance of body marking patterns in the half-banded barb, *Barbus semifasciolatus*. J. Hered. *82*, 250–251.

Friars, G. W., Bailey, J. K., and Coombs, K. A. 1990. Correlated responses to selection for grilse length in Atlantic salmon. Aquaculture *85*, 171–176.

Fujihara, M. P. and Nakatani, R. E. 1967. Cold and mild heat marking of fish. Prog. Fish-Cult. *29*, 172–174.

Fujio, Y., Nakajima, M., and Nagahama, Y. 1990. Detection of a low temperature-resistant gene in the guppy (*Poecilia reticulata*), with reference to sex-linked inheritance. Jpn. J. Genet. *65*, 201–207.

G

Gall, G. A. E. 1969. Quantitative inheritance and environmental response of rainbow trout. *In* Fish in Research, O. W. Neuhaus and J. E. Halver (Editors). Academic Press, New York.

Gall, G. A. E. 1972. Phenotypic and genetic components of body size and spawning performance. *In* Progress in Fishery and Food Science, R. W. Moore (Editor). Univ. of Washington Publications in Fisheries, New Series Vol. 5, Univ. of Washington, Seattle, WA.

Gall, G. A. E. 1974. Influence of size of eggs and age of female on hatchability and growth in rainbow trout. Calif. Fish Game *60*, 26–35.

Gall, G. A. E. 1975. Genetics of reproduction in domesticated rainbow trout. J. Anim. Sci. *40*, 19–28.

Gall, G. A. E. 1979. Two-stage trout broodstock selection program: Reproductive performance and growth rate. Calif. Dept. Fish Game Inland Fish. Admin. Rep. No. 79-2.

Gall, G. A. H. [sic]. 1986. Sexual maturation and growth rate. *In* Third World Congress on Genetics Applied to Livestock Production. X Breeding Programs for Swine, Poultry, and Fish, G. E. Dickerson and R. K. Johnson (Editors). University of Nebraska, Lincoln, NE.

Gall, G. A. E. and Gross, S. J. 1978A. Genetic studies of growth in domesticated rainbow trout. Aquaculture *13*, 225–234.

Gall, G. A. E. and Gross, S. J. 1978B. A genetics analysis of the performance of three rainbow trout broodstocks. Aquaculture *15*, 113–127.

Gall, G. A. E. and Huang, N. 1988A. Heritability and selection schemes for rainbow trout: body weight. Aquaculture *73*, 43–56.

Gall, G. A. E. and Huang, N. 1988B. Heritability and selection schemes for rainbow trout: female reproductive performance. Aquaculture *73*, 57–66.

Gall, G. A. E., Baltodano, J., and Huang, N. 1988. Heritability of age at spawning for rainbow trout. Aquaculture *68*, 93–102.

Galman, O. R., Moreau, J., Hulata, G., and Avtalion, R. R. 1988. The use of electrophoresis as a technique for the identification and control of tilapia breeding stocks in Israel. *In* The Second International Symposium on Tilapia in Aquaculture, R. S. V. Pullin, T. Bhukaswan, K. Tonguthai, and J. L.

Maclean (Editors). ICLARM Conference Proceedings 15, Department of Fisheries, Bangkok, Thailand and International Center for Living Aquatic Resources Management, Manila, Philippines.

Garcia de Leániz, C. Verspoor, E., and Hawkins, A. D. 1989. Genetic determination of the contribution of stocked and wild Atlantic salmon, *Salmo salar* L., to the angling fisheries in two Spanish rivers. J. Fish Biol. *35(Suppl. A)*, 261–270.

Garcia-Marin, J. L., Jorde, P. E., Ryman, N., Utter, F., and Pla, C. 1991. Management implications of genetic differentiation between native and hatchery populations of brown trout (*Salmo trutta*) in Spain. Aquaculture *95*, 235–249.

Garrett, G. P., Birkner, M. C. F., and Gold, J. R. 1992. Triploidy induction in largemouth bass, *Micropterus salmoides*. J. Appl. Aqua. *1(3)*, 27–34.

Garside, E. T. 1959. Some effects of oxygen in relation to temperature on the development of lake trout embryos. Can. J. Zool. *37*, 689–698.

Gervai, J., Péter, S., Nagy, A., Horváth, L., and Csányi, V. 1980A. Induced triploidy in carp, *Cyprinus carpio* L. J. Fish. Biol. *17*, 667–671.

Gervai, J., Marián, T., Krasznai, Z., Nagy, A., and Csányi, V. 1980B. Occurrence of aneuploidy in radiation gynogenesis of carp, *Cyprinus carpio* L. J. Fish Biol. *16*, 435–439.

Gharrett, A. J. and Shirley, S. M. 1985. A genetic examination of spawning methodology in a salmon hatchery. Aquaculture *47*, 245–256.

Gharrett, A. J. and Smoker, W. W. 1991. Two generations of hybrids between even-and odd-year pink salmon (*Oncorhynchus gorbuscha*): A test for outbreeding depression? Can. J. Fish. Aquat. Sci. *48*, 1744–1749.

Gile, S. R. and Ferguson, M. M. 1990. Crossing methodology and genotypic diversity in a hatchery strain of rainbow trout (*Oncorhynchus mykiss*). Can. J. Fish. Aquat. Sci. *47*, 719–724.

Giudice, J. J. 1966. Growth of a blue × channel catfish hybrid as compared to its parent species. Prog. Fish-Cult. *26*, 142–145.

Gjedrem, T. 1975. Possibilities for genetic gain in salmonids. Aquaculture *6*, 23–29.

Gjedrem, T. 1979. Selection for growth rate and domestication in Atlantic salmon. Z. Tierz. Züchtungsbiol. *96*, 56–59.

Gjedrem, T. 1983. Genetic variation in quantitative traits and selective breeding in fish and shellfish. Aquaculture *33*, 51–72.

Gjedrem, T. and Aulstad, D. 1974. Selection experiments with salmon. I. Differences in resistance to vibrio disease of salmon parr (*Salmo salar*). Aquaculture *3*, 51–59.

Gjedrem, T., Salte, R., and Gjøen, H. M. 1991A. Genetic variation in susceptibility of Atlantic salmon to furunculosis. Aquaculture *97*, 1–6.

Gjedrem, T., Gjøen, H. M., and Salte, R. 1991B. Genetic variation in red cell membrane fragility in Atlantic salmon and rainbow trout. Aquaculture *98*, 349–354.

Gjerde, B. 1984A. Response to individual selection for age at sexual maturity in Atlantic salmon. Aquaculture *38*, 229–240.

Gjerde, B. 1984B. Variation in semen production of farmed Atlantic salmon and rainbow trout. Aquaculture 40, 109–114.

Gjerde, B. 1986. Estimates of phenotypic and genetic parameters for carcass quality traits in rainbow trout. Aquaculture 57, 368–369.

Gjerde, B. 1988. Complete diallele cross between six inbred groups of rainbow trout, Salmo gairdneri. Aquaculture 75, 71–87.

Gjerde, B. 1989. Body traits in rainbow trout I. Phenotypic means and standard deviations and sex effects. Aquaculture 80, 7–24.

Gjerde, B. and Gjedrem, T. 1984. Estimates of phenotypic and genetic parameters for carcass traits in Atlantic salmon and rainbow trout. Aquaculture 36, 97–110.

Gjerde, B. and Schaeffer, L. R. 1989. Body traits in rainbow trout II. Estimates of heritabilities and of phenotypic and genetic correlations. Aquaculture 80, 25–44.

Gjerde, B., Gunnes, K., and Gjedrem, T. 1983. Effect of inbreeding on survival and growth in rainbow trout. Aquaculture 34, 327–332.

Gold, J. R., Karel, W. J., and Strand, M. R. 1980. Chromosome formulae of North American fishes. Prog. Fish-Cult. 42, 10–23.

Gold, J. R., Li, Y., Schmidt, T. R., and Tave, D. 1991. Nucleolar dominance in interspecific hybrids of cyprinid fishes. Cytobios 65, 139–147.

Goodrich, H. B. 1929. Mendelian inheritance in fish. Quart. Rev. Biol. 4, 83–99.

Goodrich, H. B., Josephson, N. D., Trinkaus, J. P., and Slate, J. M. 1944. The cellular expression and genetics of two new genes in Lebistes reticulatus. Genetics 29, 584–592.

Gordon, M. 1927. The genetics of a viviparous top-minnow platypoecilus; the inheritance of two kinds of melanophores. Genetics 12, 253–283.

Gordon, M. 1938. The genetics of Xiphophorus hellerii: Heredity in Montezuma, a Mexican swordtail fish. Copeia 1938, 19–29.

Gordon, M. 1946. Interchanging genetic mechanisms for sex determination in fishes under domestication. J. Hered. 37, 307–320.

Gordon, M. 1956. An intricate genetic system that controls nine pigment cell patterns in the platyfish. Zoologica 41, 153–162.

Goudie, C. A., Shelton, W. L., and Parker, N. C. 1986A. Tissue distribution and elimination of radiolabelled methyltestosterone fed to sexually undifferentiated blue tilapia. Aquaculture 58, 215–226.

Goudie, C. A., Shelton, W. L., and Parker, N. C. 1986B. Tissue distribution and elimination of radiolabelled methyltestosterone fed to adult blue tilapia. Aquaculture 58, 227–240.

Green, O. L., Smitherman, R. O., and Pardue, G. B. 1979. Comparisons of growth and survival of channel catfish, Ictalurus punctatus, from distinct populations. In Advances in Aquaculture, T. V. R. Pillay and W. A. Dill (Editors). Fishing News Books, Farnham, Surrey, England.

Groves, A. B. and Novotny, A. J. 1965. A thermal-marking technique for juvenile salmonids. Trans. Am. Fish. Soc. 94, 386–389.

Guerrero, R. D., III. 1975. Use of androgens for the production of all-male Tilapia aurea (Steindachner). Trans. Am. Fish. Soc. 104, 342–348.

Gunnes, K. 1980. Genetic variation in production traits between strains of Atlantic salmon. *In* Atlantic Salmon: its Future, A. E. J. Went (Editor). Fishing News Books, Farnham, Surrey, England.

Gunnes, K. and Gjedrem, T. 1978. Selection experiments with salmon. IV. Growth of Atlantic salmon during two years in the sea. Aquaculture *15*, 19–33.

Gunnes, K. and Gjedrem, T. 1981. A genetic analysis of body weight and length in rainbow trout reared in seawater for 18 months. Aquaculture *24*, 161–174.

Gunter, G. and Ward, J. W. 1961. Some fishes that survive extreme injuries, and some aspects of tenacity of life. Copeia *1961*, 456–462.

Guo, X., Hershberger, W. K., and Myers, J. M. 1990. Growth and survival of intrastrain and interstrain rainbow trout (*Oncorhynchus mykiss*) triploids. J. World Aquacul. Soc. *21*, 250–256.

Guyomard, R. 1984. High level of residual heterozygosity in gynogenetic rainbow trout, *Salmo gairdneri*, Richardson. Theor. Appl. Genet. *67*, 307–316.

Guyomard, R. 1986. Gene segregation in gynogenetic brown trout (*Salmo trutta* L.): systematically high frequencies of post-reduction. Genet. Sel. Evol. *18*, 385–392.

Guyomard, R., Chourrout, D., and Houdebine, L. 1989A. Production of stable transgenic fish by cytoplasmic injection of purified genes. *In* Gene Transfer and Gene Therapy, A. L. Beaudet, R. Mulligan, and I. M. Verma (Editors). Alan R. Liss, Inc., New York, NY.

Guyomard, R., Chourrout, D., Leroux, C., Houdebine, L. M., and Pourrain, F. 1989B. Integration and germ line transmission of foreign genes microinjected into fertilized trout eggs. Biochimie *71*, 857–863.

Gyldenholm, A. O. and Scheel, J. J. 1971. Chromosome numbers of fishes. I. J. Fish Biol. *3*, 479–486.

H

Hagan, D. W. 1973. Inheritance of numbers of lateral plates and gill rakers in *Gasterosteus aculeatus*. Heredity *30*, 303–312.

Hagen, D. W. and Blouw, D. M. 1983. Heritability of dorsal spines in the fourspine stickleback (*Apeltes quadracus*). Heredity *50*, 275–281.

Hallerman, E. M. and Kapuscinski, A. R. 1990A. Transgenic fish and public policy: Regulatory concerns. Fisheries *15(1)*, 12–20.

Hallerman, E. M. and Kapuscinski A. R. 1990B. Transgenic fish and public policy: Patenting of transgenic fish. Fisheries *15(1)*, 21–24.

Hallerman, E. M., Dunham, R. A., and Smitherman, R. O. 1986. Selection or drift—isozyme allele frequency changes among channel catfish selected for rapid growth. Trans. Am. Fish. Soc. *115*, 60–68.

Hallerman, E. M., Schneider, J. F., Gross, M., Liu, Z., Yoon, S. J., He, L., Hackett, P. B., Faras, A. J., Kapuscinski, A. R., and Guise, K. S. 1990. Gene expression promoted by the RSV long terminal repeat element in transgenic goldfish. Anim. Biotech. *1*, 79–93.

Halseth, V. 1984. En genetisk og fenotypisk analyse av eggstørrelse, rognvolum og rogantall hos Atlantisk laks. [A genetic and phenotypic analysis of egg size, egg volume, and egg number in Atlantic salmon.] Doctoral Dissertation, Agricultural University of Norway, Ås. (Cited in Refstie [1987])

Hammerman, I. S. and Avtalion, R. R. 1979. Sex determination in *Sarotherodon* (*Tilapia*). Part 2: The sex ratio as a tool for the determination of genotype—a model of autosomal and gonosomal influence. Theor. Appl. Genet. *55*, 177–187.

Happe, A., Quillet, E., and Chevassus, B. 1988. Early life history of triploid rainbow trout (*Salmo gairdneri* Richardson). Aquaculture *71*, 107–118.

Harless, J., Nairn, R. S., Svensson, R., Kallman, K. D., and Morizot, D. C. 1991. Mapping of two thyroid hormone receptor-related (*erb*A-like) DNA sequences to linkage groups U4 and XIII of *Xiphophorus* fishes (Poeciliidae). J. Hered. *82*, 256–259.

Harvey, W. D. 1990. Electrophoretic techniques in forensics and law enforcement. *In* Electrophoretic and Isoelectric Focusing Techniques in Fisheries Management, D. H. Whitmore (Editor). CRC Press, Boca Raton, FL.

Harvey, W. D. and Fries, L. T. 1987. Identification of *Morone* species and congeneric hybrids using isoelectric focusing. Proc. Southeast. Assoc. Fish. Wild Agen. *41*, 251–256.

Haus, E. O. 1984. En genetisk og fenotypisk analyse av rognstørrelse volum og antall hos regnbueaure. [A genetic and phenotypic analysis of egg size, egg volume, and number in rainbow trout.] Doctoral Dissertation, Agricultural University of Norway, Ås, Norway. (Cited in Refstie [1987])

Hayat, M., Joyce, C. P., Townes, T. M., Chen, T. T., Powers, D. A., and Dunham, R. A. 1991. Survival and integration rate of channel catfish and common carp embryos microinjected with DNA at various developmental stages. Aquaculture *99*, 249–255.

Hayford, C. O. and Embody, G. C. 1930. Further progress in the selective breeding of brook trout at the New Jersey State Hatchery. Trans. Am. Fish. Soc. *60*, 109–113.

Herbinger, C. M. and Newkirk, G. F. 1987. Atlantic salmon (*Salmo salar*) maturation timing: Relations between age at maturity and other life history traits: Implications for selective breeding. *In* Selection, Hybridization, and Genetic Engineering in Aquaculture, Vol. 1, K. Thiews (Editor). H. Heenemann GmbH and Co., Berlin, Germany.

Herke, S. W., Kornfield, I., Moran, P., and Moring, J. R. 1990. Molecular confirmation of hybridization between northern pike (*Esox lucius*) and chain pickerel (*E. niger*). Copeia *1990*, 846–850.

Hershberger, W. K. 1983. Personal communication. College of Fisheries, Univ. of Washington, Seattle, WA.

Hershberger, W. K. 1985. Personal communication. College of Fisheries, Univ. of Washington, Seattle, WA.

Hershberger, W. K. 1988. U.S. salmon breeding experience. *In* Genetics, Breeding and Domestication of Farmed Salmon Workshop, E. A. Kenney (Editor). Ministry of Agriculture and Fisheries and B.C. Salmon Farmers' Association, North Vancouver, B.C., Canada.

Hershberger, W. K. 1992. Genetic variability in rainbow trout populations. Aquaculture *100*, 51–71.

Hershberger, W. K., Myers, J. M., Iwamoto, R. N., McAuley, W. C., and Saxton, A. M. 1990A. Genetic changes in the growth of coho salmon (*Oncorhynchus kisutch*) in marine net-pens, produced by ten years of selection. Aquaculture *85*, 187–197.

Hershberger, W. K., Myers, J. M., Iwamoto, R. N., and McAuley, W. C. 1990B. Assessment of inbreeding and its implications for salmon broodstock development. *In* Genetics in Aquaculture: Proceedings of the Sixteenth U.S.-Japan Meeting on Aquaculture, Charleston, South Carolina, October 20 and 21, 1987, R. S. Svrjeck (Editor). NOAA Technical Report NMFS 92, U.S. Department of Commerce, Springfield, VA.

Hickling, C. F. 1960. The Malacca tilapia hybrids. J. Genet. *57*, 1–10.

Hildemann, W. H. 1954. Effects of sex hormones on the secondary sex characters of Lebistes reticulatus. J. Exp. Zool. *126*, 1–15.

Hill, T. K., Pardue, G. B., and Smith, B. W. 1971. An evaluation of several marks on channel catfish, *Ictalurus punctatus* (Rafinesque). Proc. Ann. Conf. Southeast. Assoc. Game Fish Comm. *24*, 304–307.

Hines, R. S., Wohlfarth, G. W., Moav, R., and Hulata, G. 1974. Genetic differences in susceptibility to two diseases among strains of the common carp. Aquaculture *3*, 187–197.

Hoffman, G. L., Dunbar, C. E., and Bradford, A. 1962. Whirling disease of trouts caused by *Myxosoma cerebralis* in the United States. U. S. Fish Wild. Serv. Spec. Sci. Rep. Fish. No. 427.

Hollebecq, M. G., Chourrout, D., Wohlfarth, G., and Billard, R. 1986. Diploid gynogenesis induced by heat shocks after activation with UV-irradiated sperm in common carp. Aquaculture *54*, 69–76.

Hollebecq, M. G., Chambeyron, F., and Chourrout, D. 1988. Triploid common carp produced by heat shock. *In* Reproduction in Fish—Basic and Applied Aspects of Endocrinology and Genetics, Y. Zohar, and B. Breton (Editors). INRA Publications, Paris, France.

Hopkins, K. D. 1979. Production of monosex tilapia fry by breeding sex-reversed fish. Doctoral Dissertation, Auburn University, AL.

Horn, J. L. 1981. Spawning success, fecundity and egg size in seven genetic groups of four-year-old channel catfish, *Ictalurus punctatus* (Rafinesque). Master's Thesis, Auburn Univ., AL.

Hörstgen-Schwark, G., Fricke, H., and Langholtz, H.-J. 1986. The effect of strain crossing on the production performance in rainbow trout. Aquaculture *57*, 141–152.

Houghton, G., Wiergertjes, G. F., Groeneveld, A., and van Muiswinkel, W. B. 1991. Differences in resistance of carp, *Cyprinus carpio* L., to atypical *Aeromonas salmonicida*. J. Fish. Dis. *14*, 333–341.

Hovey, S. J., King, D. P. F., Thompson, D., and Scott, A. 1989. Mitochondrial DNA and allozyme analysis of Atlantic salmon, *Salmo salar* L., in England and Wales. J. Fish. Biol. *35(Suppl. A)*, 253–260.

Huang, C.-M. and Liao, I.-C. 1990. Response to mass selection for growth rate in *Oreochromis niloticus*. Aquaculture *85*, 199–205.

Huang, N. and Gall, G. A. E. 1990. Correlation of body weight and reproductive characteristics in rainbow trout. Aquaculture *86*, 191–200.

Hulata, G., Moav, R., and Wohlfarth, G. 1974. The relationship of gonad and egg size to weight and age in the European and Chinese races of the common carp *Cyprinus carpio* L. J. Fish. Biol. *6*, 745–758.

Hulata, G., Moav, R., and Wohlfarth, G. 1982. Effects of crowding and availability of food on growth rate of fry in the European and Chinese races of the common carp. J. Fish Biol. *20*, 323–327.

Hulata, G., Wohlfarth, G., and Rothbard, S. 1983. Progeny-testing selection of tilapia broodstocks producing all-male hybrid progenies—preliminary results. Aquaculture *33*, 263–268.

Hulata, G., Rothbard, S., Itzkovich, J., Wohlfarth, G., and Halevy, A. 1985A. Differences in hybrid fry production between two strains of the Nile tilapia. Prog. Fish-Cult. *47*, 42–49.

Hulata, G., Wohlfarth, G., and Moav, R. 1985B. Genetic differences between the Chinese and European races of the common carp, *Cyprinus carpio* L. IV. Effects of sexual maturation on growth patterns. J. Fish. Biol. *26*, 95–103.

Hulata, G., Wohlfarth, G. W., and Halevy, A. 1986. Mass selection for growth rate in the Nile tilapia (*Oreochromis niloticus*). Aquaculture *57*, 177–184.

Hulata, G., Wohlfarth, G. W., and Halevy, A. 1988. Comparative growth tests of *Oreochromis niloticus* × *O. aureus* hybrids derived from different farms in Israel, in polyculture. *In* The Second International Symposium on Tilapia in Aquaculture, R. S. V. Pullin, T. Bhukaswan, K. Tonguthai, and J. L. Maclean (Editors). ICLARM Conference Proceedings 15, Department of Fisheries, Bangkok, Thailand and International Center for Living Aquatic Resources Management, Manila Philippines.

Hunter, G. A. and Donaldson, E. M. 1983. Hormonal sex control and its application to fish culture. *In* Fish Physiology, Vol. 9, Reproduction, Part B, Behavior and Fertility Control, W. S. Hoar, D. J. Randall, and E. M. Donaldson (Editors). Academic Press, New York.

Hunter, G. A., Donaldson, E. M., Goetz, F. W., and Edgell, P. R. 1982. Production of all-female and sterile coho salmon, and experimental evidence for male heterogamety. Trans. Am. Fish. Soc. *111*, 367–372.

Hunter, G. A., Donaldson, E. M., Stoss, J., and Baker, I. 1983. Production of monosex female groups of chinook salmon (*Oncorhynchus tshawytscha*) by the fertilization of normal ova with sperm from sex-reversed females. Aquaculture *33*, 355–364.

Hurrell, R. H., and Price, D. J. 1991. Natural hybrids between Atlantic salmon, *Salmo salar* L., and trout, *Salmo trutta* L., in juvenile salmonid populations in south-west England. J. Fish Biol. *39(Suppl. A)*, 335–341.

Hussain, M. G., Chatterji, A., McAndrew, B. J., and Johnstone, R. 1991. Triploidy induction in Nile tilapia, *Oreochromis niloticus* L. using pressure, heat and cold shocks. Theor. Appl. Genet. *81*, 6–12.

I

Ihssen, P. E. 1986. Selection of fingerling rainbow trout for high and low tolerance to high temperature. Aquaculture *57*, 370.

Ihssen, P. and Tait, J. S. 1974. Genetic differences in retention of swimbladder gas between two populations of lake trout (*Salvelinus namaycush*). J. Fish. Res. Board Can. *31*, 1351–1354.

Ijiri, K.-I. and Egami, N. 1980. Hertwig effect caused by UV-irradiation of sperm of *Oryzias latipes* (Teleost) and its photoreactivation. Mutation Res. *69*, 241–248.

Indig, F. E. and Moav, B. 1988. A prokaryotic gene is expressed in fish cells and persists in Tilapia embryos following microinjection through the micropyle. *In* Reproduction in Fish—Basic and Applied Aspects in Endocrinology and Genetics, Y. Zohar and B. Breton (Editors). INRA Publications, Paris.

Itzkovich, J., Rothbard, S., and Hulata, G. 1981. Inheritance of pink body colouration in Cichlasoma nigrofasciatum Günther (Pisces, Cichlidae). Genetica *55*, 15–16.

Iwamoto, R. N., Saxton, A. M., and Hershberger, W. K. 1982. Genetic estimates for length and weight of coho salmon during freshwater rearing. J. Hered. *73*, 187–191.

Iwamoto, R. N., Alexander, B. A., and Hershberger, W. K. 1984. Genotypic and environmental effects on the incidence of sexual precocity in coho salmon (*Oncorhynchus kisutch*). Aquaculture *43*, 105–121.

Iwamoto, R. N., Myers, J. M., and Hershberger, W. K. 1986. Genotype-environment interactions for growth of rainbow trout, *Salmo gairdneri*. Aquaculture *57*, 153–161.

Iwamoto, R. N., Myers, J. M., and Hershberger, W. K. 1990. Heritability and genetic correlations for flesh coloration in pen-reared coho salmon. Aquaculture *86*, 181–190.

J

Jalabert, B., Moreau, J., Planquette, P., and Billard, R. 1974. Déterminisme du sexe chez *Tilapia macrochir* et *Tilapia nilotica*: action de la méthyltestostérone dans l'alimentation des alevins sur la différenciation sexuelle; proportion des sexes dans la descendance des mâles "inversés." [Sex determination in *Tilapia macrochir* and *Tilapia nilotica*: Effect of methyltestosterone administered in fry feed on sex differentiation; sex ratio of the offspring produced by sex-reversed males.] Ann. Biol. Anim. Biochim. Biophys. *14*, 729–739.

Jansson, H., Holmgren, I., Wedin, K., and Anderson, T. 1991. High frequency of natural hybrids between Atlantic salmon, *Salmo salar* L., and brown trout, *S. trutta* L., in a Swedish river. J. Fish Biol. *39 (Suppl. A)*, 343–348.

Jarimopas, P. 1986. Realized response of Thai red tilapia to weight-specific selection for growth. *In* The First Asian Fisheries Forum, J. L. Maclean, L. B. Dizon, and L. V. Hosillos (Editors). Asian Fisheries Society, Manila, Philippines.

Jayaprakas, V., Tave, D., and Smitherman, R. O. 1988. Growth of two strains of *Oreochromis niloticus* and their F_1, F_2 and backcross hybrids. *In* The Second International Symposium on Tilapia in Aquaculture, R. S. V. Pullin, T. Bhukaswan, K. Tonguthai, and J. L. Maclean (Editors). ICLARM Conference Proceedings 15, Department of Fisheries, Bangkok, Thailand and International Center for Living Aquatic Resources Management, Manila, Philippines.

Johansson, N. 1981. General problems in Atlantic salmon rearing in Sweden. *In* Fish Gene Pools: Preservation of Genetic Resources in Relation to Wild Fish Stocks, N. Ryman (Editor). Ecological Bulletins No. 34, Forskningsrådsnämnden, Stockholm, Sweden.

John, G., Reddy, P. V. G. K., and Gupta, S. D. 1984. Artificial gynogenesis in two Indian major carps, *Labeo rohita* (Ham.) and *Catla catla* (Ham.). Aquaculture *42*, 161–168.

Johnson, K. R. and Wright, J. E. 1986. Female brown trout × Atlantic salmon hybrids produce gynogens and triploids when backcrossed to male Atlantic salmon. Aquaculture *57*, 345–358.

Johnson, O. W., Rabinovich, P. R., and Utter, F. M. 1984. Comparison of the reliability of a Coulter Counter with a flow cytometer in determining ploidy levels in Pacific salmon. Aquaculture *43*, 99–103.

Johnson, O. W., Dickhoff, W. W., and Utter, F. M. 1986. Comparative growth and development of diploid and triploid coho salmon, *Oncorhynchus kisutch*. Aquaculture *57*, 329–336.

Johnstone, R. 1985. Induction of triploidy in Atlantic salmon by heat shock. Aquaculture *49*, 133–139.

Johnstone, R. 1987. Survival rates and triploidy rates following heat shock in Atlantic salmon ova retained for different intervals in the body cavity after first stripping together with preliminary observations on the use of pressure. *In* Selection, Hybridization, and Genetic Engineering in Aquaculture, Vol. 2, K. Thiews (Editor). H. Heenemann GmbH and Co., Berlin, Germany.

Johnstone, R. and Youngson, A. F. 1984. The progeny of sex-inverted female Atlantic salmon (*Salmo salr* L.). Aquaculture *37*, 179–182.

Johnstone, R., Simpson, T. H., Youngson, A. F., and Whitehead, C. 1979. Sex reversal in salmonid culture Part II. The progeny of sex-reversed rainbow trout. Aquaculture *18*, 13–19.

Johnstone, R., Macintosh, D. J., and Wright, R. S. 1983. Elimination of orally administered 17 α-methyltestosterone by *Oreochromis mossambicus* (tilapia) and *Salmo gairdneri* (rainbow trout) juveniles. Aquaculture *35*, 249–257.

Johnstone, R., Knott, R. M., Macdonald, A. G., and Walsingham, M. V. 1989.

Triploidy induction in recently fertilized Atlantic salmon ova using anaesthetics. Aquaculture *78*, 229–236.

Joyce, J. A. and El-Ibiary, H. M. 1977. Persistency of hot brands and their effects on growth and survival of fingerling channel catfish. Prog. Fish-Cult. *39*, 112–114.

K

Kaastrup, P. and Hørlyck, V. 1987. Development of a simple method to optimize the conditions for producing gynogenetic offspring, using albino rainbow trout, *Salmo gairdneri* Richardson, females as an indicator for gynogenesis. J. Fish Biol. *31(Suppl. A)*, 29–33.

Kaastrup, P., Hørlyck, V., Olesen, N. J., Lorenzen, N., Vestergaard Jørgensen, P. E., and Berg, P. 1991. Paternal association of increased susceptibility to viral haemorrhagic septicaemia (VHS) in rainbow trout (*Oncorhynchus mykiss*). Can. J. Fish. Aquat. Sci. *48*, 1188–1192.

Kajishima, T. 1977. Genetic and developmental analysis of some new color mutants in the goldfish, *Carassius auratus*. Genetics *86*, 161–174.

Kallman, K. D. 1970. Different genetic basis of identical pigment patterns in two populations of platyfish, *Xiphophorus maculatus*. Copeia *1970*, 472–487.

Kallman, K. D. 1971. Inheritance of melanophore patterns and sex determination in the Montezuma swordtail, *Xiphophorus montezumae cortezi* Rosen. Zoologica *56*, 77–94.

Kallman, K. D. 1975. The platyfish, *Xiphophorus maculatus. In* Handbook of Genetics, Volume 4. Vertebrates of Genetic Interest, R. C. King (Editor). Plenum Press, New York.

Kallman, K. D. 1984. A new look at sex determination in Poeciliid fishes. *In* Evolutionary Genetics of Fishes, B. J. Turner (Editor). Plenum Press, New York.

Kallman, K. D. and Brunetti, V. 1983. Genetic basis of three mutant color varieties of *Xiphophorus maculatus:* The gray, gold and ghost platyfish. Copeia *1983*, 170–181.

Kanis, E., Refstie, T., and Gjedrem, T. 1976. A genetic analysis of egg, alevin and fry mortality in salmon (*Salmo salar*), sea trout (*Salmo trutta*) and rainbow trout (*Salmo gairdneri*). Aquaculture *8*, 259–268.

Kapuscinski, A. R. and Hallerman, E. M. 1990. Transgenic fish and public policy: Anticipating environmental impacts of transgenic fish. Fisheries *15(1)*, 2–11.

Kartavtsev, Y. F., Salmenkova, E. A., Rubtsova, G. A., and Afanes'ev, K. I. 1990. Familial analysis of allozyme variability and its interaction with body size and offspring survival in the salmon Oncorhynchus gorbuscha (Walb). Sov. Genet. *26*, 1060–1067.

Kasahara, S. 1985. Personal communication. Faculty of Applied Biological Science, Hiroshima Univ., Hiroshima-ken, Japan.

Katasonov, V. Ya. 1973. Investigation of color in hybrids of common and ornamental (Japanese) carp. Communication I. Transmission of dominant color types. Sov. Genet. *9*, 985–992.

Katasonov, V. Ya. 1974. Investigation of color in hybrids of common and ornamental (Japanese) carp. II. Pleiotropic effect of dominant color genes. Sov. Genet. *10*, 1504–1512.

Katasonov, V. Ya. 1976. Lethal action of the light color gene in carp (*Cyprinus carpio* L.). Sov. Genet. *12*, 514–516.

Katasonov, V. Ya. 1978. Color in hybrids of common and ornamental (Japanese) carp. III. Inheritance of blue and orange color types. Sov. Genet. *14*, 1522–1528.

Kelsch, S. W. and Hendricks, F. S. 1986. An electrophoretic and multivariate morphometric comparison of the American catfishes *Ictalurus lupus* and *I. punctatus*. Copeia *1986*, 646–652.

Kerby, J. H., Geiger, J. G., Harrell, R. M., Starling, C. C. and Revels, H. 1991. Relative growth and survival between triploid and diploid palmetto and sunshine striped bass to phase II. J. World. Aquacul. Soc. *22*, 32A–33A.

Khater, A. A. E. 1985. Identification and comparison of three *Tilapia nilotica* strains for selected aquacultural traits. Doctoral Dissertation, Auburn Univ., AL.

Khater, A. A. and Smitherman, R. O. 1988. Cold tolerance and growth of three strains of *Oreochromis niloticus*. *In* The Second International Symposium on Tilapia in Aquaculture, R. S. V. Pullin, T. Bhukaswan, K. Tonguthai, and J. L. Maclean (Editors). ICLARM Conference Proceedings 15, Department of Fisheries, Bangkok, Thailand and International Center for Living Aquatic Resources Management, Manila, Philippines.

Kim, D. S., Kim, I.-B., and Baik, Y. G. 1986. A report of triploid rainbow trout production in Korea. Bull. Korean Fish. Soc. *19*, 575–580.

Kim, D. S., Kim, I.-B., and Baik, Y. G. 1988. Early growth and gonadal development of triploid rainbow trout, *Salmo gairdneri*. J. Aquacul. *1*, 41–51.

Kincaid, H. L. 1975. Iridescent metallic blue color variant in rainbow trout. J. Hered. *66*, 100–102.

Kincaid, H. L. 1976A. Effects of inbreeding on rainbow trout populations. Trans. Am. Fish. Soc. *105*, 273–280.

Kincaid, H. L. 1976B. Inbreeding in rainbow trout (*Salmo gairdneri*). J. Fish. Res. Board Can. *33*, 2420–2426.

Kincaid, H. L. 1977. Rotational line crossing: An approach to the reduction of inbreeding accumulation in trout brood stocks. Prog. Fish-Cult. *39*, 179–181.

Kincaid, H. L. 1979. Development of standard reference lines of rainbow trout. Trans. Am. Fish. Soc. *108*, 457–461.

Kincaid, H. L. 1981. Trout strain registry. National Fisheries Center—Leetown, U.S. Fish and Wildlife Service, Kearneysville, WV.

Kincaid, H. L. 1983A. Results from six generations of selection for accelertaed [sic] growth rate in a rainbow trout population. Abstracts. The Future of Aquaculture in North America. Fish Culture Selection of the American Fisheries Society, 26–27.

Kincaid, H. L. 1983B. Inbreeding in fish populations used for aquaculture. Aquaculture *33*, 215–227.

Kincaid, H. L. 1983C. Personal communication. Fish Genetics Station, National Fisheries Center—Leetown, Kearneysville, WV.

Kincaid, H. L. 1987. Personal communication, National Fishery Research and Development Laboratory, Wellsboro, PA.

Kincaid, H. L., Bridges, W. R., and von Limbach, B. 1977. Three generations of selection for growth rate in fall-spawning rainbow trout. Trans. Am. Fish. Soc. *106*, 621–628.

Kindschi, G. A., Smith, C. E., and Koby, R. F, Jr. 1991. Performance of two strains of rainbow trout reared at four densities with supplemental oxygen. Prog. Fish-Cult. *53*, 203–209.

Kinghorn, B. 1983. Genetic variation in food conversion efficiency and growth in rainbow trout. Aquaculture *32*, 141–155.

Kirpichnikov, V. S. 1970. Goals and methods in carp selection. *In* Selective Breeding of Carp and Intensification of Fish Breeding in Ponds, V. S. Kirpichnikov (Editor). Israel Program for Scientific Translations, Jerusalem, Israel.

Kirpichnikov, V. S. 1972. Methods and effectiveness of breeding the Ropshian carp. Communication I. Purposes of breeding, initial forms, and system of crosses. Sov. Genet. *8*, 996–1001.

Kirpichnikov, V. S. 1981. Genetic Bases of Fish Selection. Springer-Verlag, New York.

Kirpichnikov, V. S. 1987. Selection and new breeds of pond fishes in the USSR. *In* Selection, Hybridization, and Genetic Engineering in Aquaculture, Vol. 2, K. Thiews (Editor). H. Heenemann GmbH and Co., Berlin, Germany.

Kirpichnikov, V. S. and Faktorovich, K. A. 1969. Genetische Methoden der Fischkrankheits-beitsbekaempfung. [Genetic methods for the control of fish diseases.] Z. Fisch. Hilfswiss. (NF), *17*, 227–236.

Kirpichnikov, V. S. and Faktorovich, K. A. 1972. Increase in the resistance of carp to dropsy by means of breeding. Communication II. The course of the selection and evaluation of the breeding groups selected. Sov. Genet. *8*, 592–600.

Kirpichnikov, V. S., Faktorovich, K. A., Babushkin, Y. P., Zhivotova, M. A., and Tolmacheva, N. V. 1967. Comparative resistance of different carp varieties to dropsy. Sov. Genet. *3*, 39–48.

Kirpichnikov, V. S., Faktorovich, K. A., and Suleimanyan, V. S. 1972A. Increasing the resistance of carp to dropsy by breeding. Communication I. Methods of inbreeding for resistance. Sov. Genet. *8*, 306–312.

Kirpichnikov, V. S., Ponomarenko, K. V., Tolmacheva, N. V., and Tsoi, R. M. 1972B. Methods and effectiveness of breeding Ropshian carp. Communication II. Methods of selection. Sov. Genet. *8*, 1108–1115.

Kirpichnikov, V. S., Factorovich, K. A., Ilyasov, Y. I., and Shart, L. A. 1979. Selection of common carp (*Cyprinus carpio*) for resistance to dropsy. *In* Advances in Aquaculture, T. V. R. Pillay and W. A. Dill (Editors). Fishing News Books, Farnham, Surrey, England.

Kirpitschnikow, W. S. and Faktorowitsch, K. A. 1969. See Kirpichnikov and Factorovich (1969).

Klupp, R. 1979. Genetic variance for growth in rainbow trout (*Salmo gairdneri*). Aquaculture *18*, 123–134.

Klupp, R., Heil, G., and Pirchner, F. 1978. Effects of interaction between strains and environment on growth traits in rainbow trout (*Salmo gairdneri*). Aquaculture *14*, 271–275.

Knox, D. and Verspoor, E. 1991. A mitochondrial DNA restriction fragment length polymorphism of potential use for discrimination of farmed Norwegian and wild Atlantic salmon populations in Scotland. Aquaculture *98*, 249–257.

Kolesnikov, V. A., Alimov, A. A., Barmintsev, V. A., Benyumov, A. O., Zelenia, I. A., Krasnov, A. M., Dzhabur, R., and Zelenin, A. V. 1990. High velocity mechanical injection of foreign DNA into fish eggs. Sov. Genet. *26*, 1383–1386.

Komada, N. 1977. Influence of temperature on the vertebral number of the ayu, *Plecoglossus altivelis*. Copeia *1977*, 572–573.

Komen, J., Duynhouwer, J., Richter, C. J. J., and Huisman, E. A. 1988. Gynogenesis in common carp (*Cyprinus carpio* L.) I. Effects of genetic manipulation of sexual products and incubation conditions of eggs. Aquaculture *69*, 227–239.

Komen, J., Bongers, A. B. J., Richter, C. J. J., van Muiswinkel, W. B., and Huisman, E. A. 1991. Gynogenesis in common carp (*Cyprinus carpio* L.) II. The production of homozygous gynogenetic clones and F_1 hybrids. Aquaculture *92*, 127–142.

Kosswig, C. 1964. Polygenic sex determination. Experientia *20*, 190–199.

Kowtal, G. V. 1987. Preliminary experiments in induction of polyploidy, gynogenesis and androgenesis in the white sturgeon, *Acipenser transmontanus* Richardson. *In* Selection, Hybridization, and Genetic Engineering in Aquaculture, Vol. 2, K. Thiews, (Editor). H. Heenemann GmbH and Co., Berlin, Germany.

Krasznai, Z. and Márián, T. 1986. Shock-induced triploidy and its effect on growth and gonad development of the European catfish, *Silurus glanis* L. J. Fish Biol. *29*, 519–527.

Krasznai, Z. L. and Márián, T. 1987. Induced gynogenesis on European catfish (*Silurus glanis* L.). *In* Selection, Hybridization, and Genetic Engineering in Aquaculture, Vol. 2, K. Thiews, (Editor). H. Heenemann GmbH and Co., Berlin, Germany.

Kronert, U., Hörstgen-Schwark, G., and Langholz H.-J. 1987. Investigations on selection of tilapia for late maturity. *In* Selection, Hybridization, and Genetic Engineering in Aquaculture, Vol. 1, K. Thiews (Editor). H. Heenemann GmbH and Col., Berlin, Germany.

Kronert, U., Hörstgen-Schwark, G., and Langholz, H.-J. 1989. Prospects of selecting for late maturity in tilapia (*Oreochromis niloticus*) I. Family studies under laboratory conditions. Aquaculture *77*, 113–121.

Kurokura, H., Kumai, H., and Nakamura, M. 1986. Hybridization between female red sea bream (*Pagrus major*) and male crimson sea bream (*Evynnis japonica*) by means of sperm cryopreservation. *In* The First Asian Fisheries Forum, J. L. Maclean, L. B. Dizon, and L. V. Hosillos (Editors). Asian Fisheries Society, Manila, Philippines.

Kuzema, A. I. 1971. The Ukrainian breeds of carp. Seminar/Study Tour in the U.S.S.R. on Genetic Selection and Hybridization of Cultivated Fishes, 19 April-29 May, 1968. FAO/UNDP(TA) *2926*, 228–232.

L

Lahav, M. and Lahav, E. 1990. The development of all-male tilapia hybrids in Nir David. Bamidgeh *42*, 58–61.

Langholz, H.-J. and Hörstgen-Schwark, G. 1987. Family selection in rainbow trout. *In* Selection, Hybridization, and Genetic Engineering in Aquaculture, Vol. 1, K. Thiews (Editor). H. Heenemann GmbH and Co., Berlin, Germany.

Lasley, J. F. 1978. Genetics of Livestock Improvement, 3rd edition. Prentice-Hall, Englewood Cliffs, NJ.

Latter, B. D. H. 1959. Genetic sampling in a random mating population of constant size and sex ratio. Aust. J. Biol. Sci. *12*, 500–505.

Leary, R. F., Allendorf, F. W., and Knudsen, K. L. 1984. Superior developmental stability of heterozygotes at enzyme loci in salmonid fishes. Am. Natur. *124*, 540–551.

Leary, R. F., Allendorf, F. W., and Knudsen, K. L. 1985A. Developmental instability as an indicator of reduced genetic variation in hatchery trout. Trans. Am. Fish. Soc. *114*, 230–235.

Leary, R. F., Allendorf, F. W., and Knudsen, K. L. 1985B. Inheritance of meristic variation and the evolution of developmental stability in rainbow trout. Evolution *39*, 308–314.

Leary, R. F., Allendorf, F. W., Knudsen, K. L., and Thorgaard, G. H. 1985C. Heterozygosity and developmental stability in gynogenetic diploid and triploid rainbow trout. Heredity *54*, 219–225.

Leary, R. F., Allendorf, F. W., and Knudsen, K. L. 1991. Effects of rearing density on meristics and developmental stability of rainbow trout. Copeia *1991*, 44–49.

Leberg, P. L. 1990. Influence of genetic variability on population growth: implications for conservation. J. Fish Biol. *37(Suppl A)*, 193–195.

LeGrande, W. H., Dunham, R. A., and Smitherman, R. O. 1984. Karyology of three species of catfishes (Ictaluridae: *Ictalurus*) and four hybrid combinations. Copeia. *1984*, 873–878.

Leider, S. A., Hulett, P. L., Loch, J. J., and Chilcote, M. W. 1990. Electrophoretic comparison of the reproductive success of naturally spawning transplanted and wild steelhead trout through the returning adult stage. Aquaculture *88*, 239–252.

Leong, J.-A. C., Barrie, R., Engelking, H. M., Feyereisen-Koener, J., Gilmore, R., Harry, J., Kurath, G., Manning, D. S., Mason, C. L., Oberg, L., and Wirkkula, J. 1990. Recombinant viral vaccines in aquaculture. *In* Genetics in

Aquaculture: Proceedings of the Sixteenth U.S.-Japan Meeting on Aquaculture, Charleston, South Carolina, October 20 and 21, 1987, R. S. Svrjcek (Editor). NOAA Technical Report NMFS 92, U.S. Department of Commerce, Springfield, VA.

Lester, L. J., Abella, T. A., Palada, M. S., and Keus, H. J. 1988. Genetic variation in size and sexual maturation of *Oreochromis niloticus* under hapa and cage culture conditions. *In* The Second International Symposium on Tilapia in Aquaculture, R. S. V. Pullin, T. Bhukaswan, K. Tonguthai, and J. L. Maclean (Editors). ICLARM Conference Proceedings 15, Department of Fisheries, Bangkok, Thailand and International Center for Living Aquatic Resources Management, Manila, Philippines.

Lester, L. J., Lawson, K. S., Abella, T. A., and Palada, M. S. 1989. Estimated heritability of sex ratio and sexual dimorphism in tilapia. Aquacul. Fish. Manage. *20*, 369–380.

Lewis, R. C. 1944. Selective breeding of rainbow trout at Hot Creek Hatchery. Calif. Fish Game *30*, 95–97.

Lewis, W. M. and Heidinger, R. 1971. Supplemental feeding of hybrid sunfish populations. Trans. Am. Fish. Soc. *100*, 619–623.

Li, S., Lu, W., Peng, C., and Zhao, P. 1987A. Growth performance of different populations of silver carp and big head. *In* Selection, Hybridization, and Genetic Engineering in Aquaculture, Vol. 1., K. Thiews (Editor). H. Heenemann GmbH and Co., Berlin, Germany.

Li, S., Lu, W., Peng, C., and Zhao, P. 1987B. A genetic study of the growth performance of silver carp from the Changjiang and Zhujiang Rivers. Aquaculture *65*, 93–104.

Li, Y., Gold, J. R., Tave, D., Gibson, M. D., Barnett, J., Fiegel, D. H., and Beavers, B. F. 1991. A cytogenetic analysis of the karyotypes of the golden shiner, *Notemigonus crysoleucas*, the rudd, *Scardinus erythrophthalmus*, and their reciprocal F_1 hybrids. J. Appl. Aquacul. *1(2)*, 79–87.

Lim, C. and Lovell, R. T. 1978. Pathology of the vitamin C deficiency syndrome in channel catfish (*Ictalurus punctatus*). J. Nutr. *108*, 1137–1146.

Lincoln, R. F. 1981A. The growth of female diploid and triploid plaice (*Pleuronectes plastessa*) × flounder (*Platichthys flesus*) hybrids over one spawning season. Aquaculture *25*, 259–268.

Lincoln, R. F. 1981B. Sexual maturation in triploid male plaice (*Pleuronectes platessa* and plaice × flounder (*Platichthys flesus* hybrids. J. Fish Biol. *19*, 415–426.

Lincoln, R. F. 1981C. Sexual maturation in female triploid plaice, *Pleuronectes, platessa*, and plaice × flounder, *Platichthys flesus*, hybrids. J. Fish Biol. *19*, 499–507.

Lincoln, R. F. and Scott, A. P. 1983. Production of all-female triploid rainbow trout. Aquaculture *30*, 375–380.

Lincoln, R. F. and Scott, A. P. 1984. Sexual maturation in triploid rainbow trout, *Salmo gairdneri* Richardson. J. Fish Biol. *25*, 385–392.

Linder, D., Sumari, O., Nyholm, K., and Sirkkomaa, S. 1983. Genetic and phenotypic variation in production traits in rainbow trout strains and strain crosses in Finland. Aquaculture *33*, 129–134.

Linhart, O., Kvasnička, P., Šlechtová, V., and Pokorný, J. 1986. Induced gynogenesis by retention of the second polar body in the common carp, *Cyprinus carpio* L., and heterozygosity of gynogenetic progeny in transferrin and Ldh-B[1] loci. Aquaculture *54*, 63–67.

Lodi, E. 1978. Palla: A hereditary vertebral deformity in the guppy, Poecilia reticulata Peters (Pisces, Osteichthyes). Genetica *48*, 197–200.

Lone, K. P. and Matty A. J. 1981. Uptake and disappearance of radioactivity in blood and tissues of carp (*Cyprinus carpio*) after feeding ^{3}H-testosterone. Aquaculture *24*, 315–326.

Lou, Y. D. and Purdom, C. E. 1984A. Polyploidy induced by hydrostatic pressure in rainbow trout, *Salmo gairdneri* Richardson. J. Fish. Biol. *25*, 345–351.

Lou, Y. D. and Purdom, C. E. 1984B. Diploid gynogenesis induced by hydrostatic pressure in rainbow trout, *Salmo gairdneri* Richardson. J. Fish Biol. *24*, 665–670.

Lovell, R. T. and Lim, C. 1978. Vitamin C in pond diets for channel catfish. Trans. Am. Fish. Soc. *107*, 321–325.

M

Macaranas, J. and Fujio, Y. 1990. Strain differences in cultured fish—isozymes and performance traits as indicators. Aquaculture *85*, 69–82.

Macaranas, J. M., Taniguchi, N., Pante, M. J. R., Capili, J. B., and Pullin, R. S. V. 1986. Electrophoretic evidence for extensive hybrid gene introgression into commercial *Oreochromis niloticus* (L.) stocks in the Philippines. Aquacul. Fish. Manage. *17*, 249–258.

Maclean, N. and Penman, D. 1990. The application of gene manipulation to aquaculture. Aquaculture *85*, 1–20.

Maclean, N., Penman, D., and Talwar, S. 1987A. Introduction of novel genes into the rainbow trout. *In* Selection, Hybridization, and Genetic Engineering in Aquaculture, Vol. 2, K. Thiews (Editor). H. Heeneman GmbH and Co., Berlin.

Maclean, N., Woodall, C., and Crossley, F. 1987B. Injection of the mouse MT-1 gene into rainbow trout eggs and assay of trout fry for resistance to cadmium and zinc toxicity. Experientia Suppl. *52*, 471–475.

McAndrew, B. J. and Majumdar, K. C. 1983. Tilapia stock identification using electrophoretic markers. Aquaculture *30*, 249–261.

McAndrew, B. J., Ward, R. D., and Beardmore, J. A. 1986. Growth rate and heterozygosity in the plaice, *Pleuronectes platessa*. Heredity *57*, 171–180.

McAndrew, B. J., Roubal, F. R., Roberts, R. J., Bullock, A. M., and McEwen, I. M. 1988. The genetics and histology of red, blond and associated colour variants in *Oreochromis niloticus*. Genetica *76*, 127–137.

McEvoy, T., Stack, M., Keane, B., Barry, T., Sreenan, J., and Gannon, F. 1988. The expression of a foreign gene in salmon. Aquaculture *68*, 27–37.

McGeer, J. C., Baranyi, L., and Iwama, G. K. 1991. Physiological responses to challenge tests in six stocks of coho salmon (*Oncorhynchus kisutch*). Can. J. Fish. Aquat. Sci. *48*, 1761–1771.

McGinty, A. S. 1980. Survival, growth and variation in growth of channel catfish fry and fingerlings. Doctoral Dissertation, Auburn Univ., AL.

McGinty, A. S. 1984. Suitability of communal rearing for performance testing of tilapias. Proc. Carib. Food Crops Soc. *19*, 259–266.

McGinty, A. S. 1987. Efficacy of mixed-species commonal rearing as a method for performance testing of tilapias. Prog. Fish-Cult. *49*, 17–20.

McIntyre, J. D. and Amend, D. F. 1978. Heritability of tolerance for infectious hematopoietic necrosis in sockeye salmon (*Oncorhynchus nerka*). Trans. Am. Fish. Soc. *107*, 305–308.

McIntyre, J. D. and Blanc, J.-M. 1973. A genetic analysis of hatching time in steelhead trout (*Salmo gairdneri*). J. Fish. Res. Board Can. *30*, 137–139.

McKay, L. R., Friars, G. W., and Ihssen, P. E. 1984. Genotype × temperature interactions for growth of rainbow trout. Aquaculture *41*, 131–140.

McKay, L. R., Ihssen, P. E., and Friars, G. W. 1986. Genetic parameters of growth in rainbow trout, *Salmo gairdneri*, as a function of age and maturity. Aquaculture *58*, 241–254.

McKay, L. R., Ihssen, P. E., and McMillan, I. 1992A. Early mortality of tiger trout (*Salvelinus fontinalis* × *Salmo trutta*) and the effects of triploidy. Aquaculture *102*, 43–54.

McKay, L. R., McMillan, I., Sadler, S. E., and Moccia, R. D. 1992B. Effects of mating system on inbreeding levels and selection response in salmonid aquaculture. Aquaculture *100*, 100–101.

Mair, G. C. 1992. Caudal deformity syndrome (CDS): an autosomal recessive lethal mutation in the tilapia *Oreochromis niloticus* (L.). J. Fish. Dis. *15*, 71–75.

Mair, G. C., Penman, D. J., Scott, A., Skibinski, D. O. F., and Beardmore, J. A. 1987A. Hormonal sex-reversal and the mechanisms of sex determination in *Oreochromis. In* Selection, Hybridization, and Genetic Engineering in Aquaculture, Vol. 2, K. Thiews (Editor). H. Heenemann GmbH and Co., Berlin, Germany.

Mair, G. C., Scott, A. G., Beardmore, J. A., and Skibinski, D. O. F. 1987B. A technique for induction of diploid gynogenesis in *Oreochromis niloticus* by suppression of the first mitotic division, *In* Selection, Hybridization, and Genetic Engineering in Aquaculture, Vol. 2, K. Thiews, (Editor). H. Heenemann GmbH and Co., Berlin, Germany.

Mair, G. C., Beardmore, J. A., and Skibinski, D. O. F. 1990. Experimental evidence for environmental sex determination in *Oreochromis* species. *In* The Second Asian Fisheries Forum, R. Hirano and I. Hanyu (Editors). Asian Fisheries Society, Manila, Philippines.

Mair, G. C., Scott, A. G., Penman, D. J., Beardmore, J. A., Skibinski, D. O. F. 1991A. Sex determination in the genus *Oreochromis* 1. Sex reversal, gynogenesis and triploidy in *O. niloticus* (L.). Theor. Appl. Genet. *82*, 144–152.

Mair, G. C., Scott, A. G., Penman, D. J., Skibinski, D. O. F., and Beardmore, J. A. 1991B. Sex determination in the genus *Oreochromis* 2. Sex reversal, hybridisation, gynogenesis and triploidy in *O. aureus* Steindachner. Theor. Appl. Genet. *82*, 153–160.

Majumdar, K. C. and McAndrew, B. J. 1983. Sex ratios from interspecific crosses within the tilapias. *In* International Symposium on Tilapia in Aquaculture, L. Fishelson and Z. Yaron, (Editors). Tel Aviv Univ., Tel Aviv, Israel.

Manickam, P. 1991. Triploidy induced by cold shock in the Asian catfish, *Clarius batrachus* (L.). Aquaculture *94*, 377–379.

Matsui, Y. 1934. Genetical studies on gold-fish of Japan. 2. On the Mendelian inheritance of the telescope eyes of gold-fish. J. Imp. Fish. Inst. *30*, 37–46.

May, B. and Johnson, K. R. 1990. Composite linkage map of salmonid fishes (*Salvelinus, Salmo, Oncorhynchus*). *In* Genetic Maps: Locus Maps of Complex Genomes, S. J. O'Brien (Editor). Cold Spring Harbor Laboratory Press, Cold Spring Harbor, NY.

May, B., Henley, K. J., Krueger, C. C., and Gloss, S. P. 1988. Androgenesis as a mechanism for chromosome set manipulation in brook trout (*Salvelinus fontinalis*). Aquaculture *75*, 57–70.

Meffe, G. K. 1986. Conservation genetics and the management of endangered fishes. Fisheries *11(1)*, 14–23.

Meffe, G. K. 1987. Conserving fish genomes: philosophies and practices. Env. Biol. Fishes *18*, 3–9.

Meriwether, F., II. 1980. Induction of polyploidy in Israeli carp. Proc. Southeast. Assoc. Fish Wild. Agen. *34*, 275–279.

Meyer, F. P., Sneed, K. E., and Eschmeyer, P. T. 1973. Second Report to the Fish Farmers. U.S. Bur. Sport Fish. Wild. Res. Pub. 113., Washington, DC.

Millenbach, C. 1950. Rainbow brood-stock selection and observations on its application to fishery management. Prog. Fish-Cult. *12*, 151–152.

Miller, R. R. and Walters, V. 1972. A new genus of cyprinodontid fish from Nuevo Leon, Mexico, Nat. Hist. Mus. Los Angeles Cty. Contrib. Sci. *233*.

Mires, D. 1988. The inheritance of black pigmentation in two African strains of *Oreochromis niloticus*. *In* The Second International Symposium on Tilapia in Aquaculture, R. S. V. Pullin, T. Bhukaswan, K. Tonguthai, and J. L. Maclean (Editors). ICLARM Conference Proceedings 15, Department of Fisheries, Bangkok, Thailand and International Center for Living Aquatic Resources Management, Manila, Philippines.

Mirza, J. A. and Shelton, W. L. 1988. Induction of gynogenesis and sex reversal in silver carp. Aquaculture *68*, 1–14.

Moav, R. and Wohlfarth, G. W. 1968. Genetic improvement of yield in carp. Proceedings of the World Symposium on Warm-water Pond Fish Culture. FAO Fish. Rep. No. 44, *4*, 12–29.

Moav, R. and Wohlfarth, W. G. [sic (G. W.)] 1970. Genetic correlation between seine escapability and growth capacity in carp. J. Hered. *61*, 153–157.

Moav, R. and Wohlfarth, G. W. 1974. Magnification through competition of genetic differences in yield capacity in carp. Heredity *33*, 181–202.

Moav, R. and Wohlfarth, G. 1976. Two-way selection for growth rate in the common carp (*Cyprinus carpio* L.). Genetics *82*, 83–101.

Moav, R., Wohlfarth, G., and Lahman, M. 1960A. Genetic improvement of carp II; marking fish by branding. Bamidgeh *12*, 49–53.

Moav, R., Wohlfarth, G., and Lahman, M. 1960B. An electric instrument for brandmarking fish. Bamidgeh *12*, 92–95.

Moav, R., Hulata, G., and Wohlfarth, G. 1974. The breeding potential of growth curve differences between the European and Chinese races of the common carp. First World Cong. Genet. Appl. Livestock Prod. *3*, 573–578.

Moav, R., Hulata, G., and Wohlfarth, G. 1975. Genetic differences between the Chinese and European races of the common carp. I. Analysis of genotype-environment interactions for growth rate. Heredity *34*, 323–340.

Moav, R., Soller, M., Hulata, G., and Wohlfarth, G. 1976A. Genetic aspects of the transition from traditional to modern fish farming. Theor. Appl. Genet. *47*, 285–290.

Moav, R., Brody, T., Wohlfarth, G., and Hulata, G. 1976B. Applications of electrophoretic genetic markers to fish breeding. I. Advantages and methods. Aquaculture *9*, 217–228.

Moav, R., Brody, T., and Hulata, G. 1978. Genetic improvement of wild fish populations, Science *201*, 1090–1094.

Moav, R., Brody, T., Wohlfarth, G., and Hulata, G. 1979. A proposal for the continuous production of F_1 hybrids between the European and Chinese races of the common carp in traditional fish farms of Southeast Asia. *In* Advances in Aquaculture, T. V. R. Pillay and W. A. Dill (Editors). Fishing News Books, Farnham, Surrey, England.

Møller, D., Naevdal, G., Holm, M., and Lerøy, R. 1979. Variation in growth rate and age at sexual maturity in rainbow trout. *In* Advances in Aquaculture, T. V. R. Pillay and W. A. Dill (Editors). Fishing News Books, Farnham, Surrey, England.

Monan, G. E. 1966. Aids to fish tattooing. Prog. Fish-Cult. *28*, 57–59.

Moore, W. S. 1974. A mutant affecting chromatophore proliferation in a poeciliid fish. J. Hered. *65*, 326–330.

Morán, P., Pendás, A. M., Garcia-Vázquez, E., and Izquierdo, J. 1991. Failure of a stocking policy, of hatchery reared brown trout, *Salmo trutta*, L., in Asturias, Spain, detected using *LDH-5** as a genetic marker. J. Fish Biol. *39 (Suppl. A)*, 117–121.

Moreau, J., Bambino, C., and Pauly, D. 1986. Indices of overall growth performance of 100 tilapia (Cichlidae) populations. *In* The First Asian Fisheries Forum, J. L. Maclean, L. B. Dizon, and L. V. Hosillos (Editors). Asian Fisheries Society, Manila, Philippines.

Morizot, D. C. 1990. Linkage maps of biochemical loci in non-salmonid fishes. *In* Genetic Maps: Locus Maps of Complex Genomes, S. J. O'Brien (Editor). Cold Spring Harbor Laboratory Press, Cold Spring Harbor, NY.

Morizot, D.C., Slaugenhaupt, S. A., Kallman, K. D., and Chakravarti, A. 1991. Genetic linkage map of fishes of the genus *Xiphophorus* (Teleostei: Peociliidae). Genetics *127*, 399–410.

Morkramer, S., Hörstgen-Schwark, G., and Langholz, H. J. 1985. Comparison of different European rainbow populations under intensive production conditions. Aquaculture *44*, 303–320.

Morris, J. E., D'Abramo, L. R., and Muncy, R. C. 1990. An inexpensive marking technique to assess ingestion of formulated feeds by larval fish. Prog. Fish-Cult. *52*, 120–121.

Mrakovčić, M. and Haley, L. E. 1979. Inbreeding depression in the zebra fish *Brachydanio rerio* (Hamilton Buchanan). J. Fish Biol. *15*, 323–327.

Murofushi, M., Oikawa, S., Nishikawa, S., and Yosida, T. H. 1980. Cytogenetical studies on fishes, III. Multiple sex chromosome mechanism in the filefish, *Stephanolepis cirrhifer*. Jpn. J. Genet. *55*, 127–131.

Murphy, B. R., Nielsen, L. A., and Turner, B. J. 1983. Use of genetic tags to evaluate stocking success for reservoir walleyes. Trans. Am. Fish. Soc. *112*, 457–463.

Murray, C. B. and Beacham, T. D. 1990. Marking juvenile pink and chum salmon with hot brands in the form of a binary code. Prog. Fish-Cult. *52*, 122–124.

Myers, J. M. 1986. Tetraploid induction in *Oreochromis* spp. Aquaculture *57*, 281–287.

Myers, J. M. and Hershberger, W. K. 1991A. Early growth and survival of heat-shocked and tetraploid-derived triploid rainbow trout (*Oncorhynchus mykiss*). Aquaculture *96*, 97–107.

Myers, J. M. and Hershberger, W. K. 1991B. A comparison of meiotic and interploid triploid rainbow trout (*Oncorhynchus mykiss*) incubation and growth performance. J. World. Aquacul. Soc. *22*, 44A.

Myers, J. M., Hershberger, W. K., and Iwamoto, R. N. 1986. The induction of tetraploidy in salmonids. J. World. Aquacul. Soc. *17*, 1–7.

N

Naevdal, G., Holm, M., Møller, D., and Østhus, O. D. 1975. Experiments with selective breeding of Atlantic salmon. Int. Council Exp. Sea C. M. 1975/ M:22.

Naevdal, G., Holm, M., Møller, D., and Østhus, O. D. 1976. Variation in growth rate and age at sexual maturity in Atlantic salmon. Int. Council Exp. Sea C. M. 1976/E:40.

Nagy, A. 1987. Genetic manipulations performed on warm water fish. *In* Selection, Hybridization, and Genetic Engineering in Aquaculture, Vol. 2, K. Thiews (Editor). H. Heenemann GmbH and Co., Berlin.

Nagy, A. and Csányi, V. 1982. Changes of genetic parameters in successive gynogenetic generations and some calculations for carp gynogenesis. Theor. Appl. Genet. *63*, 105–110.

Nagy, A. and Csányi, V. 1984. A new breeding system using gynogenesis and sex-reversal for fast inbreeding in carp. Theor. Appl. Genet. *67*, 485–490.

Nagy, A., Rajki, K., Horváth, L., and Csányi, V. 1978. Investigation on carp, *Cyprinus carpio* L. gynogenesis. J. Fish Biol. *13*, 215–224.

Nagy, A., Rajki, K., Bakos, J., and Csanyi, V. 1979. Genetic analysis in carp (*Cyprinus carpio*) using gynogenesis. Heredity *43*, 35–40.

Nagy, A., Csanyi, V., Bakos, J., and Horvath, L. 1980. Development of a short-term laboratory system for the evaluation of carp growth in ponds. Bamidgeh *32*, 6–15.

Nagy, A., Bercsényi, M., and Csányi, V. 1981. Sex reversal in carp (*Cyprinus carpio*) by oral administration of methyltestosterone. Can. J. Fish. Aquat. Sci. *38*, 725–728.

Nakamura, D., Wachtel, S. S., and Kallman, K. 1984. Y-Y antigen and the evolution of heterogamety. J. Heredity *75*, 353–358.

Nakamura, N. and Kasahara, S. 1955. A study on the phenomenon of the tobi-koi or shoot carp—I. On the earliest stage at which the shoot carp appears. Bull. Jpn. Soc. Sci. Fish. *21*, 73–76. [English translation in Wohlfarth (1977).]

Nakamura, N. and Kasahara, S. 1956. A study on the phenomenon of the tobi-koi or shoot carp—II. On the effect of particle size and quantity of the food. Bull. Jpn. Soc. Sci. Fish. *21*, 1022–1024. [English translation in Wohlfarth (1977).]

Nakamura, N. and Kasahara, S. 1957. A study on the phenomenon of the tobi-koi or shoot carp—III. On the result of culturing the modal group and the growth of carp fry reared individually. Bull. Jpn. Soc. Sci. Fish. *22*, 674–678. [English translation in Wohlfarth (1977).]

Nakamura, N. and Kasahara, S. 1961. A study on the phenomenon of the tobi-koi or shoot carp—IV. Effects of adding a small number of larger individuals to the experimental batches of carp fry and of culture density upon the occurrence of shoot carp. Bull. Jpn. Soc. Sci. Fish. *27*, 958–962. [English translation in Wohlfarth (1977).]

Naruse, K., Ijiri, K., Shima, A., and Egami, N. 1985. The production of cloned fish in the medaka (*Oryzias latipes*). J. Exp. Zool. *236*, 335–341.

Nei, M., Maruyama, T., and Chakraborty, R. 1975. The bottleneck effect and genetic variability in populations. Evolution *29*, 1–10.

Nenashev, G. A. 1966. Determination of the heritability of different characters in fish. Sov. Genet. *2(11)*, 39–43.

Nenashev, G. A. 1970. Heritability of some morphological (diagnostic) traits in Ropsha carp. *In* Selective Breeding of Carp and Intensification of Fish Breeding in Ponds, V. S. Kirpichnikov (Editor). Israel Program for Scientific Translations, Jerusalem, Israel.

Nilsson, J. 1990. Heritability estimates of growth-related traits in Arctic charr (*Salvelinus alpinus*). Aquaculture *84*, 211–217.

Norton, J. 1982. Angelfish genetics Part three. Fresh. Mar. Aquar. *5(7)*, 8–10, 91–92.

O

Okamoto, N., Matsumoto, T., Kato, N., Tazaki, S., Tanaka, M., Ai., N., Hanada, H., Suzuki, Y., Takamatsu, C., Tayama, T., and Sano, T. 1987. Difference in susceptibility in IPN virus among rainbow trout populations from three hatcheries in Japan. Bull. Jpn. Soc. Sci. Fish. *53*, 1121–1124.

Oldorf, W., Kronert, U., Balarin, J., Haller, R., Hörstgen-Schwark, G., and Langholz, H.-J. 1989. Prospects of selecting for late maturity in tilapia (*Oreochromis niloticus*) II. Strain comparisons under laboratory and field conditions. Aquaculture *77*, 123–133.

Onozato, H. 1982. The "Hertwig effect" and gynogenesis in chum salmon *On-*

corhynchus keta eggs fertilized with ^{60}Co γ-ray irradiated milt. Bull. Jpn. Soc. Sci. Fish. *48*, 1237–1244.

Onozato, H. 1984. Diploidization of gynogenetically activated salmonid eggs using hydrostatic pressure. Aquaculture *43*, 91–97.

Ozato, K., Kondoh, H., Inohara, H., Iwamatsu, T., Wakamatsu, Y., and Okada, T. S. 1986. Production of transgenic fish: introduction and expression of chicken δ-crystallin gene in medaka embryos. Cell Differen. *19*, 237–244.

P

Paaver, T. and Gross, R. 1990. Genetic variability of Cyprinus carpio L. stocks reared in Estonia. Sov. Genet. *26*, 839–846.

Pandian, T. J. and Varadaraj, K. 1988A. Techniques for producing all-male and all-triploid *Oreochromis mossambicus*. *In* Second International Symposium on Tilapia in Aquaculture, R. S. V. Pullin, T. Bhukaswan, K. Tonguthai, and J. L. Maclean (Editors). ICLARM Conference Proceedings 15, Department of Fisheries, Bangkok, Thailand and International Center for Living Aquatic Resources Management, Manila, Philippines.

Pandian, T. J. and K. Varadaraj. 1988B. Sterile female triploidy in *Oreochromis mossambicus*. Bull. Aquacul. Assoc. Can. *88*, 134–136.

Park, E. H. and Kang. Y. S. 1979. Karyological confirmation of conspicuous ZW sex chromosomes in two species of Pacific anguilloid fishes (Anguilliformes: Teleostomi). Cytogenet. Cell Genet. *23*, 33–38.

Parker, N. C. and Klar, G. T. 1987. Depigmented skin lesions (stripedness) in channel catfish (*Ictalurus punctatus*). Aquaculture *60*, 117–120.

Parsons, J. E. and Thorgaard, G. H. 1984. Induced androgenesis in rainbow trout. J. Exp. Zool *231*, 407–412.

Parsons, J. E. and Thorgaard, G. H. 1985. Production of androgenetic diploid rainbow trout. J. Hered. *76*, 177–181.

Parsons, J. E., Busch, R. A., Thorgaard, G. H., and Scheerer, P. D. 1986. Increased resistance of triploid rainbow trout × coho salmon hybrids to infectious hematopoietic necrosis virus. Aquaculture *57*, 337–343.

Pastene, L. A., Numachi, K., and Tsukamoto, K. 1991. Examination of reproductive success of transplanted stocks in an amphidromous fish, *Plecoglossus altivelis* (Temmink et Schlegel) using mitochondrial DNA and isozyme markers. J. Fish Biol. *39 (Suppl. A)*, 93–100.

Pauly, D., Moreau, J., and Prein, M. 1988. A comparison of overall growth performance of tilapia in open waters and aquaculture. *In* The Second International Symposium on Tilapia in Aquaculture, R. S. V. Pullin, T. Bhukaswan, K. Tonguthai, and J. L. Maclean (Editors). ICLARM Conference Proceedings 15, Department of Fisheries, Bangkok, Thailand and International Center for Living Aquatic Resources Management, Manila, Philippines.

Pella, J. J. and Milner, G. B. 1988. Use of genetic marks in stock composition analysis. *In* Population Genetics & Fishery Management, N. Ryman and F.

Utter (Editors). Washington Sea Grant Program, University of Washington Press, Seattle, WA.

Penman, D. J., Skibinski, D. O. F., and Beardmore, J. A. 1987. Survival, growth rate and maturity in triploid tilapia. *In* Selection, Hybridization, and Genetic Engineering in Aquaculture, Vol. 2, K. Thiews (Editor). H. Heenemann GmbH and Co., Berlin.

Penman, D. J., Beeching, A. J., Penn, S., and Maclean, N. 1990. Factors affecting survival and integration following microinjection of novel DNA into rainbow trout eggs. Aquaculture *85*, 35–50.

Penman, D. J., Iyengar, A., Beeching, A. J., Rahman, A., Sulaiman, Z., Bromage, N., and Maclean, N. 1992. Inheritance of the MTrGH gene in transgenic rainbow trout. Aquaculture *100*, 102.

Pezold, F. 1984. Evidence for multiple sex chromosomes in the freshwater goby, *Gobionellus shufeldti* (Pisces: Gobiidae). Copeia *1984*, 235–238.

Phang, V. P. E. and Doyle, R. W. 1989. Analysis of early growth of guppy strains, *Poecilia reticulata*, with different color patterns. Theor. Appl. Genet. *77*, 645–650.

Phang, V. P. E., Ng, L. N., and Fernando, A. A. 1989. Inheritance of the snakeskin color pattern in the guppy, *Poecilia reticulata*. J. Hered. *80*, 393–399.

Phillips, R. B. and Ihssen, P. E. 1985. Identification of sex chromosomes in lake trout (*Salvelinus namaycush*). Cytogenet. Cell Genet. *39*, 14–18.

Phillips, R. B., Zajicek, K. D., Ihssen, P. E., and Johnson, O. 1986. Application of silver staining to the identification of triploid fish cells. Aquaculture *54*, 313–319.

Pine, R. T. and Anderson, L. W. J. 1990. Blood preparation for flow cytometry to identify triploidy in grass carp. Prog. Fish-Cult. *52*, 266–268.

Piron, R. D. 1978. Spontaneous skeletal deformities in the Zebra Danio (*Brachydanio rerio*) bred for fish toxicity tests. J. Fish Biol. *13*, 79–83.

Plumb, J. A., Green, O. L., Smitherman, R. O., and Pardue, G. B. 1975. Channel catfish virus experiments with different strains of channel catfish. Trans. Am. Fish. Soc. *104*, 140–143.

Pohar, J. 1992. Estimation of heritabilities and genetic correlations for consecutive body weights of rainbow trout reared in fresh and sea water. Aquaculture *100*, 103.

Prather, E. E. 1961. A comparison of production of albino and normal channel catfish. Proc. Southeast. Assoc. Game Fish Comm. *15*, 302–303.

Pruginin, Y., Rothbard, S., Wohlfarth, G., Halevy, A., Moav, R., and Hulata, G. 1975. All-male broods of *Tilapia nilotica* × *T. aurea* hybrids. Aquaculture *6*, 11–21.

Purdom, C. E. 1969. Radiation-induced gynogenesis and androgenesis in fish. Heredity *24*, 431–444.

Purdom, C. E. 1972. Induced polyploidy in plaice (*Pleuronectes platessa*) and its hybrid with the flounder (*Platichthys flesus*). Heredity *29*, 11–24.

Purdom, C. E. 1983. Genetic engineering by the manipulation of chromosomes. Aquaculture *33*, 287–300.

Purdom, C. E., Thompson, D., and Lou, Y. D. 1985. Genetic engineering in rainbow trout, *Salmo gairdnerii* Richardson, by suppression of meiotic and mitotic metaphase. J. Fish. Biol. *27*, 73–79.

Q

Quattro, J. M., Avise, J. C., and Vrijenhoek, R. C. 1991. Molecular evidence for multiple origins of hybridogenetic fish clones (Poeciliidae: *Poeciliopsis*). Genetics *127*, 391–398.

Quillet, E., Chevassus, B., and Devaux, A. 1988A. Timing and duration of hatching in gynogenetic, triploid, tetraploid, and hybrid progenies in rainbow trout. Genet. Sel. Evol. *20*, 199–210.

Quillet, E. Chevassus, B., Blanc, J.-M., Krieg, F., and Chourrout, D. 1988B. Performances of auto and allotriploids in salmonids I. Survival and growth in fresh water farming. Aquat. Living Resour. *1*, 29–43.

R

Ramboux, A. C. R. and Dunham, R. A. 1991. Evaluation of four genetic groups of channel-blue catfish hybrids grown in earthen ponds. J. World Aquacul. Soc. *22*, 49A.

Rasmuson, M. 1981. Some aspects of available resources of genetic variation. *In* Fish Gene Pools: Preservation of Genetic Resources in Relation to Wild Fish Stocks, N. Ryman (Editor). Ecological Bulletins No. 34, Forskningsrådsnämnden, Stockholm, Sweden.

Rawstron, R. R. 1973. Harvest, mortality, and cost of three domestic strains of tagged rainbow trout stocked in large California impoundments. Calif. Fish Game *59*, 245–265.

Rawstron, R. R. 1977. Effect of a reduced bag limit and later planting date on the mortality, survival, and cost and yield of three domestic strains of rainbow trout at Lake Berryessa and Merle Collins Reservoir 1971–1974. Calif. Fish Game *63*, 219–227.

Reagan, R. E., Jr. 1979. Heritabilities and genetic correlations of desirable commercial traits in channel catfish. Res. Rep. Miss. Ag. For. Exp. Sta. *5*, No. 4.

Reagan, R. E., Jr. and Conley, C. M. 1977. Effect of egg diameter on growth of channel catfish. Prog. Fish-Cult. *39*, 133–134.

Reagan, R. E., Pardue, G. B., and Eisen, E. J. 1976. Predicting selection response for growth of channel catfish. J. Hered. *67*, 49–53.

Reddy, P. V. G. K., Kowtal, G. V., and Tantia, M. S. 1990. Preliminary observations on induced polyploidy in Indian major carps, *Labeo rohita* (Ham.) and *Catla catla* (Ham.). Aquaculture *87*, 279–287.

Refstie, T. 1980. Genetic and environmental sources of variation in body weight and length of rainbow trout fingerlings. Aquaculture *19*, 351–357.

Refstie, T. 1981. Tetraploid rainbow trout produced by cytochalasin B. Aquaculture *25*, 51–58.

Refstie, T. 1983. Induction of diploid gynogenesis in Atlantic salmon and rainbow trout using irradiated sperm and heat shock. Can. J. Zool. *61*, 2411–2416.

Refstie, T. 1987. Selective breeding and intraspecific hybridization of cold water finfish. *In* Selection, Hybridization, and Genetic Engineering in Aquaculture, Vol. 1, K. Thiews (Editor). H. Heenemann GmbH and Co., Berlin.

Refstie, T. and Steine, T. A. 1978. Selection experiments with salmon. III. Genetic and environmental sources of variation in length and weight of Atlantic salmon in the freshwater phase. Aquaculture *14*, 221–234.

Refstie, T., Steine, T. A., and Gjedrem, T. 1977. Selection experiments with salmon. II. Proportion of Atlantic salmon smoltifying at 1 year of age. Aquaculture *10*, 231–242.

Refstie, T., Stoss, J., and Donaldson, E. M. 1982. Production of all female coho salmon (*Oncorhynchus kisutch*) by diploid gynogenesis using irradiated sperm and cold shock. Aquaculture *29*, 67–82.

Reinitz, G. L., Orme, L. E., Lemm, C. A., and Hitzel, F. N. 1978. Differential performance of four strains of rainbow trout reared under standardized conditions. Prog. Fish-Cult. *40*, 21–23.

Reinitz, G. L., Orme, L. E., and Hitzel, F. N. 1979. Variations of body composition and growth among strains of rainbow trout. Trans. Am. Fish. Soc. *108*, 204–207.

Reisenbichler, R. R. and McIntyre, J. D. 1977. Genetic differences in growth and survival of juvenile hatchery and wild steelhead trout, *Salmo gairdneri*. J. Fish. Res. Board Can. *34*, 123–128.

Richter, C. J. J., Henken, A. M., Eding, E. H., Van Doesum, J. H., and DeBoer, P. 1987. Induction of triploidy by cold-shocking eggs and performance of triploids of the African catfish, *Clarias gariepinus* (Burchell, 1822). *In* Selection, Hybridization, and Genetic Engineering in Aquaculture, Vol. 2, K. Thiews (Editor). H. Heenemann GmbH and Co., Berlin.

Riddell, B. E., Leggett, W. C., and Saunders, R. L. 1981. Evidence of adaptive polygenic variation between two populations of Atlantic salmon (*Salmo salar*) native to tributaries of the S. W Miramichi River, N. B. Can. J. Fish. Aquat. Sci. *38*, 321–333.

Rinne, J. N. 1976. Coded spine clipping to identify individuals of the spiny-rayed fish *Tilapia*. J. Fish. Res. Board Can. *33*, 2626–2629.

Rishi, K. K. 1976. Karyotypic studies on four species of fishes. The Nucleus *19*, 95–98.

Robins, C. R., Bailey, R. M., Bond, C. E., Brooker, J. R., Lachner, E. A., Lea, R. N., and Scott, W. B. 1991. Common and Scientific Names of Fishes from the United States and Canada, 5th edition. American Fisheries Society Special Publication 20, American Fisheries Society, Bethesda, MD.

Robinson, G. D., Dunson, W. A., Wright, J. E., and Mamolito, G. E. 1976. Differences in low pH tolerance among strains of brook trout (*Salvelinus fontinalis*). J. Fish. Biol. *8*, 5–17.

Robison, O. W. and Luempert, L. G., III. 1984. Genetic variation in weight and survival of brook trout (*Salvelinus fontinalis*). Aquaculture *38*, 155–170.

Røed, K. H., Brun, E., Larsen, H. J., and Refstie, T. 1990. The genetic influence of serum haemolytic activity in rainbow trout. Aquaculture *85*, 109–117.

Rokkones, E., Alestrøm, P., Skjervold, H., and Gautvik, K. M. 1989. Microinjection and expression of a mouse metallothionein human growth hormone fusion gene in fertilized salmonid eggs. J. Comp. Physiol. B *158*, 751–758.

Rokkones, E., Alestrøm, P., Skjervold, H., and Gautvik, K. M. 1990. Expression of human growth hormone gene in fertilized eggs from Atlantic salmon and rainbow trout. Aquaculture *85*, 329.

Romana, M. R. R. 1988. Electrophoretic studies on induced gynogenetic diploids and triploids in tilapia (*Oreochromis niloticus* and *O. aureus*). In Proceedings of the Second International Symposium on Tilapia in Aquaculture, R. S. V. Pullin, T. Bhukaswan, K. Tonguthai, and J. L. Maclean (Editors). ICLARM Conference Proceedings 15, Department of Fisheries, Bangkok, Thailand and International Center for Living Aquatic Resources Management, Manila, Philippines.

Romanov, N. S. 1984. Effect of culture conditions on skull morphology in smolts of the masu salmon, *Oncorhynchus masou* (Brevoort). Aquaculture *41*, 147–153.

Rosenthal, H. L. and Rosenthal, R. S. 1950. Lordosis, a mutation in the guppy *Lebistes reticulatus*. J. Hered. *41*, 217–218.

Ryman, N. 1970. A genetic analysis of recapture frequencies of released young of salmon (*Salmo salar* L.). Hereditas *65*, 159–160.

Ryman, N. 1972. An attempt to estimate the magnitude of additive genetic variation of body size in the guppy-fish, *Lebistes reticulatus*. Hereditas *71*, 237–243.

Ryman, N. 1981. Conservation of genetic resources: Experiences from the brown trout (*Salmo trutta*). In Fish Gene Pools: Preservation of Genetic Resources in Relation to Wild Fish Stocks, N. Ryman (Editor). Ecological Bulletins No. 34, Forskningsr ådsnämnden, Stockholm, Sweden.

Ryman, N. 1991. Conservation genetics considerations in fishery management. J. Fish Biol. *39 (Suppl. A)*, 211–224.

Ryman, N. and Ståhl, G. 1980. Genetic changes in hatchery stocks of brown trout (*Salmo trutta*). Can. J. Fish. Aquat. Sci. *37*, 82–87.

Ryman, N. and Utter, F. 1988. Population Genetics & Fishery Management. Washington Sea Grant Program, University of Washington Press, Seattle, WA.

S

Sadler, S. E., McLeod, D., McKay, L. R., and Moccia, R. D. 1992. Selection for early spawning and repeatability of spawn date in rainbow trout. Aquaculture *100*, 103.

Sadoglu, P. and McKee, A. 1969. A second gene that affects eye and body color in Mexican blind cave fish. J. Hered. *60*, 10–14.

Sato, R. 1980. Variations in hatchability and hatching time, and estimation of the heritability for hatching time among individuals within the strain of coho salmon (*Oncorhynchus kisutch*). Bull. Nat. Res. Inst. Aqua. *1*, 21–28.

Sato, R. and Morikawa, S. 1982. Estimation of the heritability for hatchability, hatching time and duration of hatching in amago salmon (*Oncorhynchus rhodurus*). Bull. Nat. Res. Inst. Aqua. *3*, 21–30.

Saunders, R. L., Henderson, E. B., Glebe, B. D., and Loudenslager, E. J. 1983. Evidence of a major environmental component in determination of the grilse: larger salmon ratio in Atlantic salmon (*Salmo salar*). Aquaculture *33*, 107–118.

Savost'yanova, G. G. 1969. Mass selection in trout breeding. Sov. Genet. *5*, 896–902.

Saxton, A. M., Hershberger, W. K., and Iwamoto. R. N. 1984. Smoltification in the net-pen culture of coho salmon: Quantitative genetic analysis. Trans. Am. Fish. Soc. *113*, 339–347.

Scheerer, P. D. and Thorgaard, G. H. 1983. Increased survival in salmonid hybrids by induced triploidy. Can. J. Fish. Aquat. Sci. *40*, 2040–2044.

Scheerer, P. D., Thorgaard, G. H., Allendorf, F. W., and Knudsen, K. L. 1986. Androgenetic rainbow trout produced from inbred and outbred sperm sources show similar survival. Aquaculture *57*, 289–298.

Schmidt, J. 1919. Racial studies in fishes. II. Experimental investigations with *Lebistes reticulatus* (Peters) Regan. J. Genet. *8*, 147–153.

Schneider, J. F., Hallerman, E. M., Yoon, S. J., He, L., Myster, S. H., Gross, M., Liu, Z., Zhu, Z., Hackett, P. B., Guise, K. S., Kapuscinski, A. R., and Faras, A. J., 1989. Microinjection and successful transfer of the bovine growth hormone gene into the northern pike, Esox lucius. J. Cell. Biochem. Suppl. *13B*, 173.

Schom, C. B. 1986. Genetic, environmental, and maturational effects on Atlantic salmon (*Salmo salar*) survival in acute low pH trials. Can. J. Fish. Aquat. Sci. *43*, 1547–1555.

Schröder, J. H. 1969. Die Vererbung von Beflossungsmerkmalen beim Berliner Guppy (*Lebistes reticulatus* Peters). [Inheritance of fin characters in the Berlin strain of the guppy (*Lebistes reticulatus* Peters).] Theor. Appl. Genet. *39*, 73–78.

Schröder, J. H. 1976. Genetics for Aquarists. T. F. H. Publications, Neptune City, NJ.

Schultz, R. J. 1963. Stubby, a hereditary vertebral deformity in the viviparous fish, *Poeciliopsis prolifica*. Copeia *1963*, 325–330.

Schultz, R. J. 1967. Gynogenesis and triploidy in the viviparous fish Poeciliopsis. Science *157*, 1564–1567.

Scott, A. G., Mair, G. C., Skibinski, D. O. F., and Beardmore, J. A. 1987. "Blond": a useful new genetic marker in the tilapia, *Oreochromis niloticus* (L.). Aquacul. Fish. Manage. *18*, 159–165.

Scott, A. G., Penman, D. J., Beardmore, J. A., and Skibinski, D. O. F. 1989. The "YY" supermale in *Oreochromis niloticus* (L.) and its potential in aquaculture. Aquaculture *78*, 237–251.

Seeb, J., Thorgaard, G., Hershberger, W. K., and Utter, F. M. 1986. Survival in diploid and triploid Pacific salmon hybrids. Aquaculture 57, 375.

Seeb, J. E., Thorgaard, G. H., and Utter, F. M. 1988. Survival and allozyme expression in diploid and triploid hybrids between chum, chinook, and coho salmon. Aquaculture 72, 31–48.

Seyoum, S. and Kornfield, I. 1992. Identification of subspecies of Oreochromis niloticus (Pisces: Cichlidae) using restriction endonuclease analysis of mitochondrial DNA. Aquaculture 102, 29–39.

Shah, M. S. 1988. Female homogamety in tilapia (Oreochromis niloticus) revealed by gynogenesis. Asian Fish. Sci. 1, 215–219.

Shami, S. A. and Beardmore, J. A. 1978. Stabilizing selection and parental age effects on lateral line scale number in the guppy, Poecilia reticulata (Peters). Pak. J. Zool. 10, 1–15.

Shell, E. W. 1983. Fish Farming Research. Alabama Agricultural Experiment Station, Auburn Univ., AL.

Shelton C. J., Macdonald, A. G., and Johnstone, R. 1986. Induction of triploidy in rainbow trout using nitrous oxide. Aquaculture 58, 155–159.

Shelton, W. L. 1986A. Broodstock development for monosex production of grass carp. Aquaculture 57, 311–319.

Shelton, W. L. 1986B. Strategies for reducing risks from introductions of aquatic organisms: An aquaculture perspective. Fisheries 11(2), 16, 18–19.

Shelton, W. L. 1987. Genetic manipulations—sex control of exotic fish for stocking. In Selection, Hybridization, and Genetic Engineering in Aquaculture, Vol. 2, K. Thiews (Editor). H. Heenemann GmbH and Co., Berlin.

Shelton, W. L. 1990. Personal communication. Department of Zoology, University of Oklahoma, Norman, OK.

Shelton, W. L., Meriwether, F. H., Semmens, K. J., and Calhoun, W. E. 1983. Progeny sex ratios from intraspecific pair spawnings of Tilapia aurea and Tilapia nilotica. In International Symposium on Tilapia in Aquaculture, L. Fishelson and Z. Yaron (Editors). Tel Aviv Univ., Tel Aviv, Israel.

Shields, B. A., Guise, K. S., and Underhill, J. C. 1990. Chromosomal and mitochondrial DNA characterization of a population of dwarf cisco (Coregonus artedii) in Minnesota. Can. J. Fish. Aquat. Sci. 47, 1562–1569.

Shrestha, S. B. 1977. The parasites of different strains and species of catfishes (Ictalurus spp.). Master's Thesis, Auburn Univ., AL.

Shultz, F. T. 1986. Developing a commercial breeding program. Aquaculture 57, 65–76.

Siitonen, L. 1986. Factors affecting growth in rainbow trout (Salmo gairdneri) stocks. Aquaculture 57, 185–191.

Siitonen, L. and Gall, G. A. E. 1989. Response to selection for early spawn date in rainbow trout, Salmo gairdneri. Aquaculture 78, 153–161.

Siraj, S. S., Smitherman, R. O., Castillo-Gallusser, S., and Dunham, R. A. 1983. Reproductive traits for three year classes of Tilapia nilotica and maternal effects on their progeny. In International Symposium on Tilapia in Aquaculture, L. Fishelson and Z. Yaron (Editors). Tel Aviv Univ., Tel Aviv, Israel.

Smíšek, J. 1979A. The hybridization of the Vodňany and Hungarian lines of carp. Buletin VÚRH Vodňany *15*(1), 3–12. Abstract in Anim. Breed. Abs. 1980, *48(5)*, 302.

Smíšek, J. 1979B. Considerations of body conformation, heritability and biochemical characters in genetic studies in Czechoslovakia. Buletin VÚRH Vodňany. *15(2)*, 3–6. Abstract in Anim. Breed. Abs. 1980. *48*(5), 302.

Smith, L. T. and Lemoine, H. L. 1979. Colchicine-induced polyploidy in brook trout. Prog. Fish-Cult. *41*, 86–88.

Smith, R. R., Kincaid, H. L., Regenstein, J. M., and Rumsey, G. L. 1988. Growth, carcass composition, and taste of rainbow trout of different strains fed diets containing primarily plant or animal protein. Aquaculture *70*, 309–321.

Smitherman, R. O. and Dunham, R. A. 1985. Genetics and breeding. *In* Channel Catfish Culture, C. S. Tucker (Editor). Elsevier, Amsterdam, The Netherlands.

Smitherman, R. O. and Tave, D. 1987. Maintenance of genetic quality in cultured tilapia. Asian Fish. Sci. *1*, 75–82.

Smitherman, R. O., Green, O. L., and Pardue, G. B. 1974. Genetics experiments with channel catfish. Catfish Farmer and World Aquaculture News *6(3)*, 43–44.

Smitherman, R. O., Khater, A. A., Cassell, N. I., and Dunham, R. A. 1988. Reproductive performance of three strains of *Oreochromis niloticus*. Aquaculture *70*, 29–37.

Smoker, W. W. 1981. Quantitative genetics of chum salmon, *Oncorhynchus keta* (Walbaum). Doctoral Dissertation, Oregon State Univ., Corvallis, OR.

Snieszko, S. F., Dunbar, C. E., and Bullock, G. L. 1959. Resistance to ulcer disease and furunculosis in eastern brook trout, *Salvelinus fontinalis*. Prog. Fish-Cult. *21*, 111–116.

Snyder, R. J. 1991. Quantitative genetic analysis of life histories in two freshwater populations of the threespine stickleback. Copeia *1991*, 526–529.

Sodsuk, P. and McAndrew, B. J. 1991. Molecular systematics of three tilapiine genera *Tilapia, Sarotherodon,* and *Oreochromis* using allozyme data. J. Fish Biol. *39 (Suppl. A)*, 301–308.

Sola, L., Cataudella, S., and Capanna, E. 1981. New developments in vertebrate cytotaxonomy. III. Karyology of bony fishes: A review. Genetica *54*, 285–328.

Solar, I. I., Donaldson, E. M., and Hunter, G. A. 1984. Induction of triploidy in rainbow trout (*Salmo gairdneri* Richardson) by heat shock, and investigation of early growth. Aquaculture *42*, 57–67.

Solar, I. I., Baker, I. J., and Donaldson, E. M. 1987. Experimental use of female sperm in the production of monosex female stocks of chinook salmon (*Oncorhynchus tshawytscha*) at commercial fish farms. Canadian Technical Report of Fisheries and Aquatic Sciences No. 1552, Department of Fisheries and Oceans, West Vancouver, B.C., Canada.

Solar, I. I., Baker, I., and Donaldson, E. M. 1989. The development of monosex female chinook salmon stocks for the British Columbia mariculture industry. J. World. Aquacult. Soc. *20*, 72A.

Springate, J. R. C. and Bromage, N. R. 1985. Effects of egg size on early growth and survival in rainbow trout (*Salmo gairdneri* Richardson). Aquaculture *47*, 163–172.

Ståhl, G. 1983. Differences in the amount and distribution of genetic variation between natural populations and hatchery stocks of Atlantic salmon. Aquaculture *33*, 23–32.

Standal, M. and Gjerde, B. 1987. Genetic variation in survival of Atlantic salmon during the sea-ranching period. Aquaculture *66*, 197–207.

Stanley, J. G. 1976A. Production of hybrid, androgenetic, and gynogenetic grass carp and carp. Trans. Am. Fish. Soc. *105*, 10–16.

Stanley, J. G. 1976B. Female homogamety in grass carp (*Ctenopharyngodon idella*) determined by gynogenesis. J. Fish. Res. Board Can. *33*, 1372–1374.

Stanley, J. G. 1983. Gene expression in haploid embryos of Atlantic salmon. J. Hered. *74*, 19–22.

Stanley, J. G. and Jones, J. B. 1976. Morphology of androgenetic and gynogenetic grass carp, *Ctenopharyngodon idella* (Valenciennes). J. Fish. Biol. *9*, 523–528.

Stanley, J. G., Martin, J. M., and Jones, J. B. 1975. Gynogenesis as a possible method for producing monosex grass carp (*Ctenopharyngodon idella*). Prog. Fish-Cult. *37*, 25–26.

Stanley, J. G., Biggers, C. J., and Schultz, D. E. 1976. Isozymes in androgenetic and gynogenetic white amur, gynogenetic carp, and carp-amur hybrids. J. Hered. *67*, 129–134.

Streisinger, G., Walker, C., Dower, N., Knauber, D., and Singer, F. 1981. Production of clones of homozygous diploid zebra fish (*Brachydanio rerio*). Nature *291*, 293–296.

Stuart, G. W., McMurray, J. V., and Westerfield, M. 1988. Replication, integration and stable germ-line transmission of foreign sequences injected into early zebrafish embryos. Development *103*, 403–412.

Stuart, G. W., McMurray, J. V., and Westerfield, M. 1989. Germ-line transformation of the zebrafish. *In* Gene Transfer and Gene Therapy, A. L. Beaudet, R. Mulligan, and I. M. Verma (Editors). Alan R. Liss, Inc., New York, NY.

Sugama, K., Taniguchi, N., and Umeda, S. 1988. An experimental study on genetic drift in hatchery population of red sea bream. Bull. Jpn. Soc. Sci. Fish. *54*, 739–744.

Sugama, K., Taniguchi, N., Seki, S., Nabeshima, H., and Hasegawa, Y. 1990. Gynogenetic diploid production in the red sea bream using UV-irradiated sperm of black sea bream and heat shock. Bull. Jpn. Soc. Sci. Fish. *56*, 1427–1433.

Sumantadinata, K. and Taniguchi, N. 1990. Comparison of electrophoretic allele frequencies and genetic variability of common carp stocks from Indonesia and Japan. Aquaculture *88*, 263–271.

Suzuki, R., Oshiro, T., and Nakanishi, T. 1985A. Survival, growth and fertility of gynogenetic diploids induced in the cyprinid loach, *Misgurnus anguillicaudatus*. Aquaculture *48*, 45–55.

Suzuki, R., Nakanishi, T., and Oshiro, T. 1985B. Survival, growth and sterility of induced triploids in the cyprinoid loach *Misgurnus anguillicaudatus*. Bull. Jpn. Soc. Sci. Fish. *51*, 889–894.

Svärdson, G. 1945. Polygenic inheritance in Lebistes. Arkiv för Zool. *36A(6)*, 1–9.

Swain, D. P., Riddell, B. E., and Murray, C. B. 1991. Morphological differences between hatchery and wild populations of coho salmon (*Oncorhynchus kisutch*): Environmental versus genetic origin. Can. J. Fish. Aquat. Sci. *48*, 1783–1791.

Swarts, F. A., Dunson, W. A., and Wright, J. E. 1978. Genetic and environmental factors involved in increased resistance of brook trout to sulfuric acid solutions and mine acid polluted waters. Trans. Am. Fish. Soc. *107*, 651–677.

Swarup, H. 1959. Production of triploidy in *Gasterosteus aculeatus* (L.). J. Genet. *56*, 129–142.

T

Taggart, J. B. and Ferguson, A. 1984. An electrophoretically-detectable genetic tag for hatchery-reared brown trout (*Salmo trutta* L.). Aquaculture *41*, 119–130.

Taniguchi, N., Hamada, R., and Fujiwara, H. 1981. Genetic difference in growth between full-sib groups of a maternal half-sib family observed in the juvenile stage of red seabream and nibe-croaker. Bull. Jpn. Soc. Sci. Fish. *47*, 731–734.

Taniguchi, N., Sumantadinata, K., and Iyama, S. 1983. Genetic change in the first and second generations of hatchery stock of black seabream. Aquaculture *35*, 309–320.

Taniguchi, N., Macaranas, J. M., and Pullin, R. S. V. 1985. Introgressive hybridization in cultured tilapia stocks in the Philippines. Bull. Jpn. Soc. Sci. Fish *51*, 1219–1224.

Taniguchi, N., Kijima, A., Fukai, J., and Inada, Y. 1986A. Conditions to induce triploid and gynogenetic diploid in ayu, *Plecoglossus altivelis*. Bull. Jpn. Soc. Sci. Fish. *52*, 49–53.

Taniguchi, N., Kijima, A., Tamura, T., Takegami, K., and Yamasaki, I. 1986B. Color, growth and maturation in ploidy-manipulated fancy carp. Aquaculture *57*, 321–328.

Taniguchi, N., Hatanaka, H., and Seki, S. 1990. Genetic variation in quantitative characters of meiotic- and mitotic-gynogenetic diploid ayu, *Plecoglossus altivelis*. Aquaculture *85*, 223–233.

Tave, D. 1984A. Genetics of dorsal fin ray number in the guppy, *Poecilia reticulata*. Copeia *1984*, 140–144.

Tave, D. 1984B. Quantitative genetics of vertebrae number and position of dorsal fin spines in the velvet belly shark, *Etmopterus spinax*. Copeia *1984*, 794–797.

Tave, D. 1984C. Effective breeding efficiency: An index to quantify the effects that different breeding programs and sex ratios have on inbreeding and genetic drift. Prog. Fish-Cult. *46*, 262–268.

Tave, D. 1986. A quantitative genetic analysis of 19 phenotypes in *Tilapia nilotica*. Copeia *1986*, 672–679.

Tave, D. 1987. Improving productivity in catfish farming by selection. Aquacul. Mag. *13(5)*, 53–55.

Tave, D. 1988. Effective breeding number and broodstock management. *In* Genetics, Breeding and Domestication of Farmed Salmon Workshop, E. A. Kenney (Editor). Ministry of Agriculture and Fisheries and B. C. Salmon Farmers' Association, North Vancouver, B.C., Canada.

Tave, D. 1989. Channel catfish yield trial centers: An idea whose time has come. Aquacul. Mag. *15(1)*, 50–52.

Tave, D. 1990A. Effective breeding number and broodstock management: I. How to minimize inbreeding. *In* Proceedings Auburn Symposium on Fisheries and Aquaculture, R. O. Smitherman and D. Tave (Editors). Alabama Agricultural Experiment Station, Auburn University, AL.

Tave, D. 1990B. Effective breeding number and broodstock management: II. How to minimize genetic drift. *In* Proceedings Auburn Symposium on Fisheries and Aquaculture, R. O. Smitherman and D. Tave (Editors). Alabama Agricultural Experiment Station, Auburn University, AL.

Tave, D. 1990C. Chromosomal manipulation. Aquacul. Mag. *16(1)*, 62–65.

Tave, D. and Smitherman, R. O. 1980. Predicted response to selection for early growth in *Tilapia nilotica*. Trans. Am. Fish. Soc. *109*, 439–445.

Tave, D. and Smitherman, R. O. 1982. Spawning success of reciprocal hybrid pairings between blue and channel catfishes with and without hormone injection. Prog. Fish-Cult. *44*, 73–74.

Tave, D., McGinty, A. S., Chappell, J. A., and Smitherman, R. O. 1981. Relative harvestability by angling of blue catfish, channel catfish, and their reciprocal hybrids. N. Am. J. Fish. Manage. *1*, 73–76.

Tave, D., Bartels, J. E., and Smitherman, R. O. 1982. Stumpbody *Sarotherodon aureus* (Steindachner) (= *Tilapia aurea*) and tail-less *S. niloticus* (L.) (= *T. nilotica*): two vertebral anomalies and their effects on body length. J. Fish Dis. *5*, 487–494.

Tave, D., Bartels, J. E., and Smitherman, R. O. 1983. Saddleback: a dominant, lethal gene in *Sarotherodon aureus* (Steindachner) (= *Tilapia aurea*). J. Fish Dis. *6*, 59–73.

Tave, D., Rezk, M., and Smitherman, R. O. 1989A. Genetics of body color in *Tilapia mossambica*. J. World. Aquacul. Soc. *20*, 214–222.

Tave, D., Smitherman, R. O., and Jayaprakas, V. 1989B. Estimates of additive genetic effects, maternal effects, specific combining ability, maternal heterosis, and egg cytoplasm effects for cold tolerance in *Oreochromis niloticus* (L.). Aquacul. Fish. Manage. *20*, 159–166.

Tave, D., Jayaprakas, V., and Smitherman, R. O. 1990A. Effects of intraspecific hybridization in *Tilapia nilotica* on survival under ambient winter temperature in Alabama. J. World Aquacul. Soc. *21*, 201–204.

Tave, D., Lovell, R. T., Smitherman, R. O., and Rezk, M. 1990B. Flesh and peritoneal lining color of gold, bronze, and black *Tilapia mossambica*. J. Food Sci. *55*, 255–256.

Tave, D., Rezk, M., and Smitherman, R. O. 1990C. Effect of body colour of *Oreochromis mossambicus* (Peters) on predation by dragonfly nymphs. Aquacul. Fish. Manage. *21*, 157–161.

Tave, D., Smitherman, R. O., Jayaprakas, V., and Kuhlers, D. L. 1990D. Estimates of additive genetic effects, maternal genetic effects, individual heterosis, maternal heterosis, and egg cytoplasmic effects for growth in *Tilapia nilotica*. J. World Aquacul. Soc. *21*, 263–270.

Tave, D., Rezk, M., and Smitherman, R. O. 1991. Effect of body colour of *Oreochromis mossambicus* (Peters) on predation by largemouth bass, *Micropterus salmoides* (Lacepède). Aquacul. Fish. Manage. *22*, 149–153.

Teichert-Coddington, D. R., and Smitherman, R. O. 1988. Lack of response by *Tilapia nilotica* to mass selection for rapidly early growth. Trans. Am. Fish. Soc. *117*, 297–300.

Thomas, A. E. 1975. Marking channel catfish with silver nitrate. Prog. Fish-Cult. *37*, 250–252.

Thompson, B. Z., Wattendorf, R. J., Hestand, R. S., and Underwood, J. L. 1987. Triploid grass carp production. Prog. Fish-Cult. *49*, 213–217.

Thompson, D. 1983. The efficiency of induced diploid gynogenesis in inbreeding. Aquaculture *33*, 237–244.

Thorgaard, G. H. 1977. Heteromorphic sex chromosomes in male rainbow trout. Science *196*, 900–902.

Thorgaard, G. H. 1978. Sex chromosomes in the sockeye salmon: A Y-autosome fusion. Can. J. Genet. Cytol. *20*, 349–354.

Thorgaard, G. H. 1986. Ploidy manipulation and performance. Aquaculture *57*, 57–64.

Thorgaard, G. H. and Allen, S. K., Jr. 1988. Chromosome manipulation and markers in fishery management. *In* Population Genetics & Fishery Management, N. Ryman and F. Utter (Editors). Washington Sea Grant Program, University of Washington Press, Seattle, WA.

Thorgaard, G. H., Jazwin, M. E., and Stier, A. R. 1981. Polyploidy induced by heat shock in rainbow trout. Trans. Am. Fish. Soc. *110*, 546–550.

Thorgaard, G. H., Rabinovitch, P. S., Shen, M. W., Gall, G. A. E., Propp, J., and Utter, F. M. 1982. Triploid rainbow trout identified by flow cytometry. Aquaculture *29*, 305–309.

Thorgaard, G. H., Scheerer, P. D., and Parsons, J. E. 1985. Residual paternal inheritance in gynogenetic rainbow trout: implications for gene transfer. Theor. Appl. Genet. *71*, 119–121.

Thorgaard, G. H., Scheerer, P. D., Hershberger, W. K., and Myers, J. M. 1990. Androgenetic rainbow trout produced using sperm from tetraploid males show improved survival. Aquaculture *85*, 215–221.

Thorpe, J. E., Morgan, R. I. G., Talbot, C., and Miles, M. S. 1983. Inheritance of developmental rates in Atlantic salmon, *Salmo salar* L. Aquaculture *33*, 119–128.

Tipping J. M. 1991. Heritability of age at maturity in steelhead. N. Am. J. Fish. Manage. *11*, 105–108.

Tofteberg, P. and Hansen, T. 1987. Relationship between age at maturity and growth rate in farmed rainbow trout *Salmo gairdneri*. *In* Selection, Hybridization, and Genetic Engineering in Aquaculture, Vol. 1, K. Thiews (Editor). H. Heenemann GmbH and Co., Berlin.

Tomasso, J. R. and Carmichael, G. J. 1991. Differential resistance among channel catfish strains and intraspecific hybrids to environmental nitrite. J. Aquat. Anim. Health *3*, 51–54.

Tomita, H. and Matsuda, N. 1961. Deformity of vertebrae induced by lathyrogenic agents and phenylthiourea in the medaka (*Oryzias latipes*). Embryologia *5*, 413–422.

Tsoi, R. M. 1981. Artificial mutagenesis and gynogenesis in carp breeding practice. Sov. Genet. *17*, 734–738.

Tsoi, R. M. 1987. Elevage des especes productives de carpe par la methode de selection par mutation. [Creation of productive lines of carp by selection and mutation.] *In* Selection, Hybridization, and Genetic Engineering in Aquaculture, Vol. 2., K. Thiews (Editor). H. Henemann GmbH and Co., Berlin.

U

Ueda, T., Sato, R., and Kobayashi, J. 1984A. Triploid rainbow trout induced by high-pH · high-calcium. Bull. Jpn. Soc. Sci. Fish. *54*, 2045.

Ueda, T., Sato, R., and Kobayashi, J. 1988B. The origin of the genome of haploid masu salmon and rainbow trout recognized in abnormal embryos. Bull. Jpn. Soc. Sci. Fish. *54*, 619–625.

Ueno, K. 1984. Induction of triploid carp and their haematological characteristics. Jpn. J. Genet. *59*, 585–591.

Ueno, K., Ikenaga, Y., and Kariya, H. 1986. Potentiality of application of triploidy to the culture of ayu, *Plecoglossus altivelis* temminck [sic] et Schlegel. Jpn. J. Genet. *61*, 71–77.

Ulla, O. and Gjedrem, T. 1985. Number and length of pyloric caeca and their relationship to fat and protein digestibility in rainbow trout. Aquaculture *47*, 105–111.

Uraiwan, S. 1988. Direct and indirect responses to selection for age at first maturation of *Oreochromis niloticus*. *In* The Second International Symposium on Tilapia in Aquaculture, R. S. V. Pullin, T. Bhukaswan, K. Tonguthai, and J. L. Maclean (Editors). ICLARM Conference Proceedings 15, Department of Fisheries, Bangkok, Thailand and International Center for Living Aquatic Resources Management, Manila, Philippines.

Uraiwan, S. and Doyle, R. W. 1986. Replicate variance and the choice of selection procedures for tilapia (*Oreochromis niloticus*) stock improvement in Thailand. Aquaculture *57*, 93–98.

U.S. Fish and Wildlife Service. 1982. Loss of genetic variability in strains of lake trout stocked in Lakes Michigan and Ontario. *In* Fisheries and Wildlife

Research 1981, T. G. Scott and P. H. Eschmeyer (Editors). U.S. Fish and Wildlife Service, U.S. Government Printing Office, Denver, CO.

U.S. Fish and Wildlife Service. 1984. Minimum number of parents needed to protect genetic stability in fish brood stocks. *In* Fisheries and Wildlife Research and Development 1983. P. H. Eschmeyer and D. K. Harris (Editors). U.S. Fish and Wildlife Service, U.S. Government Printing Office, Denver, CO.

Utter, F. M. 1991. Biochemical genetics and fishery management: an historical perspective. J. Fish Biol. *39 (Suppl. A)*, 1–20.

Utter, F. W., Allendorf, F. W., and May, B. 1976. The use of protein variation in the management of salmonid populations. Trans. N. Am. Wild. Natur. Resour. Conf. *41*, 373–384.

Utter, F. M., Johnson, O. W., Thorgaard, G. H., and Rabinovitch, P. S. 1983. Measurement and potential applications of induced triploidy in Pacific salmon. Aquaculture *35*, 125–135.

Utter, F., Aebersold, P., and Winans, G. 1988. Interpreting genetic variation detected by electrophoresis. *In* Population Genetics & Fishery Management, N. Ryman and F. Utter (Editors). Washington Sea Grant Program, University of Washington Press, Seattle, WA.

Uwa, H. 1965. Gynogenetic haploid embryos of the medaka (*Oryzias latipes*). Embryologia *9*, 40–48.

Uyeno, T. and Miller, R. R. 1971. Multiple sex chromosomes in a Mexican cyprinodontid fish. Nature *231*, 452–453.

V

Valenti, R. J. 1975. Induced polyploidy in *Tilapia aurea* (Steindachner) by means of temperature shock treatment. J. Fish. Biol. *7*, 519–528.

Vanelli, M. L., Pancaldi, C., Alicchio, R., and Palenzona, D. 1981. Genetic control of body traits and growth pattern in *Lebistes*. Can. J. Genet. Cytol. *23*, 141–149.

Varadaraj, K. and Pandian, T. J. 1989. First report on production of supermale tilapia by integrating endocrine sex reversal with gynogenetic technique. Curr. Sci. *58*, 434–441.

Varadaraj, K. and Pandian, T. J. 1990. Production of all-female sterile-triploid *Oreochromis mossambicus*. Aquaculture *84*, 117–123.

Verspoor, E. 1988. Reduced genetic variability in first-generation hatchery populations of Atlantic salmon (*Salmo salar*). Can. J. Fish. Aquat. Sci. *45*, 1686–1690.

Verspoor, E. and Hammar, J. 1991. Introgressive hybridization in fishes: the biochemical evidence. J. Fish. Biol. *39 (Suppl. A)*, 309–334.

Vincent, R. E. 1960. Some influences of domestication upon three stocks of brook trout (*Salvelinus fontinalis* Mitchill). Trans. Am. Fish. Soc. *89*, 35–52.

Volz, C. D. and Wheeler, C. O. 1966. A portable fish-tattooing device. Prog. Fish-Cult. *28*, 54–56.

von Limbach, B. 1970. Fish genetics laboratory. Progress in Sport Fishery Research 1970. U.S. Fish Wild. Serv. Res. Pub. *106*, 153–160.

von Hertell, U., Hörstgen-Schwark, G., Langholz, H.-J., and Jung, B. 1990. Family studies on genetic variability in growth and reproductive performance between and within test fish populations of the zebrafish, *Brachydanio rerio.* Aquaculture *85*, 307–315.

W

Wahl, R. W. 1974. Heat tolerance in strains of brook trout (*Salvelinus fontinalis*). Doctoral Dissertation, Pennsylvania State Univ., State College, PA.

Wales, J. H. and Berrian, W. 1937. The relative susceptibility of various strains of trout to furunculosis. Calif. Fish Game *23*, 147–148.

Wales, J. H. and Evins, D. 1937. Sestonosis, a gill irritation in trout. Calif. Fish Game *23*, 144–146.

Wallbrunn, H. M. 1958. Genetics of the Siamese fighting fish, Betta splendens. Genetics *43*, 289–298.

Wangila, B. C. C. and Dick, T. A. 1988. Influence of genotype and temperature on the relationship between specific growth rate and size in rainbow trout. Trans. Am. Fish. Soc. *117*, 560–564.

Waples, R. S. 1990A. Conservation genetics of Pacific salmon. II. Effective population size and the rate of loss of genetic variability. J. Hered. *81*, 267–276.

Waples, R. S. 1990B. Conservation genetics of Pacific salmon. III. Estimating effective population size. J. Hered. *81*, 277–289.

Waples, R. S. and Teel, D. J. 1990. Conservation genetics of Pacific salmon. I. Temporal changes in allele frequency. Conservation Biol. *4*, 144–156.

Waples, R. S., Winans, G. A., Utter, F. M., and Mahnken, C. 1990. Genetic approaches to the management of Pacific salmon. Fisheries *15(5)*, 19–25.

Wattendorf, R. J. 1986. Rapid identification of triploid grass carp with a Coulter Counter and channelyzer. Prog. Fish-Cult. *48*, 125–132.

Webster, D. A. and Flick, W. A. 1981. Performance of indigenous, exotic, and hybrid strains of brook trout (*Salvelinus fontinalis*) in waters of the Adirondack Mountains, New York. Can. J. Fish. Aquat. Sci. *38*, 1701–1707.

Wedekind, H., Hörstgen-Schwark, G., and Langholz, H. J. 1990. Investigations on sex ratio in *Oreochromis niloticus.* Aquaculture *85*, 321–322.

Weigand, M. D., Hataley, J. M., Kitchen, C. L. and Buchanan, L. G. 1989. Induction of developmental abnormalities in larval goldfish. *Carassius auratus* L., under cool incubation conditions. J. Fish. Biol. *35*, 85–95.

Welch, H. E. and Mills, K. H. 1981. Marking fish by scarring soft fin rays. Can. J. Fish. Aquat. Sci. *38*, 1168–1170.

Whitmore, D. H. 1990. Electrophoretic and Isoelectric Focusing Techniques in Fisheries Management. CRC Press, Boca Raton, FL.

Williamson, J. H. and Carmichael, G. J. 1990. An aquacultural evaluation of Florida, northern, and hybrid largemouth bass, *Micropterus salmoides.* Aquaculture *85*, 247–257.

Winemiller, K. O. and Taylor, D. H. 1982. Inbreeding depression in the convict cichlid, *Cichlasoma nigrofasciatum* (Baird and Girard). J. Fish Biol. *21*, 399–402.

Winge, Ö. 1922. One-sided masculine and sex-linked inheritance in *Lebistes reticulatus*. J. Genet *12*, 145–162.

Winge, Ö. 1923. Crossing-over between the X- and the Y-chromosome in *Lebistes*. J. Genet. *13*, 201–217.

Winge Ö. 1927. The location of eighteen genes in *Lebistes reticulatus*. J. Genet. *18*, 1–43.

Winge, Ø. and Ditlevsen, E. 1947. Colour inheritance and sex determination in *Lebistes*. Heredity *1*, 65–83.

Withler, R. E. 1986. Genetic variation in carotenoid pigment deposition in the red-fleshed and white-fleshed chinook salmon (*Oncorhynchus tshawytscha*) of Quesnel River, British Columbia. Can. J. Genet. Cytol. *28*, 587–594.

Withler, R. E. 1988. Genetic consequences of fertilizing chinook salmon (*Oncorhynchus tshawytscha*) eggs with pooled milt. Aquaculture *68*, 15–25.

Withler, R. E. 1990. Genetic consequences of salmonid egg fertilization techniques. Aquaculture *85*, 326.

Withler, R. E., Clarke, W. C., Riddell, B. E., and Kreiberg, H. 1987. Genetic variation in freshwater survival and growth of chinook salmon (*Oncorhynchus tshawytscha*). Aquaculture *64*, 85–96.

Wlodek, J. M. 1968. Studies on the breeding of carp (*Cyprinus carpio*) at the experimental pond farms of the Polish Academy of Sciences in southern Silesia, Poland. Proceedings of the World Symposium on Warm-water Pond Fish Culture. FAO Fish. Rep. No. 44, *4*, 93–116.

Wohlfarth, G. W. 1977. Shoot carp. Bamidgeh *29*, 35–56.

Wohlfarth, G. W. 1986. Decline in natural fisheries—a genetic analysis and suggestion for recovery. Can. J. Fish. Aquat. Sci. *43*, 1298–1306.

Wohlfarth, G. W. and Milstein, A. 1987. Predicting correction factors for differences in initial weight among genetic groups of common carp in communal testing. Aquaculture *60*, 13–25.

Wohlfarth, G. W. and Moav, R. 1969. The genetic correlation of growth rate with and without competition in carp. Int. Assoc. Theor. Appl. Limnol. Proc. *17*, 702–704.

Wohlfarth, G. and Moav, R. 1970. The effects of variation in spawning time on subsequent relative growth rate and viability in carp. Bamidgeh *22*, 42–47.

Wohlfarth, G. W. and Moav, R. 1972. The regression of weight gain on initial weight in carp. I. Methods and results. Aquaculture *1*, 7–28.

Wohlfarth, G. W. and Moav, R. 1991. Genetic testing of common carp in cages 1. Communal versus separate testing. Aquaculture *95*, 215–223.

Wohlfarth, G. W. and Wedekind, H. 1991. The heredity of sex determination in tilapias. Aquaculture *92*, 143–156.

Wohlfarth, G., Lahman, M., and Moav, R. 1963. Genetic improvement of carp. IV. Leather and line carp in fish ponds of Israel. Bamidgeh *15*, 3–8.

Wohlfarth, G., Moav, R., and Hulata, G. 1975A. Genetic differences between the Chinese and European races of the common carp. II. Multi-character variation—a response to the diverse methods of fish cultivation in Europe and China. Heredity *34*, 341–350.

Wohlfarth, G., Moav, R., Hulata, G., and Beiles, A. 1975B. Genetic variation in seine escapability of the common carp. Aquaculture *5*, 375–387.

Wohlfarth, G. W., Moav, R., and Hulata, G. 1983. A genotype-environment interaction for growth rate in the common carp, growing in intensively manured ponds. Aquaculture *33*, 187–195.

Wohlfarth, G. W., Moav, R., and Hulata, G. 1987. Breeding programs in Israeli aquaculture. *In* Selection, Hybridization, and Genetic Engineering in Aquaculture, Vol. 2, K. Thiews (Editor). H. Heenemann GmbH and Co., Berlin.

Wolf, L. E. 1941. Further observations on ulcer disease of trout. Trans. Am. Fish. Soc. *70*(1940), 369–381.

Wolf, L. E. 1942. Fish-diet disease of trout. A vitamin deficiency produced by diets containing raw fish. N.Y. State Cons. Dep. Fish. Res. Bull. No. 2.

Wolf, L. E. 1954. Development of disease-resistant strains of fish. Trans. Am. Fish. Soc. *83*(1953), 342–349.

Wolters, W. R., Libey, G. S., and Chrisman, C. L. 1981. Induction of triploidy in channel catfish. Trans. Am. Fish. Soc. *110*, 310–312.

Wolters, W. R., Chrisman, C. L., and Libey, G. S. 1982A. Erythrocyte nuclear measurements of diploid and triploid channel catfish, *Ictalurus punctatus* (Rafinesque). J. Fish Biol. *20*, 253–258.

Wolters, W. R., Libey, G. S., and Chrisman, L. 1982B. Effect of triploidy on growth and gonad development of channel catfish. Trans. Am. Fish. Soc. *111*, 102–105.

Wolters, W. R., Lilyestrom, C. G., and Craig, R. J. 1991. Growth, yield, and dress-out percentage of diploid and triploid channel catfish in earthen ponds. Prog. Fish-Cult. *53*, 33–36.

Wright, J. E., Jr. 1972. The palomino rainbow trout. Penn. Angler *41*, 8–9, 26.

Wu, C. 1990. Retrospects and prospects of fish genetics and breeding research in China. Aquaculture *85*, 61–68.

Wu, C., Ye, Y., and Chen. R. 1986. Genome manipulation in carp (*Cyprinus carpio* L.). Aquaculture *54*, 57–61.

Wu, C., Chen, R., Ye, Y., and Huang, W. 1990. Production of all-female carp and its application in fish cultivation. Aquaculture *85*, 327.

X

Xia, D., Liu, A., Wu, T., and Sun, Y. 1990. Study of clones of commercial fish. Aquaculture *85*, 327–328.

Y

Yamamoto, T. 1955. Progeny of artificially induced sex-reversals of male genotype (XY) in the medaka (Oryzias latipes) with special reference to YY-male. Genetics *40*, 406–419.

Yamamoto, T. 1958. Artificial induction of functional sex-reversal in genotypic females of the medaka (Oryzias latipes). J. Exp. Zool. *137*, 227–263.

Yamamoto, T. 1961. Progenies of sex-reversal females mated with sex-reversal males in the medaka, *Oryzias latipes*. J. Exp. Zool. *146*, 163–179.

Yamamoto, T. 1963. Induction of reversal in sex differentiation of YY zygotes in the medaka, Oryzias latipes. Genetics *48*, 293–306.

Yamamoto, T. 1964A. Linkage map of sex chromosomes in the medaka, Oryzias latipes. Genetics *50*, 59–64.

Yamamoto, T. 1964B. The problem of viability of YY zygotes in the medaka, Oryzias latipes. Genetics *50*, 45–58.

Yamamoto, T. 1969A. Inheritance of albinism in the medaka, *Oryzias latipes*, with special reference to gene interaction. Genetics *62*, 797–809.

Yamamoto, T. 1969B. Sex differentiation. *In* Fish Physiology, Vol. 3, Reproduction and Growth, Bioluminescence, Pigments, and Poisons, W. S. Hoar and D. J. Randall (Editors). Academic Press, New York, NY.

Yamamoto, T. 1973. Inheritance of albinism in the goldfish, *Carassius auratus*. Jpn. J. Genet. *48*, 53–64.

Yamamoto, T. 1975A. The medaka, *Oryzias latipes*, and the guppy, *Lebistes reticularis* [sic]. *In* Handbook of Genetics. Vol. 4. Vertebrates of Genetic Interest, R. C. King (Editor). Plenum Press, New York, NY.

Yamamoto, T. 1975B. Linkage map of sex chromosomes. *In* Medaka (Killifish): Biology and Strains, T. Yamamoto. Keigaku Publishing Co., Tokyo, Japan.

Yamamoto, T. 1975C. A YY male goldfish from mating estrone-induced XY female and normal male. J. Hered. *66*, 2–4.

Yamamoto, T. 1977. Inheritance of nacreous-like scaleness in the ginbuna, *Carassius auratus langsdorfii*. Jpn. J. Genet. *52*, 373–377.

Yamamoto, T. and Kajishima, T. 1968. Sex hormone induction of sex reversal in the goldfish and evidence for male heterogamity. J. Exp. Zool. *168*, 215–221.

Yamazaki, F. 1983. Sex control and manipulation in fish. Aquaculture *33*, 329–354.

Yant, D. R., Smitherman, R. O., and Green, O. L. 1976. Production of hybrid (blue × channel) catfish and channel catfish in ponds. Proc. Southeast. Assoc. Game Fish Comm. *29*, 82–86.

Yoon, S. J., Liu, Z., Kapuscinski, A. R., Hackett, P. B., Faras, A., and Guise, K. S. 1989. Successful gene transfer in fish. *In* Gene Transfer and Gene Therapy, A. L. Beaudet, R. Mulligan, and I. M. Verma (Editors). Alan R. Liss, Inc., New York, NY.

Yoon, S. J., Hallerman, E. M., Gross, M. L., Liu, Z., Schneider, J. F. Faras, A. J. Hackett, P. B., Kapuscinski, A. R., and Guise, K. S. 1990A. Transfer of the gene for neomycin resistance into goldfish, *Carassius auratus*. Aquaculture *85*, 21–33.

Yoon, S. J., Liu, Z., Kapuscinski, A. R., Hackett, P. B., Faras, A., and Guise, K. S. 1990B. Successful gene transfer in fish. *In* Genetics in Aquaculture: Proceedings of the Sixteenth U.S.-Japan Meeting on Aquaculture, Charleston, South Carolina, October 20 and 21, 1987, R. S. Svrjeck (Editor). NOAA

Technical Report NMFS 92, U.S. Department of Commerce, Springfield, VA.

Youngblood, P. N. 1980. Growth and feed conversion of six genetic groups of adult channel catfish selected as broodstock. Master's Thesis, Auburn Univ., AL.

Z

Zhang, H., Gall, G. A. E., and Hung, S. S. O. 1990. Effect of sire, diet and week of spawning on volume of eggs retained by artificially spawned rainbow trout. Aquaculture *87*, 23–33.

Zhang, P., Hayat, M., Joyce, C., Gonzalez-Villaseñor, L. I., Lin, C. M., Dunham, R. A., Chen, T. T., and Powers, D. A. 1990. Gene transfer, expression and inheritance of pRSV-rainbow trout-GH cDNA in the common carp, *Cyprinus carpio* (Linnaeus). Mol. Repro. Develop. *25*, 3–13.

Zhu, Z., Li, G., He, L., and Chen, S. 1985. Novel gene transfer into the fertilized eggs of gold fish (*Carassius auratus* L. 1758) Z. Angew. Ichthyol. *1*, 31–34.

Zhu, Z., Xu, K., Li, G., Xie, Y., and He, L. 1986. Biological effects of human growth hormone gene microinjected into the fertilized eggs of loach *Misgurnus anguillicaudatus*. Kexue Tongbao *81*, 988–990. (cited in Wu [1990]).

Index